T0332921

Darwin Mythology

Many historical figures have their lives and works shrouded in myth, both in life and long after their deaths. Charles Darwin (1809–82) is no exception to this phenomenon and his hero-worship has become an accepted narrative. This concise, accessible and engaging collection unpacks this narrative to rehumanize Darwin's story and establish what it meant to be a "genius" in the Victorian context. Leading Darwin scholars have come together to argue that, far from being a lonely genius in an ivory tower, Darwin had fortune, diligence and – crucially – community behind him. The aims of this essential work are twofold. First, to set the historical record straight, debunking the most pervasive myths and correcting falsehoods. Second, to provide a deeper understanding of the nature of science itself, relevant to historians, scientists and the public alike.

KOSTAS KAMPOURAKIS is the author and editor of several books about evolution, genetics, philosophy and history of science, as well as the editor of the Cambridge University Press book series *Understanding Life*. He teaches biology and science education courses at the University of Geneva.

Darwin Mythology

Debunking Myths, Correcting Falsehoods

Edited by

Kostas Kampourakis
University of Geneva

CAMBRIDGE
UNIVERSITY PRESS

CAMBRIDGE
UNIVERSITY PRESS

Shaftesbury Road, Cambridge CB2 8EA, United Kingdom

One Liberty Plaza, 20th Floor, New York, NY 10006, USA

477 Williamstown Road, Port Melbourne, VIC 3207, Australia

314–321, 3rd Floor, Plot 3, Splendor Forum, Jasola District Centre,
New Delhi – 110025, India

103 Penang Road, #05–06/07, Visioncrest Commercial, Singapore 238467

Cambridge University Press is part of Cambridge University Press & Assessment,
a department of the University of Cambridge.

We share the University's mission to contribute to society through the pursuit of
education, learning and research at the highest international levels of excellence.

www.cambridge.org
Information on this title: www.cambridge.org/9781009375702

DOI: 10.1017/9781009375719

First published 2024

A catalogue record for this publication is available from the British Library

Library of Congress Cataloging-in-Publication Data
Names: Kampourakis, Kostas, editor.
Title: Darwin mythology : debunking myths, correcting falsehoods / edited by
Kostas Kampourakis, Université de Genève.
Description: Cambridge, United Kingdom ; New York, NY, USA : Cambridge
University Press, 2024. | Includes bibliographical references and index.
Identifiers: LCCN 2023046941 (print) | LCCN 2023046942 (ebook) |
ISBN 9781009375702 (hardback) | ISBN 9781009375689 (paperback) |
ISBN 9781009375719 (epub)
Subjects: LCSH: Darwin, Charles, 1809–1882. | Natural selection. |
Evolution (Biology) – Philosophy. | Evolution (Biology) – History.
Classification: LCC QH375 .D355 2024 (print) | LCC QH375 (ebook) |
DDC 576.8/2–dc23/eng/20240112
LC record available at https://lccn.loc.gov/2023046941
LC ebook record available at https://lccn.loc.gov/2023046942

ISBN 978-1-009-37570-2 Hardback
ISBN 978-1-009-37568-9 Paperback

In memory of Ronald L. Numbers, who taught us the value of mythoclastic endeavors, and of John L. Heilbron who taught us their limitations

Contents

Figures

Contributors

PETER J. BOWLER is Professor Emeritus of the History of Science at Queen's University, Belfast. He is a Fellow of the British Academy and a Member of the Royal Irish Academy. He has an MA from Cambridge University and a PhD from the University of Toronto and has taught at universities in Canada, Malaysia and the United Kingdom. He has published a number of books on the history of evolutionism, including *The Eclipse of Darwinism* (Johns Hopkins University Press, 1983), *Theories of Human Evolution* (John Hopkins University Press, 1986), *The Non-Darwinian Revolution* (John Hopkins University Press, 1988), *Life's Splendid Drama* (University of Chicago Press, 1996) and *Evolution: The History of an Idea* (University of California Press, 25th anniversary ed., 2009). Recent books include *Darwin Deleted: Imagining a World without Darwin* (University of Chicago Press, 2013) and *Progress Unchained: Ideas on Evolution, Human History and the Future* (Cambridge University Press, 2021).

JOHN HEDLEY BROOKE taught the history of science at Lancaster University before becoming the first Andreas Idreos Professor of Science & Religion at Oxford (1999–2006), where he was Director of the Ian Ramsey Centre and Fellow of Harris Manchester College. A Gifford Lecturer (Glasgow University 1994–5) and member of the International Academy of the History of Science, he was designated a "Distinguished Fellow" by the Institute of Advanced Study, Durham University, in 2007. He has lectured worldwide and been President of the British Society for the History of Science, the Historical Section of the British Science Association, the International Society for Science and Religion and the UK Forum for Science & Religion. Books include his prize-winning *Science and Religion: Some Historical Perspectives* (Cambridge University Press, 1991) republished as a "Canto Classic" in 2014; *Thinking About Matter* (Routledge, 1995); *Reconstructing Nature* (Oxford University Press, 2000, with Geoffrey Cantor); and *Science and Religion around the World* (Oxford University Press, 2011, coedited with Ronald Numbers).

RICHARD W. BURKHARDT, JR. is Professor of History Emeritus at the University of Illinois Urbana-Champaign. He is the author of *The Spirit of System: Lamarck and Evolutionary Biology* (Harvard University Press, 1st ed. 1977, 2nd ed. 1995) and *Patterns of Behavior: Konrad Lorenz, Niko Tinbergen, and the Founding of Ethology* (University of Chicago Press, 2005). He continues to write on the history of biology, ethology and zoos.

PIETRO CORSI is Professor Emeritus of the History of Science at the University of Oxford, and Directeur d'études honoraire at the E.H.E.S.S., Paris. He has published monographs on science and religion in nineteenth-century England and on Jean-Baptiste Lamarck. He has also produced articles and the critical edition of a scientific correspondence concerning the earth sciences in nineteenth-century Italy and the failed project of the Geological Map of the Kingdom of Italy. He has been the coeditor of *KOS*, and editor-in-chief of the Italian language edition of *The New York Review of Books*. Corsi has taken part in several exhibition projects, the last to date being *Les Origines du Monde*, Paris, Musée d'Orsay, curated by Laura Bossi.

DAVID DEPEW is Professor Emeritus of Rhetoric of Inquiry at the University of Iowa. Previously, he was Professor of Philosophy at California State University–Fullerton. His work is largely on the history, philosophy and public understanding of evolutionary biology, with special interest in the history, conceptual foundations and current challenges to the Modern Evolutionary Synthesis. He is the coauthor, with Bruce H. Weber, of *Darwinism Evolving: Systems Dynamics and the Genealogy of Natural Selection* (MIT Press, 1995); with the late Marjorie Grene of *Philosophy of Biology: An Episodic History* (Cambridge University Press, 2005); and with John P. Jackson of *Darwinism, Democracy and Race* (Routledge, 2017). He has also coedited several collections on these themes with Bruce H. Weber.

PATRICIA FARA is an Emeritus Fellow of Clare College, Cambridge. Currently President of the Antiquarian Horological Society and a Fellow of the Royal Historical Society, she was President of the British Society for the History of Science from 2016 to 2018 and won the 2022 Abraham Pais prize of the American Physical Society. A regular contributor to *In Our Time* and other radio and TV programs, she is especially interested in the Enlightenment period, with a particular emphasis on scientific imagery and women in science. In addition to many articles, her books include the prize-winning *Science: A Four Thousand Year History* (Oxford University Press, 2009) (which has been translated into nine languages), as well as *Erasmus Darwin: Sex, Science and Serendipity* (Oxford University Press, 2012) and *A Lab of One's Own: Science and Suffrage in the First World War* (Oxford University

Press, 2018). Her most recent book is *Life after Gravity: Isaac Newton's London Career* (Oxford University Press, 2021).

LIV GRJEBINE is a postdoctoral research fellow in the Department of the History of Science at Harvard University. She did her PhD at the Sorbonne on the dissemination of Darwinism in nineteenth-century France. She studies the relations between science and society from the nineteenth century to the present. She has published a dozen articles in France and the United States in newspapers and scholarly journals, including *The Conversation, Isis: A Journal of the History of Science Society, The Huffington Post* and *Scientific American*. She has received several awards, including the Prix de la Chancellerie des Universités de Paris, the Harvard University Arthur Sachs Scholarship and the Harvard University Q Award.

KIMBERLY A. HAMLIN is the James and Beth Lewis Professor of History at Miami University in Oxford, Ohio. She teaches classes on the history of women, gender, sex, science and medicine in the USA. She is the author of *From Eve to Evolution: Darwin, Science, and Women's Rights in Gilded Age America* (University of Chicago Press, 2014), *Free Thinker: Sex, Suffrage, and the Extraordinary Life of Helen Hamilton Gardener* (W.W. Norton, 2020) and several articles. Hamlin's article "The 'Case of a Bearded Woman': Hypertrichosis and the Construction of Gender in the Age of Darwin" (*American Quarterly*, 2011) received both the Margaret Rossiter Prize from the History of Science Society and the Emerging Scholar Award from the Nineteenth Century Studies Association. Hamlin is the founder and past chair of the American Studies Association's Science and Technology Caucus and past cochair of the History of Science Society's Women's Caucus.

JOHN L. HEILBRON was Professor of History Emeritus at the University of California–Berkeley, who wrote about early modern science in its cultural setting for almost fifty years. His most recent work in this genre relates to Galileo: a biography of the man (*Galileo*, Oxford University Press, 2010), of a painting made during the English Civil War mysteriously referring to the man (*The Ghost of Galileo*, Oxford University Press, 2021) and of a champion second-generation Galilean prized by popes and princes (*The Incomparable Monsignor*, Oxford University Press, 2022). During his long career he had many opportunities for debunking myths and even more for creating them.

KOSTAS KAMPOURAKIS is the author and editor of several books about evolution, genetics, philosophy and history of science, his most recent ones being *Understanding Evolution* (Cambridge University Press, 2020), *Understanding Genes* (Cambridge University Press, 2021), *Ancestry*

Reimagined: Dismantling the Myth of Genetic Ethnicities (Oxford University Press, 2023), and *How We Get Mendel Wrong and Why It Matters: Challenging the Narrative of Mendelian Genetics* (CRC Press/ Routledge, 2024). He teaches biology and science education courses at the University of Geneva. He is the editor of the book series *Understanding Life*, published by Cambridge University Press. In the past, he was the editor-in-chief of the Springer journal *Science & Education*, the founding editor of the Springer book series *Science: Philosophy, History and Education* and the founding coeditor of the Springer book series *Contributions from Biology Education Research*.

EMILY M. KERN is Assistant Professor in History of Science at the University of Chicago in the Department of History and the Committee on Conceptual and Historical Studies of Science. Previously, she was a postdoctoral research associate in the Program in New Earth Histories at the University of New South Wales and coeditor of the volume *New Earth Histories: Geo-Cosmologies and the Making of the Modern World* with Alison Bashford and Adam Bobbette (University of Chicago Press, 2023). She is currently at work on her first book, a global history of the search for the "cradle of humankind" from the Enlightenment to the late twentieth century.

JAMES G. LENNOX is Professor Emeritus of History and Philosophy of Science, University of Pittsburgh. He has published widely on the history and philosophy of biology with a special focus on Aristotle, Charles Darwin and evolutionary biology. His books include *Aristotle: On the Parts of Animals I–IV*, a translation and commentary (Oxford University Press, 2001), *Aristotle's Philosophy of Biology* (Cambridge University Press, 2001), *Aristotle on Inquiry* (Cambridge University Press, 2021), as well as many coedited volumes, including *Philosophical Issues in Aristotle's Biology* (Cambridge University Press, 1987), *Self-motion from Aristotle to Newton* (Princeton University Press, 1994) and *Concepts and Their Role in Knowledge: Reflections on Objectivist Epistemology* (University of Pittsburgh Press, 2013).

ALAN C. LOVE is Distinguished McKnight University Professor in the Department of Philosophy at the University of Minnesota, as well as Winton Chair in the Liberal Arts and Director of the Minnesota Center for Philosophy of Science. He is the author of *Evolution and Development: Conceptual Issues* (Cambridge University Press, 2024) and editor of *Conceptual Change in Biology: Scientific and Philosophical Perspectives on Evolution and Development* (Springer, 2015) and *Beyond the Meme: Development and Structure in Cultural Evolution* (University of Minnesota Press, 2019, with William Wimsatt). His research concentrates on a variety of conceptual issues in the life sciences, including function in genomics, reductionism in protein folding and development, criteria of explanatory adequacy for explaining complex evolutionary phenomena and how concepts accomplish more than categorization in research programs. He uses a combination of approaches, including collaboration with scientific practitioners, to study developmental biology, evolutionary biology, evo-devo, functional morphology, molecular biology and genetics and paleontology.

ALISON M. PEARN was the Associate Director of the Darwin Correspondence Project at Cambridge University and collaborated on twenty of the thirty volumes of *The Correspondence of Charles Darwin* (Cambridge University Press, 1985–2023). She is one of the founders of Epsilon (epsilon.ac.uk) a digital platform recreating the international networks of nineteenth-century scientific correspondence, author of *Darwin: All that Matters* (John Murray Publishing, 2015) and has written on Darwin's life and work for both general and academic audiences. In 2009, Darwin's bicentenary year, she curated an exhibition about the *Beagle* expedition, A Voyage Round the World, and edited a companion volume of essays (Cambridge University Press). Most recently she has curated exhibitions in Cambridge and in New York celebrating the completion of the Darwin Correspondence Project.

ERIK L. PETERSON is Associate Professor of the History of Science & Medicine and Associate Provost for General Education at the University of Alabama. Erik also cohosts "Speaking of Race," a featured podcast of the American Anthropological Association. He researches the impact of genetics and evolutionary biology on society, and vice versa, and is especially interested in the persistence of race science. His books include *The Life Organic* (University of Pittsburgh Press, 2017), *A Deeper Sickness* (Beacon 2022), *Understanding Charles Darwin* (Cambridge University Press, 2023) and his forthcoming *Shortest History of Eugenics* (The Experiment, 2023), which reframes this often-told story as a fast-paced 150-year-long narrative of vigilante medicine gone bad. Erik's authored and coauthored essays appear in *The Journal of the History of Biology, The British Journal for the History of Science, Science & Education, History & Philosophy of the Life Sciences, Theoretical Biology Forum* and *Genetics*.

ANYA PLUTYNSKI is Professor of Philosophy and affiliate faculty with the Division of Biology and Biomedical Sciences at Washington University–St. Louis. She received her PhD in Philosophy and MA in Biology at the University of Pennsylvania. Her research is on the history and philosophy of science, with a special focus on twentieth-century biology and medicine. She is coeditor with Sahotra Sarkar of the *Blackwell's Companion to Philosophy of Biology* (Blackwell, 2007) and coeditor with Justin Garson and Sahotra Sarkar of the *Routledge Handbook of Philosophy of Biodiversity* (Routledge, 2017). Her most recent book is *Explaining Cancer: Finding Order in Disorder* (Oxford University Press, 2018).

GREGORY RADICK is Professor of History and Philosophy of Science at the University of Leeds. Educated at Rutgers University and Cambridge University, he has published widely in the history of the life and human sciences. His latest book is *Disputed Inheritance: The Battle over Mendel and the Future of Biology* (University of Chicago Press, 2023). Earlier books include *The Simian Tongue: The Long Debate about Animal Language* (winner of the 2010 Suzanne J. Levinson Prize of the History of Science Society for best book in the history of the life sciences and natural history, University of Chicago Press, 2007), *Darwin in Ilkley* (with Mike Dixon, The History Press, 2009), and, as coeditor, *The Cambridge Companion to Darwin* (with Jonathan Hodge, Cambridge University Press, 2003, 2009). He has served as President of the British Society for the History of Science (2014–16) and the International Society for the History, Philosophy, and Social Studies of Biology (2019–21). In 2022 he was appointed to the Board of Trustees of the Science Museum Group.

ROBERT J. RICHARDS is Morris Fishbein Distinguished Service Professor of the History of Science at the University of Chicago, where he is a professor in the Departments of History, Philosophy and Psychology, and a member of the Committee on Conceptual and Historical Studies of Science. He also directs the Fishbein Center for the History of Science and Medicine. His primary areas of research are German Romanticism and the history of evolutionary theory. He is the author of several books, including *Darwin and the Emergence of Evolutionary Theories of Mind and Behavior* (University of Chicago Press, 1987), which won the Pfizer Prize from the History of Science Society, *The Romantic Conception of Life: Science and Philosophy in the Age of Goethe* (University of Chicago Press, 2002) and *The Tragic Sense of Life: Ernst Haeckel and the Struggle over Evolutionary Thought* (University of Chicago Press, 2008).

NICOLAAS RUPKE is Professor Emeritus of the History of Science at Göttingen University and Johnson Professor in the College at Washington & Lee. Educated at Groningen and Princeton, he has held research fellowships at the Smithsonian, Oxford, Tübingen, NIAS, the Wellcome Institute, the NHC, the Institute of Advanced Studies in Canberra, the Institute for Advanced Study in Princeton, and he was the initial occupant of the Tyrone Professorship of Medical History at Vanderbilt University. His books include scientific biographies of William Buckland, Richard Owen and Alexander von Humboldt. Rupke's metabiographical approach to Humboldt has received considerable acclaim. Currently, Rupke works on Johann Friedrich Blumenbach and the non-Darwinian tradition in evolutionary theory. He is a fellow of the German Academy of Sciences Leopoldina and the Göttingen Academy of Sciences (recently renamed the Lower Saxony Academy of Sciences at Göttingen).

MICHAEL RUSE is the retired Lucyle T. Werkmeister Professor of Philosophy and Director of the Program in the History and Philosophy of Science at Florida State University. He is also Professor Emeritus at the University of Guelph where he taught from 1965 to 2000. He is the founding editor (1985) of the journal *Biology and Philosophy*. Today, he is coeditor of the Cambridge *Elements* series in the Philosophy of Biology. His seventy books include *The Gaia Hypothesis: Science on a Pagan Planet* (Chicago University Press, 2013), *Darwinism as Religion: What Literature Tells Us About Evolution* (Oxford University Press, 2016), *On Purpose* (Princeton University Press, 2019), *On Faith and Science* (coauthored with Edward Larson, Yale University Press, 2017) and, most recently, *Why We Hate: Understanding the Roots of Human Conflict* (Oxford University Press, 2022), *Understanding the Christianity-Evolution Relationship* (Cambridge University Press, 2023) and *The New Biology: The Battle between Mechanism and Organicism* (coauthored with Michael Reiss, Harvard University Press, 2023).

SHRUTI SANTOSH is a graduate student at Washington University in St Louis working on the History and Philosophy of Science. She is interested in questions about scientific objectivity and is currently investigating the role sociopolitical values play in the classification of women's health disorders.

FRANK J. SULLOWAY is Adjunct Professor in the Department of Psychology at the University of California–Berkeley. He received his PhD in the history of science from Harvard University (1978) and is a recipient of a MacArthur Award (1984–9). His book *Freud, Biologist of the Mind* (Harvard University Press, 1979) provides a radical reanalysis of the origins and validity of psychoanalysis and received the Pfizer Award of the History of Science Society. Dr Sulloway has published extensively on the life and theories of Charles Darwin. His research also includes evolutionary psychology, where he has applied a Darwinian perspective to the understanding of sibling competition and family dynamics, and has studied the evolution of personality differences among animal species. Among his additional research interests are the behavior and evolution of Darwin's Galápagos finches, and avian strategies against nest predation and brood parasitism in Australian fairywrens.

JOHN VAN WYHE is a historian of science at the National University of Singapore who has written extensively on the history of phrenology, evolutionary theories and scientific travelers, specializing in Charles Darwin and Alfred Russel Wallace. He is the founder and director of The Complete Work of Charles Darwin Online and has published fifteen books and numerous articles. His work has overturned many traditional, and often academically orthodox, beliefs and myths about Darwin and Wallace. He lectures and broadcasts around the world.

Acknowledgments

I am indebted to all contributors for turning the making of the present book into a delightful experience. Words are not enough to express my gratitude to them for their diligent work as authors and reviewers, as well as for the inspiring discussions we had over three days during July 2022. I am also additionally indebted to John van Wyhe, Frank Sulloway and John Brooke who read through the whole manuscript and sent me very useful suggestions, as well as to John Heilbron for a meticulous editing of my introduction for idiom and style.

I am also very grateful to our editor Lucy Rhymer for her support all the way since the inception and until the publication of the present book, which exists also thanks to the diligent work of Rosa Martin, Laheba Alam, Geethanjali Rangaraj, and Helen Kitto.

Last, but not least, I would like to express my deepest gratitude to Ronald L. Numbers and John L. Heilbron for their insights and influence on my thinking. Ron's book *Galileo goes to Jail and other Myths about Science and Religion* (Harvard University Press, 2009) was an inspiration for me, and editing together *Newton's Apple and Other Myths about Science* (Harvard University Press, 2015) was a fascinating and formative experience. I hope that with the present book I successfully complete a useful trilogy on "mythoclasting" books that Ron started. John has also influenced my thinking about "mythoclasting," starting with his 2014 talk and the subsequent discussions we had in Lexington, Virginia, about what exactly historical myths are. Because of this influence and in spite of being a historian of physics, not Darwinism, I invited him to contribute the very first chapter in the present volume, which stems from that 2014 talk. I am very glad and grateful that he accepted.

Unfortunately, neither Ron nor John will see this book in print, as they both sadly passed away within a few months in 2023 while this book was in production. It is with great sadness, but also with great admiration for their personalities and scholarships, that I dedicate the present book to their memory.

Kostas Kampourakis

Introduction: Myths and Darwin

Kostas Kampourakis

I.1   Introduction

"Great is the power of steady misrepresentation; but the history of science shows that fortunately this power does not long endure." This is what Charles Darwin wrote in the sixth edition of his book *On the Origin of Species by Means of Natural Selection, or the Preservation of Favoured Races in the Struggle for Life* (hereafter, *Origin of Species*).[1] It should be obvious from the title of the present book that its contributors do not agree with Darwin's conclusion that however powerful misrepresentation is, it does not persist. The aim of the present book is to argue that misrepresentation often does persist, and to explain the reasons why this happens. Some of these misrepresentations have existed since Darwin's time. Others are more recent. Some are attributed to Darwin himself, others to his contemporaries, both friends and foes. The aim of the present book is to analyze myths and falsehoods related to Darwin's life and work, explain their origins and suggest how we should think about them.

Charles Darwin is a scientific hero; his image can be found in every biology textbook, and it was on the 10GBP banknote between 2000 and 2018 (his image was replaced in 2017 by that of Jane Austen).[2] Darwin is credited with changing our perception of humanity's place in nature. And he is also an exemplar case of hero-worshipping, which sometimes ends up being naïve. This in turn leads to stereotypes about big discoveries, eureka moments, lonely genius work and more. On his twitter account, popular science writer Richard Dawkins quoted Michael Ghiselin stating that "[Darwin] was a genius of the first rank and a man of immense learning, and he often worked a century or more ahead of his time. What a pleasure it is to bask in the sunlight of his intellect."[3] Evolutionary biologist Jerry Coyne agrees:

---

[1] Darwin, C. (1872) *On the Origin of Species by Means of Natural Selection* (6th ed). London: John Murray, p. 421.

[2] www.bankofengland.co.uk/news/2000/may/new-10-banknote-design; www.bankofengland.co.uk/news/2018/february/the-darwin-10-is-almost-extinct-just-one-week-to-go (accessed December 14, 2022)

[3] Richard Dawkins on Twitter (@RichardDawkins, 11:16 AM · Aug 7, 2020)

Those of us who admire him [Darwin] do so not because he was morally perfect, but because he proposed, in one huge go, a theory that gave pretty much dispositive evidence that organisms evolved, did so slowly, that lineages split, creating a common ancestry between all species, and that the mechanism of adaptive change was natural selection. Those are four or five huge theories, and all have, with time, been shown to be correct. At one go, in one huge book (supplemented by a spate of other ones, including *The Descent of Man and Selection in Relation to Sex*), Darwin rid the world of centuries of incorrect creationist thought and gave rise to a fertile new field with a million new questions to explore. Now that is an accomplishment worth celebrating![4]

I must note that I share Richard Dawkins's and Jerry Coyne's admiration for Charles Darwin. This is why I have invested time and effort, along with the many stellar scholars contributing to this book, to consider myths and false-hoods about his life and work. But what does being a genius, and working ahead of one's time, mean? Should this be taken to imply that most of Darwin's contemporaries were not brilliant, or at least not as brilliant as Darwin was? Does this in turn mean that Darwin worked alone? Didn't he get any insights at all from anybody else? Was his theory stemmed from his own mind alone, as Athena stemmed from the head of Zeus according to the myth? Did Darwin indeed "in one huge go" change forever our thinking about nature and our place in it, at the same time dismantling thoughts that stood unchallenged for centuries? And did he single-handedly give rise to a new research field?

Darwin was without question a brilliant naturalist, observer and experimentalist and scholar. But this kind of hero-worshipping should be avoided because it is misleading – science is not done, and does not advance, by individuals who make big breakthroughs in one go. Science is done by communities, which consist of individuals many of whom have something important to contribute to the overall advancement. Even when some individuals happen to see something that others do not, the validation of a novel perspective or findings by the community is absolutely necessary. Most importantly, coming up with anything novel takes time and effort – it took Darwin twenty years of painstaking work – while one works in a particular context and with particular resources to hand – and Darwin had experiences and resources that most others lacked. This kind of hero-worshipping is also better avoided because it dehumanizes science; in the last chapter of the present book, I explain how the stories in its twenty-four chapters can help us better understand science as a human activity. My aim is to humanize Darwin and to emphasize a number of points about how science is done.

At the same time, it is interesting how many authors have tried to show in book-length accounts that not only Darwin was not original, but also – and even

---

[4] Jerry Coyne, whyevolutionistrue.com (2021/06/05) (accessed October 16, 2022)

worse – that he stole the main idea of natural selection from others and presented it as his own. For instance, Arnold Brackman has argued that Darwin stole priority from Alfred Russel Wallace, by suggesting that "Wallace, not Darwin, first wrote out the *complete* theory of the origin and divergence of species by natural selection – the theory which is today universally ascribed to Darwin." According to Brackman, Wallace "was the victim of a conspiracy by the scientific aristocracy of the day and was robbed in 1858 of his priority in the proclaiming of the theory," something also facilitated by his "lowly social status" and "the nature of his character."[5] D. J. Dempster has argued that it was the Scottish breeder Patrick Matthew who first made public the idea of natural selection in 1831, by writing about "a natural process of selection." As Dempster wrote, "That *The Origin of Species* stimulated interest in evolution and natural selection throughout the world is historical fact. But Darwin claimed that the theory was entirely original and his own and that he owed nothing to his predecessors. This was clearly false, but it was and is still largely believed."[6] And more recently, A. N. Wilson has argued that Darwin stole the idea of natural selection from Edward Blyth, a pharmacist from Tooting, who had published it in magazine articles in 1837. "Darwin stole these ideas, and covered the evidence of his plagiarism by slicing the relevant pages from his notebooks." Wilson added: "Whatever his [Darwin's] motives were in cutting out the opening pages of his Notebook B, his rival in the field of transmutational theory, Edward Blyth, was conveniently removed from the scene."[7] This is in fact a claim that was first made many years ago by Loren Eiseley, who wrote:

But let the world not forget that Edward Blyth, a man of poverty and bad fortune, shaped a key that dropped half-used from his hands when he set forth hastily on his own ill-fated voyage. That key, which was picked up and reforged by a far greater and more cunning hand, was no less than natural selection. At that moment, probably in 1837, the *Origin* was born.[8]

It thus seems that Darwin is a big hero for some and a big cheater for others. A primary aim of the present book is to show that Darwin was a brilliant scholar who was neither a lonely genius ahead of his time, nor a cheater; rather he was a man of his time, somehow obsessed with the study of nature and fortunate to be able to devote his life to it as, thanks to his family's wealth, he never had to earn his living. Darwin is to be celebrated among many others; but he also

---

[5] Brackman, A.C. (1980) *A Delicate Arrangement: The Strange Story of Charles Darwin and Alfred Russel Wallace.* New York: Times Books, p. xi.

[6] Dempster, W.J. (1983) *Patrick Matthews and Natural Selection.* Edinburgh: Paul Harris Publishing, pp. 8–9.

[7] Wilson, A.N. (2017) *Charles Darwin: Victorian Mythmaker.* London: John Murray pp.104, 282.

[8] Eiseley, L. (1979) *Darwin and the Mysterious Mr. X: New Light on the Evolutionists.* New York: EP.Dutton, pp. 79–80.

needs to be depicted as what he was: a diligent and meticulous worker who was operating within a broad scientific community. A second main aim of the present book is to debunk myths and correct falsehoods about his life and work. So let us now consider what a myth is.

I.2   What Is a Myth?

The word *mythology* (μυθολογία, from the Greek words "mythos" for myth, and "logos" for account) denotes the study of myths. So, "Darwin Mythology," the title of the present book, indicates that it is about the study of myths related to Darwin. There are so many prevalent myths about Darwin that filling in a whole book with chapters debunking those was not difficult. These myths are so prevalent that several readers might be surprised to read that the photo of the cover of the present book is not an original, but an edited one. I chose this photo because it represents one of the most popular myths about Darwin: that he kept his theory secret for 20 years because he was scared. John van Wyhe analyzes this myth in Chapter 10. For a book on Darwin mythology what is better than an image illustrating a myth?

But what exactly is a myth? The Cambridge Dictionary defines myth as "an ancient story or set of stories, especially explaining the early history of a group of people or about natural events and facts," as well as "a commonly believed but false idea."[9] So a myth can simply be a false idea, but can it also be something more? As John Heilbron explained in the first chapter of the present book, myths can be more than simple falsehoods. When it comes to historical myths, it can be important to correct them in order to set the historical record straight. But it can also be important to understand why they exist and what role(s) they may have served. Perhaps you are wondering now how a false belief may have served a role. But it is exactly this service that makes myths more than just falsehoods.

What role can a myth serve? In his study of myths in ancient Greece, philosopher Cornelius Castoriadis defined a myth as "a representation by means of a narrative of the meaning which a given society endows the world with." Castoriadis then explained that a myth can have many meanings: some of these meanings can be manifest in the narrative and understanding of a particular story; and others can be latent, and related to coming to know what the myth entails. He also emphasized an important feature of myths: that a myth is carrier of an essential and universal meaning for the society in which it emerges, but also beyond that. In this sense, a myth is more than a tale exactly because of this essential and universal meaning it carries, although the

---

[9] https://dictionary.cambridge.org/dictionary/english/myth (accessed September 1, 2022)

distinction between myth and tale is not always simple.[10] With respect to historical myths, some of them are often more than false tales.

But are all myths the same? Or are there different kinds of myths? Historian of science Costas Gavroglu has made an interesting distinction among three kinds of myths.[11] The first kind can be described as *anodyne*. In these cases, there is always an element of truth, which is however emphasized more than what it is worth. For instance, whereas Galileo conducted many experiments to test the predictions of Aristotelian theory, he does not seem to have performed any experiments at the tower of Pisa.[12] Or whereas during his trial he conceded that he had perhaps overemphasized the advantages of the heliocentric system, he did criticize the geocentric system in his writings. Such episodes get more attention than they deserve, and eventually become myths. Gavroglu argues that the reason for this emphasis is often that we tend to admire heroical figures who break away from the past, and so focus on characteristic episodes that illustrate this. The second kind of myths according to Gavroglu are the *dangerous* myths. These are myths that represent deeply held conceptions, ideological commitments and political views. One such myth is that there was no intellectual activity or development during the Middle Ages.[13] Another such myth is that alchemy and astrology were superstitious pursuits that made no contributions to science.[14] And last but not least that science and religion have always been in conflict, with the latter constantly impeding the progress of the former.[15] Finally, there is a third kind of myth that Gavroglu calls *historiographical myths*. These are the myths that the first historians of science created in their attempt to correct the myths of the first and second kind. Simply put, the third type concerns features of science and its history. One such myth is that social, cultural and ideological factors are irrelevant to the study of the development of science.

The last question that will concern us is why do myths matter, or why should we care at all about myths? With respect to falsehoods, the aim can simply be to

[10] Castoriadis, C. (2004) *Ce Qui Fait la Grèce, 1. D'Homère à Héraclite*. Paris: Seuil, pp. 164–166 (translations from French are my own).
[11] Γαβρόγλου, Κ. (2004) *Το Παρελθόν των Επιστημών ως Ιστορία*. Πανεπιστημιακές Εκδόσεις Κρήτης, pp. 23–26.
[12] Heilbron, J. L. (2015). Myth 5. That Galileo publicly refuted Aristotle's conclusions about motion by repeated experiments made from the campanile of Pisa. In Numbers R.L. and Kampourakis K. (Eds) *Newton's Apple and Other Myths about Science* (pp. 40–47). Harvard University Press.
[13] Shank, M.H. (2015). Myth 1. That there was no scientific activity between Greek antiquity and the scientific revolution. In Numbers R.L. and Kampourakis K. (Eds) *Newton's Apple and Other Myths about Science* (pp. 7–15). Harvard University Press.
[14] Principe L.M. (2015). Myth 4. That alchemy and astrology were superstitious pursuits that did not contribute to science and scientific understanding In Numbers R.L. and Kampourakis K. (Eds) *Newton's Apple and Other Myths about Science* (pp. 32–39). Harvard University Press.
[15] Harrison, P. (2015). Myth 24. That religion has typically impeded the progress of science. In Numbers R.L. and Kampourakis K. (Eds) *Newton's Apple and Other Myths about Science* (pp. 195–201). Harvard University Press.

set the historical record straight. With respect to myths, no matter of what kind, I think that addressing them and analyzing them can have at least two important valuable outcomes. The first one is to understand why they emerged in the first place, and why they have persisted across time. This in turn can provide useful insights and understanding of the respective societies and cultures. The second outcome, which is also the main aim of the present book, is to refrain from perpetuating notions that oversimplify the complex social, political, cultural, historical (and so on) contexts in which science has been, and is being done, and thus provide an authentic view – or at least as authentic as possible – of the nature of science: what science is and how it is done. The notion of heroic figures (usually white and male), of "fathers" of disciplines who made the big discoveries and saved us from our ignorance, is naïve to say the least. Everyone involved in science has always had to rely on predecessors and contemporaries to achieve whatever was achieved.

Science writer Philip Ball has argued that there exist modern myths, not just old myths retold in modern times. As he explains, what distinguishes modern myths is that "these stories *could not have been told* in earlier time, because their themes did not yet exist. Modern myths explore dilemmas, obsessions, and anxieties specific to the condition of modernity."[16] Such myths stem from the stories of Robinson Crusoe or Victor Frankenstein, and according to Ball this is where we go "to explore questions that do not have definitive answers, to seek purpose and meaning in a world beyond our power to control and comprehend." We can thus "reveal some of the dilemmas and the anxieties of our age: what we dream, what we fear."[17] I will not claim that the myths analyzed in the present book are myths in the sense that Ball describes. Yet I find similarities in the uses of the modern myths in Ball's sense and the hero-worshipping myths about Darwin, Mendel and other great men of science. Could they serve the same needs as modern myths? As Ball points out, there is a key difference: modern myths are explicitly fictional, and yet encode deep truths. In contrast, "great men" stories about scientists such as Darwin are thought (by many) to be true, yet they are not. However, he agrees that there is some overlap, especially when they become storified (like the trial of Galileo, as Brecht used it, or the Scopes trial), because they both fill a perceived (if unconscious) need! (Philip Ball, personal communication). What is the need in our case? I am inclined to think that it is our need for scientific heroes who will save us from our ignorance and who will see what we cannot see. This provides the base for hero-worshiping at its best, which in turn makes some of the myths persist.

---

[16] Ball, P. (2021) *Modern Myths: Adventures in the Machinery of the Popular Imagination.* Chicago: University of Chicago Press, p. 18.

[17] Ball, P. (2021) *Modern Myths: Adventures in the Machinery of the Popular Imagination.* Chicago: University of Chicago Press, p. 19.

As quotations in each chapter in the present book show, the myths analyzed are found in textbooks or webpages to which lay people may look for information. In each chapter a detailed analysis of the respective myth provides an understanding of how it came to exist, what message it transmitted, and why addressing and challenging it is useful for a better understanding of the nature of science. As the understanding is sometimes implicit in the various chapters, in my concluding chapter I bring all these messages together to develop an account of the nature of science as it emerges from the myths related to the life and work of Charles Darwin.

Which may make you wonder: Why him?

I.3 Why Charles Darwin?

At first thought the answer may seem obvious: Charles Darwin is one of the most celebrated naturalists of all time. But this is not all. Darwin is at the same time a person that most people have heard of, and one whose actual contributions are not widely understood – and sometimes are quite misunderstood. What is also special about Darwin is that the written record, both by him and about him, is so vast that we can represent his life and his work in an astonishing level of detail. Darwin left an enormous correspondence and numerous published and unpublished writings, all of which are now available online for scholars to study.[18] Also numerous are the books that have already been written about Charles Darwin's life and work (see the Further Reading section at the end of the present book). Let us then briefly consider his life and work up to the time he started to develop his theory of evolution.[19]

Charles Robert Darwin (Figure I.1) was born on February 12, 1809, in his family house, The Mount, in Shrewsbury, Shropshire, England. He was the fifth child of the wealthy physician Robert Waring Darwin and his wife Susannah Wedgwood. His grandfathers, the physician and poet Erasmus Darwin and the potter Josiah Wedgwood I, were close friends, radical nonconformists and founders of the Lunar Society. With his mother having passed away in 1817, his three older sisters took on maternal responsibility. In 1818, Charles boarded the Shrewsbury School, even though it was less than a mile away from his family home. Being unsuccessful in school, in 1825 Charles was sent by his

---

[18] His writings can be found at the Darwin Online website (http://darwin-online.org.uk); his correspondence can be found at the Darwin Correspondence Project website (www.darwinproject.ac.uk).

[19] Although there exist numerous sources about Darwin's life and work, my account in what follows is based on information from Desmond, D., Moore, J., and Browne, J. (2007). *Charles Darwin*. Oxford and New York: Oxford University Press, which is the shortest biography available written by all three Darwin's main biographers, with some additions from Van Helvert, P., & Van Wyhe, J. (2021). *Darwin: A Companion-With Iconographies By John van Wyhe*. World Scientific, which is the most recent and the most detailed reference book available.

Figure I.1 Charles Darwin in 1857 (Photograph by Maull & Fox, c. 1857.
(DAR 225:175). With permission from the syndics of Cambridge University
Library.

father to the University of Edinburgh to study medicine, where his older brother
Erasmus had been studying medicine since 1822. Erasmus was trained as
a physician in 1825, but never practiced medicine.

Neither did Charles. His dislike for anatomy and his professors, but above all
his revulsion for surgery that at the time was done without anesthesia, led him to
neglect his medical studies. Instead, he learned to stuff birds from John
Edmonstone, a freed Black servant whom he paid a guinea an hour every day
for two months. He also became a student of zoologist and physician Robert
Edmond Grant, who was a proponent of the theories of Jean Lamarck concerning
evolution. In 1826–27, Charles took part in his studies and collections of marine
animals in the seashore close to Edinburgh. In March 1827, Darwin made his
debut presentation to the Plinian society of his own discovery that the black spores
often found in oyster shells were the eggs of a skate leech. He also attended Robert
Jameson's lectures, learning about rocks as sedimentary precipitates, while par-
ticipating in practical sessions three times a week. Edinburgh was where Darwin
became a naturalist and engaged in activities that really interested him.

In 1827, Charles enrolled in the University of Cambridge to qualify as a clergyman, his father being unhappy that his younger son had no interest in becoming a physician, and also concerned that he would waste his life and family fortune. Although several members of his family were Freethinkers, openly lacking conventional religious beliefs, Charles did not initially doubt the literal truth of the Bible. He studied for an ordinary BA degree, which was a prerequisite to doing theology training. During his studies he read William Paley and accepted his argument that design in nature proved the existence of God. There, he became good friends with his cousin William Darwin Fox, who introduced him to the competitive collecting of beetles. He also came to know John Stevens Henslow, professor of botany, whose lectures he subsequently attended. In March 1830, Charles passed his first major exams and started to work closely with Henslow, becoming his walking companion and joining him in field trips. Henslow had a major influence on Darwin, making him see plants under a different light than before. In the final BA examination of January 1831, Charles ranked tenth of the 178 poll candidates.[20] Inspired by Alexander von Humboldt's *Personal Narrative*, he planned a month's visit to Tenerife with some classmates after graduation. To prepare himself for this project, he also attended the geology lectures of Adam Sedgwick in spring 1831, and in August of the same year he accompanied him (at Henslow's request) on a fortnight of fieldwork in North Wales. This was a formative experience for Darwin, as his work with Henslow had been.

Henslow eventually recommended Darwin to Robert FitzRoy, the captain of HMS *Beagle*, to join him on an expedition to chart the coastline of South America, which would give Darwin valuable opportunities to develop his career as a naturalist.[21] His father initially objected to the voyage and Darwin refused the offer. However, in the end Robert was persuaded by his brother-in-law, Josiah Wedgwood II, to agree to his son's participation. Charles accepted the offer in September 1831 and returned to Cambridge to consult Henslow. Eventually, after many delays, the *Beagle* sailed on 27 December 1831. Charles was the captain's closest companion during the trip, despite their different views on various topics, including slavery. The *Beagle* visited various places in South America and Charles had the opportunity to study a rich variety of geological features, fossils and living organisms, and meet a wide range of

---

[20] The poll (an abbreviation of the Greek "Hoi Polloi" for the crowd) consisted of those students who took an ordinary pass degree rather than an honours degree. See van Wyhe, J. (2009). Charles Darwin's Cambridge Life 1828–1831. *Journal of Cambridge Studies*, *4*, 2–13.

[21] Although it has been claimed that Darwin initially joined as Fitzroy's companion, and not as the official naturalist of the ship, it seems that Darwin was indeed the official naturalist of the *Beagle*. See van Wyhe, J. (2013). "My appointment received the sanction of the Admiralty": Why Charles Darwin really was the naturalist on HMS Beagle. *Studies in History and Philosophy of Science Part C: Studies in History and Philosophy of Biological and Biomedical Sciences*, *44*(3), 316–326.

people. He also methodically collected an enormous number of specimens, which established his reputation as a naturalist. During the voyage, he read Charles Lyell's *Principles of Geology*, which explained geological features as the outcome of gradual processes over huge periods of time. This provided him with a theoretical scheme for his observations. After experiencing an earthquake near Concepcion in 1835, he noticed mussel-beds lifted out of the sea showing that the land had been raised about ten feet. Most famously, in 1835 he also visited the Galápagos Islands. He returned to England in October 1836, after a five-year voyage around the globe.

Between 1836 and 1842 Charles lived in London, where he set the foundations of his theory in his Notebooks. It was then and there that he put together his understanding of breeding practices, his experiences from the *Beagle* voyage and his conclusions from the theory of population by Thomas Malthus. It was in September 1838, when he read Malthus's *An Essay on the Principle of Population*, that Darwin first came up with the idea of natural selection, which was nevertheless incomplete. He spent the rest of his life in Down House, in Kent (Figure I.2), where he wrote most of his important works. That was also the place where Charles raised his children with Emma Wedgwood, whom he married in 1839. If the name sounds familiar, you are right. Emma was the daughter of Josiah Wedgwood II, Charles' uncle who had convinced Robert Darwin to give Charles his permission to travel aboard the *Beagle*. Josiah Wedgwood II was the brother of Charles's mother Susannah and the son of Josiah Wedgwood I, who in turn was Charles's maternal grandfather and Emma's paternal grandfather. So, Emma and Charles were first cousins. They spent their whole lives together, having ten children, seven of whom reached adulthood, until Charles's death on April 19, 1882.

In 1842 Darwin wrote a short "Pencil Sketch" of his theory, which was by 1844 expanded to a 240-page "Essay." His friend, the botanist Joseph Dalton Hooker, read the "Essay" and sent notes that provided Darwin with the feedback that he needed. Darwin was concerned about publishing his theory in an incomplete form, as his ideas about evolution would be highly controversial. Older ideas about evolution – especially the work of Jean Lamarck – had been soundly dismissed by the British scientific community and were associated with political radicalism. The anonymous publication of the book *Vestiges of the Natural History of Creation* in 1844, written by Robert Chambers, created another controversy over radicalism and evolution, and was severely attacked by Darwin's friends who dismissed it as unfit for reputable scientists. Darwin embarked on his study of barnacles in 1846, which earned him the Royal Society's Royal Medal in 1853. Then he continued to further develop his theory by introducing the principle of divergence. He also began writing what would be a big book called *Natural Selection*. But when in 1858 he received a letter from Alfred Russel Wallace describing a similar evolutionary process,

(a)

1. The Exterior from the Garden.—2. Mr. Darwin's Study.
THE HOME OF THE LATE CHARLES DARWIN, DOWN, KENT

Figure I.2 The house of Charles Darwin (Down House) in Kent: Darwin's study (above) and the exterior (next page). Wood engraving by J. R. Brown, 1882. Credit: Wellcome Collection. Public Domain Mark.

Hooker and Lyell arranged for a joint presentation in the Linnean society – where neither Wallace nor Darwin were present – and started writing an abstract of his theory. This was published as the *Origin of Species* in November 1859.

As is well known, there were several kinds of reactions to the publication of Darwin's theory. The most famous confrontation is said to have taken place at a meeting of the British Association for the Advancement of Science in Oxford, on June 30, 1860. According to legend, the Bishop of Oxford, Samuel Wilberforce, attempted to pour scorn on Darwin's *Origin*, receiving a forceful response from Thomas Henry Huxley. The story is that when asked by Wilberforce whether he was descended from apes on his grandfather's or his grandmother's side of the family, Huxley replied to Wilberforce that he would rather be descended from an ape than from a cultivated man who used his gifts

(b)

Figure I.2 (cont.)

of culture and eloquence in the service of prejudice and falsehood, or something like that. No written records of the meeting exist, so we cannot know what really happened. But the sensation characterizing this story has made it a famous one, and indeed one of the most cited instances of the conflict between science and religion.

In the *Origin of Species*, Darwin mentioned almost nothing about the origin of humans, which became the topic of another book *The Descent of Man*, published in 1871. There he proposed the theory of sexual selection and clearly expressed his views about human races. He also speculated about the possible origins of humans in Africa. This obviously had implications for how people understood the relation among humans at the time; being related is not the same as being equal. The philosophical implications of Darwin's theory are also important. He is considered to have banished teleology and essentialism from biology and so to have brought about a scientific revolution. For some, Darwin's theory did for humanity what Kepler's theory did for the Earth: removed it from the center of the universe to the periphery. Humans were no longer the pinnacle of creation, but only one species among others that has evolved through natural processes from common ancestors.

The chapters in the present book provide ample historical information about what Darwin did, and what influence his work had. So, I will refrain from writing anything else about him and his work, and let you explore the fascinating contributions that follow.

Myth 1   That Myths Are Simple Falsehoods

John L. Heilbron

1.1   Introduction

During the last decade, energetic debunkers have identified and destroyed dozens and dozens of myths in the history of science. They have been able to dispatch so many so easily because they include mere errors in their quarry. But an error, mistake, or false belief that has lasted only because no one has bothered to check its correctness is not *ipso facto* a myth.

To qualify as myth, a false belief should be persistent and widespread and have a plausible, assignable reason for its endurance. The myths of greatest interest have the additional attribute of transmitting a useful caricature or an inspiring illusion. Although erroneous or fabulous, such myths are not entirely wrong, and their exaggerations bring out aspects of a situation, relationship, or project that might otherwise be ignored. In support of this assertion, I summon Francesco Bianchini, the author of a history of the first thirty-two centuries of human existence and the most important astronomer in Italy during his time. Bianchini had to rely on his euhemerist interpretations of ancient myths, which, he believed, "are mostly only histories somewhat corrupted by equivocations and additions, from which, however, they are easily purged."[1]

Most of my chapter concerns three myths that meet my specifications and that I will try to purge of their distracting elements: *The Greatest Myth of Science and Religion*, *The Greatest Myth of Science*, and *The Greatest Myth of History of Science*.

1.2   The Greatest Myth of Science and Religion

According to a leading mythoclast, "the greatest myth in the history of science and religion" is the claim, made noisily by two American professors in the late

---

This chapter originated in a talk given in May 2014 at a meeting at Washington & Lee University in Lexington, Virginia. I am grateful to the editor for the invitation to publish it here and to allow it to retain the informal style of the original.
[1] F. Bianchini (1697). *La istoria universal provata con monumenti*. Rome. For the latest news about this polymath see J. L. Heilbron (2022). *The Incomparable Monsignor: Francesco Bianchini's World of Science, History, and Court Intrigue*. Oxford: Oxford University Press.

nineteenth century, that science and religion have been in a state of constant conflict.[2] Taken literally, the claim is certainly false: there are countless cases where individuals practiced their religion and their science without conflict, sometimes symbiotically. The American professors, one of whom was a historian and the other a chemist, chose to ignore periods of peaceful coexistence and coprosperity. The historian, Andrew Dickson White, blasted away with military metaphors; science was locked in a "sacred struggle" against religion, "with battles fiercer, with sieges more persistent, with strategy more vigorous than in any of the comparatively petty encounters of Alexander or Caesar or Napoleon." The chemist, John William Draper, saw the struggle as a problem in gas dynamics, a "narrative of the conflict of two contending powers, the expansive force of the human intellect on the one side, and the compression arising from tradition[al] faith and human interests on the other."[3]

It does not follow from the sometime collaboration of science and religion and the vacuity of the charge of perpetual enmity that science and religion have almost always been friends, or should or can work fruitfully together. But it has been fashionable to say so. That is to propagate a grosser and more dangerous myth than the terrible tale told by White and Draper. Their story and its Enlightenment version as delivered by Voltaire and Diderot make an enduring myth because, like a good caricature, the story contains an element of truth usefully exaggerated. Science and religion are always potentially at odds. When they are not opposed *in potentia*, neither is living up to its fundamental commitments. They must be perpetually in disagreement because they are different belief systems anchored in different and conflicting authorities. Many educated people have believed and still believe that Scripture rules out various views about the nature and structure of the physical world. Still more, perhaps, believe as their religious leaders direct.

Even where religion has abandoned natural knowledge to science, it has reserved the right to interfere in its practice. Several Christian sects, and not just fundamentalists, have sought to limit research using human stem cells. They believe that this research raises grave questions about the nature of humanity, the rights of the fetus, and the risks and ethics of medical applications. These questions lie within territory claimed by religion. Since aborted and superfluous test-tube fetuses are a primary source of important stem cells, the ethics of their use touches the great sociopolitical issue of abortion. Churches opposed to abortion ruled it out altogether, although some Catholic theologians have found

[2] R. L. Numbers (2009). *Galileo Goes to Jail, and Other Myths about Science and Religion.* Cambridge, MA: Harvard University Press, p.1.
[3] First of the Course of Scientific Lectures-Prof. White on "The Battlefields of Science," New York Daily Tribune, December 18, 1869, p. 4; J. W. Draper (1875). *History of the Conflict between Religion and Science.* London: Henry S. King & Company, p. vi. (http://moses.law.umn.edu/d arrow/documents/Draper_History_Conflict_Religion_Science.pdf).

ways that might moderate this teaching. The US Federal government during George W. Bush's presidency at first would not fund human stem-cell research and then, under pressure, agreed to support projects that used only established cell lines.[4]

I think that this is as it should be in a democracy. Organized religions should speak out against scientific work, especially if publicly funded, when they deem it contrary to their ethical commitments. The consequences are by no means always negative. Churches served as centers of opposition to the continuing improvement of nuclear weapons and helped to promote test-ban treaties. Some of them decried eugenics and unregulated cloning. Many advocate green causes and consequently challenge science policy that may result in increasing pollution, decreasing nonrenewable resources, and accelerating climate change. It will not do to argue that these examples miss the point because they concern not science but its applications. No more than applied religion, that is, social pressure from faith-based ethics, can applied science be omitted from the reckoning.

It is particularly important in today's world, when fundamentalism is on the rise and illiberal theocracies are ready to exploit every weakness in democratic regimes, that we recognize the perpetual potential warfare between science and religion. This potential is often actualized in battles over school curricula. Promoting legislation against teaching evolution is a familiar American example. Pressure to bring teachings of Islam into schools in Britain is increasing in proportion to the growth of Muslim populations there. Let us not replace the myth of constant warfare with one of inevitable peace in the relations between the two parties.

The Greatest Myth of Science and Religion inspires conflict and argument because it is supposed to apply in general. Most of the myths considered by the debunkers are not universal propositions, however, but assertions about a particular episode or fact. Some belong to the class of golden-mean myths identified by Claude Lévy-Strauss.[5] These are stories designed to resolve tensions between equally strong but opposite claims. Two familiar stories from the history of science fulfill this function to a sufficient approximation.

The first is the tale of the torture of Galileo by the Inquisition during his trial in 1633. This is certainly false if torture signifies inflicting only physical pain. There is no doubt, however, that Galileo was tortured psychologically by his interrogators' references to methods of dealing with recalcitrant sinners; and he certainly suffered from confinement for life to his small villa and from prohibitions against talking and writing about his favorite subjects. So, though

---

[4] See the essays in S. Holland, K. Lebacqz, and L. Zoloth., eds. (2001). *The Human Embryonic Stem Cell Debate: Science, Ethics, and Public Policy.* Cambridge, MA: MIT Press.

[5] G. S. Kirk (1970). *Myth: Its Meaning & Functions in Ancient & Other Cultures.* London: Cambridge University Press and Berkeley: University of California Press, pp. 44, 48–9.

literally wrong, the story of Galileo's torture is psychologically correct. Thus interpreted, it is a golden-mean myth. It has survived in its cruder form as a useful weapon in polemics against the Roman Catholic Church.

My second example is the story that Columbus was the first to teach that the world is round and that learned churchmen attacked him for this fresh insight. This fabrication has this much truth in it, however, that Columbus was the first and perhaps the last person to think that the Earth is half the size that judicious evaluators of previous measurements supposed it to be. To argue the practicability of his proposed voyage, Columbus claimed against wiser heads that the ocean between Spain and Japan could be crossed in thirty days. The event proved both him and his adversaries wrong: Columbus had not got halfway to Japan before his thirty days were up, just as Isabella's advisors predicted; but neither had Columbus perished, because in thirty days he hit the West Indies. The myth that Columbus was the first to assert that the Earth is a globe reconciles the truths that his claims were bold and unprecedented, that Isabella's advisors were conservative, and that, as we might represent it to the National Science Foundation, venture capital is sometimes needed to break the mold of accepted scientific theories.

1.3   The Greatest Myth of Science

We must not overlook that science and religion have their fair shares of myths. Choose for yourself among such splendid ones as Eve and the Serpent, Noah and his Ark, Moses and the Ten Commandments, the Burning Bush, Lot's Wife, the teachings of Freud and Jung, weather forecasts, and Theories of Everything. From this cornucopia I shall choose the Theory of Everything or, as it is known to its intimates, the TOE. It is not to be confused with a Grand Unified Theory, or GUT, which, in the peculiar anatomy of physics, is a part of a TOE. The beliefs that there exists a full and rational account of the universe and that human beings, or anyway mathematicians, can discover it, is the greatest and most inspiring myth of science.

The pedigree of this belief was long and noble before Thales proclaimed that all is water, and others, on equally solid ground, declared for air or fire. Their very distant descendants today work in the more sublime fields of mathematical physics. Should these gifted investigators succeed in finding a TOE, they almost certainly would not be able to convey the good news to ordinary human beings. Instead, we would be offered the super-myth that the world is entirely comprehensible to people unable to communicate their knowledge to anyone else. I am reminded of Thomas Carlyle's observation that people exist for whom devising a new universe is no harder than cooking an apple dumpling; although, he conceded, "there have been minds to whom the question, how the apples were got in, presented difficulties."[6] I'll mention two discarded dumplings before venturing a few words about current cookery in quantum cosmology.

My first TOE is the system of weightless fluids that constituted the basis of physics and chemistry around 1800. There were eight of them – two for electricity, two for magnetism, one each for heat and light, and fire and phlogiston for those who still believed in them. The fluids in this eightfold way were composed of particles that attracted or repelled one another in the manner of electric charges, that is, by the sort of action-at-a-distance presupposed in the theory of universal gravitation. Its most important advocate, the Marquis de Laplace, made much of the transfer of distance forces from the Heavens to the Earth, from planetary gravity to electrical charges. The system of imponderables thus was a TOE in appearance, although scarcely a GUT in practice, for the only common property of the forces was their algebraic form.

Advances in the theory of heat soon evaporated most of the weightless fluids and eliminated the distance forces they carried. Physics duly grew a new TOE, based on the conservation of energy and incorporating, as its GUT, Maxwell's synthesis of the theories of electricity, magnetism, and light. This theory of everything aimed to represent all the phenomena in the physical world in terms of matter and motion. In its most ambitious and far-fetched form, it foresaw a universe composed of primitive vortex rings (atoms) swimming about in an omnipresent incompressible fluid, just as smoke rings would do in the air if air were frictionless and inelastic. On this model, a mathematician with infinite cleverness and no end of time would be able to calculate the course of the universe.

Physicists soon discovered that even the ordinary concepts of mechanics did not suffice to describe the goings-on in the microworld, which opened its doors to them around 1900 with the discoveries of x-rays, radioactivity, and the electron. The program of mechanical reduction fell victim to quantum theory just as mechanical reduction had wiped out the system of imponderables. Many brilliant people have extended the dominion of the quantum from the microworld to the cosmos. In the process they discovered a few fundamental particles and made many more. A Standard Model (as it is called) reduced this particle zoo to a relatively few primary particles or progenitors; in its first draft, the scheme required five pairs of primaries and their antiparticles, a factoid I shall return to soon.

The Standard Model incorporates two nuclear forces and electricity, that is, three of the four forces physicists recognize, the odd one out being gravity. The most promising approach for integrating gravity with the other forces in a Theory of Everything is Superstring Theory. This scheme lodges an infinitesimal elastic

---

[6] T. Carlyle (2nd ed., 1841). *Sartor resartus*. London: J. Fraser, p. 2.

string within every fundamental particle. Calculating the wriggles of these worms and relating the results to anything observable make a task of the same exquisite difficulty and elusive glory as calculating a universe of vortex atoms. One version of the theory operates in eleven dimensions, ten of space and one of time. The tightly curled space dimensions we do not experience might be detected if we had a particle accelerator the size of the galaxy, or maybe the universe (estimates are a little wobbly in string theory), to unroll them. The little strings have no problem entering these tight dimensions, however, as they are rather small, one millionth of a trillionth of a trillionth of a trillionth of a centimeter.

Are we not in the realm of myth? I am reminded of the story of Atlantis as delivered by Plato. According to him, five pairs of twin brothers, fathered by the god Neptune on a mortal woman, ruled the island. They exercised their power to create an infinite abundance of everything under a single omnipotent law laid down by Neptune.[7] Who cannot see in this story a prophetic anticipation of the Standard Model? Like the first generation of Atlantis, the first version of the Standard Model derived infinite abundance from the activity of five sets of twins – the neutrino, electron, three sorts of quarks, and their antiparticles.

Although Atlantis did not exist, it had a place on medieval maps; and although these maps had little to do with reality, they helped to inspire the voyage of Columbus. Building on myth and mismeasurement, he sailed beyond Atlantis and discovered America. The man who first proposed a state-supported institution for scientific research, Francis Bacon, called it The New Atlantis and staffed it with Solomons whose goal was to make a TOE. Faith in the myth that everything is understandable keeps fundamental research fresh and funded.

1.4   The Greatest Myth in the History of Science

No plausible concept of science can cover the many sorts of natural knowledge developed since Thales pronounced that everything is water. Histories of the various sorts of natural knowledge are not sufficiently parallel in content or method to fall into a single narrative. Thus, the concept of the scientific revolution of the seventeenth century, during which all natural knowledge went through the same or similar transformations at roughly the same time, cannot capture the situation, any more than a history of the German-speaking peoples can be reduced to the history of Prussia.

The Scientific Revolution, which is the main episode in the history of Western science, has been a particularly fruitful historiographical concept

---

[7] Plato, *Critias*, in *The Dialogues of Plato*, ed. B. Jowett (5 vols., 1892). London: Macmillan, vol. 3, pp. 535–43.

because it is a metaphor as well as a myth. It has inspired historians to look for radical changes, sharp breaks, revolutionary heroes, martyrs, slogans: "eppure si muove," "cogito ergo sum," "nullius in verba," "plus ultra," "hypotheses non fingo." The metaphor of revolution goes well with the heroic view of history, which, as Carlyle put it in his off-putting lectures on heroes and hero worship, is the belief that progress in this world is the work of a very few great men. There is no better nursery for myth than the hothouse of hero worship; the heroes of the Scientific Revolution have become the stuff that myths are made of. The widest and most useful collection of such myths concerns the life and work of Galileo, a modern Archimedes for his mathematical approach to physical problems, a second Columbus for his discovery of a new world in the Heavens, and a new David for his slaying of the Goliath Aristotle.

According to the eminent historian of science Paolo Rossi, the Galileos devised by modern historians divide into eight major and five minor types.[8] From this abundance I have chosen two for further examination, Galileo the Faithful Platonist and Galileo the Modern Physicist.

Galileo the Platonist is largely the creation of Alexandre Koyré, who taught the world that the history of science is more than a chronicle of the discoveries and inventions of unprejudiced minds. Not that Koyré looked beyond ideas for motivations; he began life as a historian of philosophy and wrote history as if novel ideas interact only with other ideas as they propagate through resistant minds. The work of historians of science is to tease out prejudicial preconceptions and show how they evolved into painfully won discoveries.

The prejudice that filled the mind of Koyré's Galileo was the idea that nature is constructed on mathematical principles. Consequently, he tried to transform natural knowledge from its preoccupation with causes, of which Aristotle had prescribed a superfluity, to the more modest task of mathematical description. By luck and logic, he hit on his law of free fall, from which, in time, he worked out the geometrical trajectory of a projectile flying without resistance. This recondite information pleased mathematicians, as it extended the scope of their subject; irritated philosophers, as it diminished theirs; and made no difference to gunners, who discovered nothing useful in Galileo's flight of fancy. As for Galileo's famous experiments with free fall (Figure 1.1), whether from towers or ship masts, or of oblique fall along inclined planes, most were either thought experiments, that is, pure fabrications, or after-the-fact demonstrations of conclusions Galileo had reached by reason and intuition alone.

Koyré's caricature captured much of importance about Galileo's ideas about motion, without, however, betraying much about what moved Galileo. The purely intellectual Platonic Galileo quickly became a master myth with profound consequences for the development of much of our field. In addition to its

---

[8] P. Rossi (1994). "Immagini di Galileo," *Nuncius*, 9(1), 3–14, at 5–9.

Figure 1.1 Galileo's demonstration at the Tower of Pisa.[9] Source: Print
Collector/Contributor/Hulton Fine Art Collection/Getty Images

extension by Koyré and others to Descartes and Newton, it inspired the
research and even the choice of profession of some of the founders of the
academic study of the history of science in the United States. Koyré's caricature
appealed to people trained in physics who valued theory above experiment and
science above society.

Empiricists preferred Galileo the Modern Physicist as imagined by Stillman
Drake. This version of the hero did physics in a manner differing only slightly
from the way Ernest Lawrence and his cyclotroneers practiced it at the
University of California when Drake was a student there in the 1930s. His
main study then taught him to despise academic philosophizing. The way to
knowledge was to follow Lawrence, experiment with an open mind, confirm
and reinforce the results, and try to subsume them under a general law. Finding
in Galileo's writings a clear statement of this method and believing that Galileo

[9] The kernel of truth in this enduring myth is extracted in J. L. Heilbron (2015). Myth 5. That
Galileo publicly refuted Aristotle's conclusions about motion by repeated experiments made
from the campanile of Pisa. In R. L. Numbers and K. Kampourakis (Eds.) *Newton's Apple and
Other Myths about Science* (pp. 40–7). Cambridge, MA: Harvard University Press.

had achieved his results by its rigorous practice, Drake recognized in him the first manifestation of the scientist in the history of civilization.

This recognition implied a belief in the reliability of Galileo's reports of experimental results. Since the inventor of the tale of the Leaning Tower, Vincenzo Viviani, claimed to have heard it from Galileo, Drake was inclined to accept it. More importantly, he accepted Galileo's reports of trials on inclined planes that confirmed the law of fall to a hair's breadth and a heart's beat. The possibility of this exactness could be tested. Several resourceful historians did so and, much to the surprise of Koyré's followers, showed that Galileo could have found his famous law by induction from experiments on inclined planes. According to another anecdote transmitted by the practiced mythologizer Viviani, Galileo discovered the isochronism of the pendulum by counting his pulse while observing swings of a chandelier in the Cathedral of Pisa. An elaboration of the story ascribes the swinging to an earthquake and the discovery to Galileo's courage to pursue new knowledge in the face of grave danger.

Drake's efforts to establish his caricature of Galileo became a lifelong productive obsession. He turned himself into the chief authority on Galileo's manuscripts, translated Galileo's major works into good English to such effect that Anglophone readers may easily mistake the father of physics for a modern thinker, and wrote many informative articles and monographs to confirm this impression. His abiding interest, as he put it in the title of one of these monographs, was *Galileo at Work*. Drake's Galileo always follows the scientific method, never claiming anything that he could not support by "sensate experience and necessary demonstration." So armed, he won all his disputes, especially with Jesuits, and when he allowed himself to speculate, he anticipated Newtonian physics.

The propriety of the concept of "the Scientific Revolution" is often attacked while its parent myth, that there is a single history of science, escapes almost unscathed. Indeed, the parent myth has prospered and enlarged as historians of the social sciences and of non-European natural knowledge shelter under its umbrella. Perhaps it is a good thing that our fundamental myth remains vital. It is a Grand Unifying Theory. It keeps the profession together, increases resources, augments impact, and inspires excellent work, although the writing of a coherent and universal history of science is as remote as a recreation of the Big Bang.

1.5   An Ovidian Twist

Errors that serve or qualify as myths imply something significant about the history they traduce. Indeed, they have helped to make history in so far as historical actors believed in them. The grand myth of incessant warfare between science and religion helped underpin the still-powerful opposing myths of value-free science and creation science. The grand myth of the Big Bang Theory of Everything, recently reaffirmed by the detection of gravitational waves, continues to drive invasive investigations into the birth and growth of the universe. The great myth of the history of science has produced, in the scholarship devoted to the Scientific Revolution, the most thorough and competent historiography in our field.

Ovid's *Metamorphoses* teaches that things are not what they seem. What you see as a cow might be Io, one of those fresh country lasses that Jove delighted to ravish. Jove's wife, Juno, who, shamefully, was also his sister, had her suspicions about poor Io. "'If I'm not wrong,' she thought, 'I'm being wronged.'" She flew down from Olympus to investigate. As she approached his trysting ground, Jove transformed Io into a cow. Juno took a fancy to the animal and asked to have it. What should Jove do? Ovid describes his dilemma: It would be "Too cruel to give his darling! Not to give / Suspicious; shame persuades but love dissuades / Love would have won; but then–if he refused / His wife (his sister too) so slight a gift / A cow, it might well seem no cow at all!"[10]

Ovid's accounts of transformations of men and women into animals are full of details calculated, as W. S. Gilbert makes Pooh-Bah say in the *Mikado*, "to give verisimilitude to an otherwise bald and unconvincing narrative." It is just this sort of detail that, according to the eminent historian of antiquity, Paul Veyne, persuaded the Greeks to ascribe some truth to their myths.[11] But Ovid appears to have had a higher reason than decorating myths for his step-by-step descriptions of the transformation of people into animals. The last book of the *Metamorphoses* contains a long account of the philosophy of Pythagoras. Ovid did not include it to advertise Pythagoras' views about "The great world's origin, the cause of things / What nature is, what god, and whence the snow."[12] He had in mind the relevance to myth of the Pythagorean doctrine of the transmigration of souls.

"There is no death," Ovid has Pythagoras say, "but only change ... what we call birth is but a different new beginning." Our souls may return in animals: the ox, the farmer's brother in toil at the plough, may have been his brother in fact. Therefore, Ovid's Pythagoras thunders, Abstain! Eat not the flesh of your fellow creatures! Heed my message! Abstain![13] Some unimaginative people have interpreted this injunction as a call for a vegetarian diet. They may be right. But it is more certainly a warning against the wanton slaying of myths. Therefore, take care! They might contain something of value. Heed my message! Things are not always what they seem. The myth you slay today may contain a truth you need tomorrow.

[10] Ovid. (1632). *Metamorphoses*, transl. George Sandys. London: J. Lichfield, p. 19.
[11] P. Veyne (1983). *Les grecs ont-ils cru à leurs myths?* Paris: Seuil.
[12] Ovid, *Metamorphoses*, 354.   [13] Ovid, *Metamorphoses*, 357, 359.

Myth 2   That Most European Naturalists Before Darwin Did Not Think That Species Change Was Possible

Pietro Corsi

2.1   Introduction

It is commonly thought that almost nobody before Darwin had thought that species change was possible. For instance, in an article about species in *The Conversation*, we read that "Before Charles Darwin, nearly all scientists believed that life on earth, including humans, was created by God thousands of years earlier and had remained unchanged over time."[1] This should be no surprise, however, as this is a view that Darwin himself instigated. The opening sentence of the Historical Sketch Darwin prefaced to the third edition of the *Origin of Species* sums up the commonly shared conviction that nothing much had happened before 1859, and what did, was not really worth talking about:

> I will here attempt to give a brief, but imperfect sketch of the progress of opinion on the Origin of Species. The great majority of naturalists believe that species are immutable productions, and have been separately created. This view has been ably maintained by many authors. Some few naturalists, on the other hand, believe that species undergo modification, and that the existing forms of life have descended by true generation from pre-existing forms.[2]

Stuff for fastidious historians, in other words. Darwin's account of the "few" who did not believe in separate creations has contributed, and is contributing, to the perpetuation of this myth. His "Historical Sketch," by the way, was not that sophisticated or accurate – the one produced by Isidore Geoffroy Saint-Hilaire (Figure 2.1) in volume two of his *Histoire Naturelle Générale des Règnes Organiques* that Darwin himself perused, for instance, was a philologically accurate and perceptive survey of standpoints on the species question from Linnaeus to the 1850s.[3] The shorter survey offered by Alexandre Godron in *De*

---

[1] T. M. Crowe, M. Muasya, and T. G. Mandiwan-Neudani (2016). The long struggle to understand species: from pre-Darwin to the present day. *The Conversation*, 6.

[2] C. Darwin (1861). *On the Origin of Species by Means of Natural Selection* (3rd ed.). London: John Murray.

[3] Isidore Geoffroy Saint-Hilaire (1859). *Histoire Naturelle Générale des Règnes Organiques*. Paris: Masson, 3 vols., 1854–62, vol. 2, ch. VI, pp. 365–446.

Figure 2.1 Isidore Geoffroy Saint-Hilaire. Credit: Photograph by
Franck. Wellcome Collection. Public Domain Mark

l'Espèce (1859), a further source for Darwin, was equally remarkable, though
less informed than the one compiled by Isidore.[4] Darwin might also have
forgotten that in late Spring 1845 he had read the entry "Espèce," written by
Frédéric Gérard for the *Dictionnaire Universel d'Histoire Naturelle*, which
started with a survey of the question of species, listing authors, works and
numbers of pages. One that even Isidore appreciated and quoted.

We are of course grateful to Curtis Johnson for his painstaking reconstruction of the when and how Darwin had crossed path with each of the thirty-five
people mentioned in the Sketch – I say people, because only some of the
authors referred to were or had been naturalists.[5] Even though the incipit of
the Sketch gives the impression that Darwin was going to comment on the

4 Alexandre Godron (1859). *De l'Espèce et des Races dans les Êtres Organisés et Spécialement de
   l'Unité de l'Espèce Humaine*. Paris: J.B. Baillière, 2 vols., vol. 1, pp. 1–14.
5 Curtis N. Johnson (2019). *Darwin's Historical Sketch: An Examination of the Preface to the
   Origin of Species*. Oxford: Oxford University Press.

"few" naturalists who had argued for descent, the final choice offered a spectrum of opinion put forward by a varied population of actors, from naturalists like Jean-Baptiste Lamarck (Burkhardt, Chapter 8, this volume) or Richard Owen (Rupke, Chapter 14, this volume), and indeed his grandfather Erasmus (Fara, Chapter 3, this volume), to theological and philosophical polemists like Baden Powell (whom Johnson promoted to the rank of Bishop) and amateur commentators like Henry Freke.

Though it would be unfair to accuse Darwin of reticence, his Sketch gives the impression that here and there views were occasionally put forward casting doubt on the direct creation of every single species, or favoring more or less sophisticated alternative explanations – some even heralding the concept, if not the term, of natural selection. According to Darwin, there had been no previous substantial debate on species, their origin and possible changes over time or geography, worth talking about. A view that even a few historians share to this day. Once again, reticence if perhaps too strong a word, but the way in which Darwin referred to, or omitted authors he knew quite well, calls for some form of explanation. I will say more below on the case of Gérard and I will presently limit my remarks to the case of Jean-Baptiste Bory de Saint-Vincent. Darwin mentioned his name in a footnote, stating that apparently, according to Godron, the French naturalist did believe that "new species are continually being produced" (1985, p. xviii). Mislead by Darwin's elusive reference, Johnson concluded a dense discussion of sources Darwin could have been familiar with concerning Bory, by stating: "If one had asked Darwin in 1860 what he knew about Bory's published writings, he would have had to confess: 'nothing at all'" (p. 324). Which is of course an overstatement.

In the short survey that follows, I argue that since the last two decades of the eighteenth century, with different timing and emphasis, various authors active in the European cultural space engaged in conversations on the origin of life and the possibility, and extent, of change organisms might undergo throughout time and space. As Darwin himself had done, it is impossible to limit such a survey to the "professional" naturalists (however defined) and to the perusal of specialist publications in several branches of natural history. Reading works of natural history, especially Buffon's *Histoire Naturelle*, and offering comments on the one or the other feature of the work, was a kind of collective undertaking within the European Republic of Letters, involving a spectrum of authors ranging from theologians and men of letters to natural philosophers and members of the medical profession. In fact, one may argue that Darwin was most probably right at least on one point: a good number of "professional naturalists" (collectors, authors of local or global floras and faunas) kept at a safe distance from contentious and potentially dangerous topics and on the whole assumed species immutability. Many also wished to distance themselves from the advancing tide of popular works they found scientifically inacceptable and yet attracted the attention of the reading public, as Robert Chambers' *Victorian Sensation* abundantly proved.[6]

2.2 Beginning with France

If a number of "naturalists" were silent or fiercely critical, generalist periodicals, dailies, dictionaries, encyclopedias, ambitious narratives of natural phenomena addressed to the educated public, as well as an equally consistent number of naturalists, were not, and engaged in extensive exchanges on the question of species. In the country I am more familiar with, France, the persistence with which major editorial enterprises, dictionaries and encyclopedias in particular, kept addressing the issue of species indicates that readers were prepared to subscribe to major works that offered – among other – philosophically satisfying accounts of living nature and its endless transformations. In other words, the demand very much conditioned the offer.

This claim counters a further, persistently shared view, that the country of Lamarck proved hostile to Darwin and to evolution: as Liv Grjebine demonstrates in her chapter (Chapter 19) in the present volume, academic resistance to Darwin and to evolution after 1859 represented a fraction of the public debate. Many different voices made themselves heard, debating the *Origin of Species*, but also works by Herbert Spencer (Depew, Chapter 15, this volume) or Ernst Haeckel (Richards, Chapter 20, this volume) were made easily available in French language translations. There is sufficient evidence to substantiate the claim that a varied and rich public debate on species, constantly adapting to changes in natural history disciplines, in philosophy and in the relationship between scientific cultures and politics, constituted a continuum in nineteenth century France. I will thus mainly deal with France, before very briefly arguing that the situation was not much different in the German-speaking countries, Italy, Spain and the British Isles.

I do not discuss Lamarck in any meaningful detail, a task Chip Burkhardt has taken on in his chapter.[7] I rather refer to the ways in which Lamarck's work and name were used, by summarily concentrating on four popular encyclopaedias published from the early 1800s until the 1850s. The authors of successive entries often made explicit or implicit reference to articles published elsewhere,

---

[6] James A. Secord (2003). *Victorian Sensation: The Extraordinary Publication, Reception, and Secret Authorship of Vestiges of the Natural History of Creation*. Chicago IL: The University of Chicago Press.

[7] I have discussed Lamarck at length in Pietro Corsi (1989). *The Age of Lamarck: evolutionary theories in France 1790–1830*. Berkeley CA: University of California Press, and in Pietro Corsi (2001). *Lamarck. Genèse et enjeux du transformisme, 1770–1830* (French language expanded ed.) Paris: CNRS Editions.

thus reinforcing the impression that we are not faced with occasional pronouncements, but with purposeful involvement in an ongoing conversation. Since the mid-1840s, some of the participants to this debate even felt the need to monitor its historical development, to write their own "historical sketch." It is finally to be recollected that, due to the marketing strategy for expensive multi-volumes encyclopedias, which were sold in instalments, key entries were often reprinted as separate pamphlets, thus generating additional income for the publishers and additional circulation. Darwin and Hooker obtained their copies of Frédéric Gérard's 1845 entry "Espèce," which we discuss below, as a separate pamphlet. Particularly long entries were also published in book format, as for instance the *Principes d'Organogénie* by Serres, originally printed in the *Encyclopédie Nouvelle*; Isidore Geoffroy Saint-Hilaire printed his contribution to encyclopaedias as a separate volume of essays.[8]

2.3   Julien-Joseph Virey's "Corps Organisés" (1803)

I begin my diachronic survey with Julien-Joseph Virey, one of the most prolific science writers and popularizers of the first half of the nineteenth century. Out of the many entries he contributed to the *Nouveau Dictionnaire d'Histoire Naturelle* (1803–04, 2nd ed., 1816–19), I have selected a particularly poignant quotation from the article "Corps Organisés":

A single germ, by developing successively and creating a great number of similar individuals, in the long span of centuries will have seen them modify themselves little by little, and turn into species more or less similar to each other due to the influence of climates, temperatures, &c. These species were further modified by the succession of ages as they experienced the long and profound influence of all that surrounds them, and as they mixed with each other. These mixtures, these variations, these species, will go on becoming subdivided; for one day, let there be no doubt, what we regard as varieties, will become a species which will still have its varieties.[9]

Virey was not preceding Darwin with the claim that varieties were incipient species. He was instead commenting on works published in the previous few decades, though neither names, nor works were mentioned. The passage quoted above ended Virey's discussion of Buffon's and Diderot's tenets concerning the possibility of considering all living beings as the successive transformations of an initial prototype. In the entry "L'âne" of his *Histoire naturelle*, Buffon had formulated the hypothesis that the web of similarities joining species within a genus could be further investigated for higher classification categories,

---

[8] Étienne-Renaud-Augustin Serres (1842). *Principes d'Organogénie*. Paris: Charles Gosselin; Isidore Geoffroy Saint-Hilaire (1841). *Essais de Zoologie Générale*. Paris: Roret.

[9] Julien-Joseph Virey (1803). Corps organisés, *Nouveau Dictionnaire d'histoire Naturelle*, 6, p. 268.

authorizing the conclusion that all forms of life originated from a primordial organism.[10] Whereas Buffon immediately rejected the hypothesis, Diderot capitalized on it to theorize the descent of all organisms from a handful of protypes.[11]

In the cultural and political climate of the early 1800s, Virey could not quote Diderot (he himself had been accused of being a Spinozist, that is, a fellow traveler of atheists) and avoided mentioning Buffon, whose heritage was at the center of a tense scientific and political debate. It is also possible that Virey, who was very familiar with natural history and medical publications in English, had read Erasmus Darwin, a likely source for the idea that the single germs giving rise to all plants and animals could be in fact reduced to a single primordial filament. He also implicitly endorsed the Linnaean acknowledgment that the huge number of species being discovered throughout the globe may be due to a natural cause: a small number of originally created forms could have produced viable hybrids when confronted with constant changes in geo-climatic conditions. The view that varieties were incipient species was Virey's personal addition to the debate, one that entries in later French encyclopaedic compilations took up again and again.[12]

2.4   Achille Requin's "Animal" (1834)

Virey was also the first author writing for an educated audience to summarize the early evolutionary writings by Lamarck, the *Recherches sur l'Organisation des Corps Vivants*, published in 1802, in particular. Increasingly and inevitably, Lamarck became an author many encyclopedia articles referred to, almost invariably with the utmost respect, even when the authors did not share his views. Starting in the mid-1820s, Étienne Geoffroy Saint-Hilaire's polemical exchanges with Cuvier, which included reference to the question of species, also became material for comment. One single author, belonging to a new generation of science writers, will retain our attention as reflecting the new priorities in natural history debates: Achille Requin. A medical doctor, in 1834 he contributed the entry "Animal" to the *Encyclopédie Pittoresque à Deux Sous*, later on *Encyclopédie Nouvelle*, edited by Jean Reynaud and Pierre Leroux. Requin was an impressive medical student who had fallen under the charm of Geoffroy Saint-Hilaire and of Etienne Serres. Concerning the origin and transformations of life, Requin sided with Lamarck even after 1833, the year in which Geoffroy started to take his distance from the naturalist he once publicly declared

---

[10] Georges Louis Leclerc de Buffon (1753). *"L'âne", IV. Histoire Naturelle, Générale et Particulière*. Paris: Imprimerie Royale, 36 vols., 1749–89, pp. 381–3.

[11] Denis Diderot (1754). *Pensées sur l'interprétation de la nature*, pp. 33–7.

[12] Achille Requin (1834). "Animal," *Encyclopédie Pittoresque à Deux Sous*, 1, pp. 554–64; Frédéric Gérard (1845). "Espèce," *Dictionnaire Universel d'Histoire Naturelle*, 5, pp. 428–52.

far superior to Cuvier.[13] Geoffroy, a good friend of the editors, did not appreciate and Requin ceased his collaboration to the *Encyclopédie*.

In his 1834 article "Animal" Requin reproduced, without acknowledgment, several passages from Lamarck's *Philosophie Zoologique*. He reminded readers that changes in the environment induced changes in habits, which in turn modified organs. Expanding upon Lamarck's passing reference to domestication as evidence of species mutability, Requin also asserted that there was a further law of organic change in action, following which "variations accidentally acquired by individuals of one species are transmitted through heredity, if these individuals mate among themselves. To this law we owe the multitude of domesticated races we have produced thanks to the diversity of climate, nourishment, education, etc."[14]

2.5   Frédéric Gérard's "Espèce" (1845)

A third article deserves our attention, as representing the state of affairs during the 1840s: the entry "Espèce" Frédéric Gérard published in 1845 in the *Dictionnaire Universel d'Histoire Naturelle* (1839–49), edited by Charles Dessalines d'Orbigny. The article was the occasion for an exchange of letters between Darwin and Joseph Dalton Hooker to which we refer below. The notion Virey had sketchily put forward in 1803 – that varieties were incipient species, and that given enough time organisms would keep splitting up – was fully endorsed by Gérard, who took it for granted: "I am thus convinced, with Lamarck, Poiret, and Geoffroy, that varieties become Species, and that it is in this way that new Species are formed which throw so much hesitation and uncertainty in the science."[15]

Whereas Virey had indulged in verbose theoretical flights, Gérard reflected new standards of natural history writing for the educated public. Much of his article was taken up by a long discussion of the difficulty naturalists experienced to distinguish between species and varieties, listing case upon case. He also surveyed the literature on hybrids. His commitment to evolution was an original combination of elements taken from Lamarck, Geoffroy Saint-Hilaire, Bory de Saint-Vincent and Jean Louis Poiret, a botanist who had collaborated with Lamarck on the completion of the botany volumes of the *Encyclopédie Méthodique*. Many examples of the uncertainties of species identification in the vegetable kingdom were taken from Poiret, though Augustin Pyramus de

---

[13] Pietro Corsi (2011). "The Revolutions of Evolution: Geoffroy and Lamarck, 1825–1840." *Bulletin du Musée d'Anthropologie Préhistorique de Monaco*, 51, pp. 97–122.

[14] Jean-Baptiste Lamarck (1809). *Philosophie Zoologique*. Paris: Dentu, 2 vols., vol. 1, p. 236; Requin, "Animal," p. 563.

[15] Gérard, "Espèce," p. 447. Marie Huiban (2007). *Frédéric Gérard (1806–1857) Transformisme et républicanisme dans la France des années 1840 et 1850*. Paris: Université Paris 1, Mémoire de licence; Goulven Laurent (1987). *Paléontologie et évolution en France de 1800 à 1860*. Paris: Éditions du Comité des travaux historiques et scientifiques.

Candolle was also a favorite source. Like several authors before him, Gérard accepted the hypothesis formulated by Linnaeus that crossings among existing organisms were an additional source of speciation.

He concluded his dense and well-informed contribution by stating his theoretical commitment:

> The facts, far from confirming the criterion established by naturalists for the determination of the Species, agree in demonstrating that Species are neither eternal nor immutable, but essentially mobile; that the organic forms, corresponding to the different degrees of evolution of living bodies, are susceptible of modifications, the limits of which are unknown to us, and which are produced by the influence of the environments, the transmission by way of generation of the qualities acquired, and the crossing of neighbouring species.[16]

Gérard also offered an account of definitions of species put forward by a variety of naturalists since the time of Linnaeus, with the intention of sketching the history of the subject. Indeed, Gérard did not feel that the discussion he was taking part in was made up of rhapsodic outburst of theoretical fervor, but represented a consistent, albeit highly diversified tradition of theoretical engagement in natural history.

When in 1853 the Catholic priest Louis-François Jéhan contributed a *Dictionnaire d'Anthropologie* to the grandiose apologetic editorial enterprise started by the Abbé Jacques-Paul Migne, he reprinted a body of French, and German texts translated into French, guilty of propagating infidel science. Since space was not an issue in a volume spanning 1593 columns, Jéhan printed *verbatim* "Génération Spontanée," Gérard's contribution to the *Dictionnaire Universel*. Since the article run for pages and pages, a reader happening to open the volume in the midst of Gérard's text, would have been taken by surprise at reading elated professions of materialist faith in a conservative Catholic publication.[17] In a previous, even more ponderous work, the 3 volumes, 2665 page-long *Dictionnaire de Zoologie* (1852–3), Jéhan wrote a lengthy preface to volume 2 discussing in detail the state of affairs in the debates on species.[18] Much of the information he was relying upon came from the dictionaries we have reviewed, including the *Nouveau Dictionnaire d'Histoire Naturelle* edited by Virey and the *Encyclopédie Nouvelle*. The fact that members of the scientific establishment kept on the whole silent concerning the complex discussion on species change, meant nothing to a minister of the Church determined to provide the clergy with means to defending Revealed Truth against aggressions in the name of science.

---

[16] Gérard, "Espèce," p. 452.

[17] Louis-François Jéhan (1853). *Dictionnaire d'Anthropologie: ou Histoire naturelle de l'Homme et des Races Humaines.* Paris: J-P. Migne, pp. 389–540.

[18] Louis-François Jéhan (1852). "De l'origine des êtres organisés," *Dictionnaire de Zoologie.* Paris: J.-P. Migne, 3 vol., 1852–3, vol. 2, pp. 10–98.

2.6   The Omissions in Darwin's "Historical Sketch"

To finish this survey where we started, when Darwin looked at Isidore Geoffroy Saint-Hilaire's and Godron's historical accounts of debates on species he must have felt rather unequal to the task of writing his own historic sketch. The historical detail and bibliographic proficiency shown by Isidore were indeed remarkable. As we stated above, Isidore praised Gérard's historical sketch, but found him overzealous. At times, Isidore's critique reads like the remarks Hooker sent to Darwin concerning the French author: a compilator rather than a professional.[19]

Darwin's attitude to Gérard and to Bory deserves a short comment, in light of the fact that in the Historical Sketch he dismissed Bory and simply omitted Gérard. Darwin had asked Hooker about Bory's contention that volcanic islands isolated in the middle of the oceans showed a very high number of "polymorphous" plants – to him, the indication of recent spontaneous generation activity going on to stock new lands with life forms. To protect himself from appearing to find something of interest in a foreign amateur, Darwin defined Bory an idiot, but he still asked the question. Even Hooker found Darwin's attitude against Bory a bit extreme. With respect to Gérard, Darwin was even more ambiguous.

Hooker had lent his own copy of the article "Espèce" reprinted as a pamphlet.[20] Darwin had marked it and sent it back to his friend. Soon afterwards, and for two times, Darwin asked Hooker to send the text back to him, so that he could transfer the marks and annotations he had made on the copy he had now acquired – this would have spared him the task of perusing again such a worthless publication. Bory was stupid, Gérard worthless, and yet the amount of attention bestowed on their texts appears to contradict the harsh judgement. No space, needless to say, to further discuss this fascinating encounter. What is equally interesting is the speed with which Gérard's text reached the British Isles and Darwin. Camille Montagne, a leading amateur Parisian botanist, had sent Gérard's pamphlet soon after publication, in late spring 1845, to William Jackson Hooker, who forwarded it to his son Joseph Dalton, who in his turn mailed it to Darwin, knowing the latter's interest in the species question.

In spite of what Darwin said of the article, he expressed the wish to obtain a copy for himself, and Montagne was happy to oblige. Darwin also bought the separate printouts of Gérard's entries "Génération spontanée" and "Géographie." The articles were rich in accurate bibliographic references, and clearly contradicted Darwin's later claim that very few naturalists doubted

---

[19] For all the letters referred to, see The Darwin Correspondence Project, www.darwinproject.a
c.uk.

[20] Janet Browne (1995). *Charles Darwin: Voyaging*. London: Jonathan Cope, Ch. 19.

the stability of species.[21] Worthless as he was deemed to be, Gérard managed to attract quite a share of attention on the two sides of the Channel, without mentioning the money spent in buying copies and sending them around in the space of a few weeks.

2.7   Beyond France

I hinted in the opening remarks that the same method of systematic perusal of a wide spectrum of publications may bring about significant results also for other European countries. To use a euphemism, I am less familiar with the German scene, and more generally with publications in the German language: I find it safer to rely on Bob Richards' work on the legacy of *Naturphilosophie* (broadly speaking) on Darwin.[22] Nicolaas Rupke also adds new critical appreciation of the German legacy to contemporary evolutionism in his chapter in the present volume. My research has been limited to representatives of amateur engagement with systems of nature and evolutionary speculations such as Johann Christian Rödig, or the Italian Giuseppe Gautieri, working in Jena and interacting with Goethe and Schelling.[23] I have also been particularly interested in the wave of translations of German medical and physiological works into French, mainly undertaken by the publishers Charles Pancoucke and the Baillière brothers, and entrusted to Antoine Jourdan during the period 1810–40. Works by Tiedemann, Treviranus, Burdach, Carus introduced French and worldwide readers to prominent representatives of German life sciences and their views on spontaneous generation or the limits and quality of species change. Not to multiply names – which would be rather easy – it suffices to say that Gérard quoted Burdach from the translation by Jourdan.[24]

My conviction that a systematic perusal of encyclopaedias and periodicals in German would yield results comparable to the ones produced by reading through French sources has found confirmation in the work of Albrecht Rengger, a Swiss diplomat and amateur geologist writing in German, whose article on the progress of geology published in 1826 in the *Edinburgh New Philosophical Journal* has been attributed in turn to Robert Edmund Grant and

[21] Frédéric Gérard (1845). *De la Zoogénie et de la distribution des êtres organisés à la surface du Globe*. Paris: Privately printed.

[22] Robert J. Richards (1992). *The Meaning of Evolution: The Morphological Construction and Ideological Reconstruction of Darwin's Theory*. Chicago IL: Chicago University Press; Robert J. Richards (2002). *The Romantic Conception of Life: Science and Philosophy in the Age of Goethe*. Chicago IL: Chicago University Press; Sander Gliboff (2008). *H. G. Bronn, Ernst Haeckel, and the Origins of German Darwinism*. Cambridge MA: MIT Press.

[23] Pietro Corsi (2005). "Before Darwin: Transformist Concepts in European Natural History." *Journal of the History of Biology*, 38, pp. 67–83.

[24] P. Pietro Corsi (2018). "'Systèmes de la nature' and Theories of Life. Bridging the Eighteenth and the Nineteenth Centuries," in J. Riskin (ed.), *Early Adventures in Evolution, Republics of Letters*, 6 (1). https://arcade.stanford.edu/rofl_issue/volume-6-issue-1.

more recently to Robert Jameson, the journal editor. Rengger, a conservative liberal politician provided interesting information of the kind of authors and works an educated German (and French) speaking reader considered as relevant to the question of species.[25] Thus, for instance, in 1824 Rengger, but also Hermann Schaafhausen in 1853, considered Gian Battista Brocchi's theory of species senescence a proposition to be evaluated, as Darwin and Lyell had done in the 1830s.[26]

Much has been written on the debate on species in the British Isles before Darwin. Many years ago, I called attention upon Baden Powell, the Professor of Geometry at Oxford who spent his life in trying to conciliate scientific progress and Anglican theology. Already in 1838 Powell referred to the "anxious debate" on species change, and enthusiastically endorsed *Vestiges* and *Origin of Species*.[27] As far as the French authors and publications I have briefly discussed in this chapter, it is appropriate to note that Baden Powel and many others, Richard Owen *in primis*, knew Geoffroy Saint-Hilaire well, French still being a language every educated reader was familiar with. An author who clearly made use of French encyclopaedias and of German authors in French translation was Sir Richard Vyvyan. The fact that his *Harmony of the Comprehensible World*, published in 1845, was privately printed and enjoyed very limited circulation is irrelevant with respect to the testimony the work offers as to the reading list available to a rich, admittedly idiosyncratic reader. Vyvyan even quoted an article that Virey had published in the *Nouveau Dictionnaire d'Histoire Naturelle* on the influence of the Moon on mankind. Finally, in my recent paper on the so-called "Edinburgh Lamarckians" I have shown the persistence and promptness with which the *New Edinburgh Philosophical Journal* translated articles from French publications.[28]

A recent excellent work on pre-1859 debates on species in Turin, shows the extent to which authors and themes we touched upon in this paper circulated in an important capital of pre-Unity Italy.[29] The article "Animal" Requin had published in 1834 was even translated in full (without acknowledgment) in a natural history textbook authored by Francesco

[25] P. Corsi (2021). "Edinburgh Lamarckians? The authorship of three anonymous papers (1826–1829)." *Journal of the History of Biology*, 54(3), 345–74.

[26] Albrecht Rengger (1824). *Beyträge zur Geognosie, besonders zu derjenigen der Schweiz und ihrer Umgebungen*. Stuttgart and Tübingen: J. G. Cotta; Hermann Schaafhausen (1835). "Ueber Beständigkeit und Umwandlung der Arten," *Verhandlungen des Naturhistorischen Vereins der preussischen Rheinlande und Westphalens*, 10, pp. 420–51.

[27] Baden Powell (1838). *The Connexion of Natural and Divine Truth: or, the Study of the Inductive Philosophy Considered as Subservient to Theology*. London: J.W. Parker, pp. 150–1; P. Corsi (1988). *Science and Religion: Baden Powell and the Anglican Debate*. Cambridge: Cambridge University Press.

[28] Corsi, "Edinburgh Lamarckians?"

[29] Fabio Forgione (2018). *Il potere dell'evoluzione. Il dibattito sulla variabilità delle specie nella Torino dell'Ottocento*. Milan: Franco Angeli.

Costantino Marmocchi.[30] Finally, a book just published by Agustí Camós Cabeceram on debates on evolution in Spain during the nineteenth century has provided evidence of similar unacknowledged translations of pro-Lamarckian entries from French encyclopaedias in the Spanish nonspecialist literature.[31]

2.8   Conclusion

Far from being negligible, let alone nonexistent, debates on the origin and stability of species, on spontaneous generation, on the history of life on Earth were common currency within European literate audiences before the publication of the *Origin of Species*. The publication of Darwin's work persuaded many that the convictions concerning the transformations of life they had read about during the previous years or decades were finally vindicated by an authoritative and much respected naturalist. Almost invariably, authors and readers engaged in long-standing debates were not interested in the "details" of Darwin's work. A great number of "converted" did not even bother to read the *Origin of Species*, nor were they that interested in understanding what exactly natural selection, let alone the principle of the divergence of characters, really meant. Pre-Darwinian transformist traditions happily adapted to a world in which "evolution" was spelled "Darwin." Not a few, as we well know, wrote to Darwin claiming they had preceded him: by which they meant that they believed since a long time in the existence of natural laws accounting for the change life forms could and had undergone. In several countries, many "Darwinians" held very un-Darwinian views, often persisting in endorsing the one or the other of the transformist traditions they had adhered to before 1859.

---

[30] Francesco Costantino Marmocchi (1844). *Prodromo della storia naturale generale e comparata d'Italia*. Florence: Società editrice fiorentina, 2 vols., vol. 1, pp. 723–76.

[31] Agusti Camos Cabeceram (2021). *La huella de Lamarck en España en el siglo XIX*. Madrid: Consejo Superior de Investigationes Cientificas.

Myth 3   That Charles Darwin Was Not Directly Influenced by the Evolutionary Views of His Grandfather Erasmus

Patricia Fara

3.1   Introduction

The Danish philosopher Søren Kierkegaard once quipped that although life can only be understood backwards, it must be lived forwards. His neat aphorism encapsulates my own feelings about learning from the past. History does, of course, entail sticking to the facts – but inevitably, it also involves personal interpretation. Any narrative that I write is colored by my identity as a white, female, middle-class European, as well as by my personal desire to understand how we have reached the present and hence – even more important for me – how to improve the future. Similarly, Charles Darwin was influenced by his need to defend his theory and protect his reputation, while subsequent historians have been swayed by various agendas stemming from their own inclinations and shifting historiographical fashions.

Although it has often been said that Charles Darwin was not directly influenced by the evolutionary views of his grandfather Erasmus (Figure 3.1), such a negative contention is too weak to have acquired the mythical status of other familiar anecdotes. It has scarcely dented the propaganda of Darwinian enthusiasts, who preach that during his voyage on the *Beagle* Charles suddenly acquired an insight that empowered him to revolutionize natural human history and demolish biblical accounts. Perhaps the most gratifying myth of such sudden (verging on divine) inspiration is that he conceived the theory of evolution by natural selection when he noticed the differing shapes of birds' beaks across the Galapagos Islands (see Sulloway, Chapter 5, this volume); despite having been disproved, this story's simplicity and clarity have enabled it to survive as an easily understandable insight into a complex theory. In contrast, no key moment of creativity, no specific event, is introduced by the notion that Charles thought independently of his far less famous grandfather – there is no satisfying mental clunk of an apple falling to the ground or a cloud of steam lifting a kettle's lid. Instead, this is a question of judgment based on

Figure 3.1 Erasmus Darwin, Charles' grandfather. Stipple engraving by
J. Heath, 1804, after J. Rawlinson. Credit: Wellcome Collection. Public
Domain Mark

weighing up the pros and cons of historical evidence. The claim is, therefore,
open to challenge and reassessment.

Intuitively, it feels obvious that as a leading exponent of evolutionary ideas
in the late eighteenth century, Erasmus must have influenced his grandson's
thoughts about the natural world. Of course, that is not proof – but there are
several supporting strands of evidence-based argument. The assertion that
Erasmus did not influence Charles arose from at least two separate sources.
The first and more important was Charles himself, who was concerned both to
distance himself from his disreputable grandfather and to safeguard the integ-
rity of his controversial theory by ruling out inconvenient alternatives. Since
then, in a prolonged second phase, many historians and scientists have found it
advantageous to consolidate this view by celebrating Charles as a heroic genius
whose flash of inspiration single-handedly changed human thought forever.

Furthermore, rooting evolution in the nineteenth century confirmed that scientific ideas are not eternal, free-floating descriptions of the natural world, but are products of their cultural context.

These complementary versions of the past provided differing justifications for what Kierkegaard called living forwards; on the other hand, they mutually reinforced one another to establish a unified retrospective understanding.

3.2   Family Matters

Although they never met – Erasmus died seven years before Charles was born – Charles was familiar both with *Zoonomia*, the medical text that included Erasmus's first tentative suggestions in print about the possibility of evolution, and with *The Temple of Nature*, his long, posthumously published poem sketching out millennia of progress from an initial sub-marine organism to human beings. Charles accused his grandfather of over-reliance on speculation, yet in broad terms their conclusions are close. Decades before Charles discussed sexual selection, Erasmus wrote that "The final cause of this contest among males seems to be, that the strongest and most active animal should propagate the species which thus should become improved."[1] Charles even quoted the following words of his grandfather, commenting approvingly that "they are interesting as forecasting the progress of modern thought": "The stronger locomotive animals devour the weaker ones without mercy. Such is the condition of organic nature! whose first law might be expressed in the words, 'eat or be eaten,' and which would seem to be one great slaughter-house, one universal scene of rapacity and injustice."[2] How reminiscent of Charles's famous last paragraph about warring nature, famine and death!

Even so, Charles repeatedly minimized his grandfather's impact. Perhaps determined to demonstrate his intellectual independence, he criticized Erasmus for concocting grand theories based on a paucity of solid facts. He acknowledged the possibility of being influenced by Erasmus as a young man, but also took pains to distance himself from any such early enthusiasm, reminiscing that at Edinburgh, the biologist Robert Grant

burst forth in high admiration of Lamarck and his views on evolution. I listened in silent astonishment and as far as I can judge, without any effect on my mind. I had previously read the Zoönomia of my grandfather, in which similar views [to Lamarck's] are maintained, but without producing any effect on me. Nevertheless it is probable that the hearing rather early in life such views maintained and praised may have favoured my upholding them under a different form in my Origin of Species. At this time I admired greatly the Zoönomia; but on reading it a second time after an interval of ten or fifteen

---

[1]   Quoted Nora Barlow (1958). *The Autobiography of Charles Darwin*. London: Collins, p. 151.
[2]   Charles Darwin (1879). "Preliminary Notice." In Ernst Krause, *Erasmus Darwin*. London: John Murray, pp. 1–127, p. 113.

years, I was much disappointed, the proportion of speculation being so large to the facts given.

Despite such protests that Erasmus was guilty of too much speculation, Charles took the initiative to compile a biography paying tribute to his repudiated ancestor for possessing "the true spirit of the philosopher."[3]

Born nearly a century apart, these two proponents of evolution were tied together by family bonds, and on further consideration, Charles's denial of Erasmus's significance seems less well-founded. Unable to escape being a Darwin, he was inevitably affected by inherited characteristics and acquired beliefs that had been passed down through the intervening generation. Moreover, the notes that he scribbled in his grandfather's books and various other comments provide indisputable evidence of close study.

Charles was conditioned into thinking about inheritance from an early age. Named after Erasmus's oldest son Charles, who had died as an Edinburgh medical student, he must have worried about being destined for the same fate. In addition, he felt afflicted by his grandfather's large nose and stammer, while the extended family perpetuated Erasmus's aversion to alcohol so forcefully that Charles felt guilty whenever he drank a glass of wine.[4]

One major ethical inheritance was the condemnation of slavery. Both of Charles's grandfathers, Erasmus Darwin and Josiah Wedgwood, championed abolition, and their campaign was actively continued by his sisters and cousins. As Charles read for himself, Erasmus scattered protests against slavery throughout his poems, while in his agricultural book *Phytologia*, he voiced an impassioned plea for British farmers to produce sugar at home: "Great God of Justice! grant that it may soon be cultivated only by the hands of freedom."[5] Similarly, Charles responded angrily to slavery throughout his life (Peterson, Chapter 22, this volume). Landing in Brazil, he was appalled by the "extent to which the trade is carried on … the ferocity with which it is defended; the respectable (!) people who are concerned in it are far from being exaggerated at home."[6] Erasmus had poetically imagined the despair of mothers watching their children being transported overseas,[7] while Charles exhorted fathers to contemplate "your wife and your little children … being torn from you and sold like beasts to the first bidder."[8]

---

[3] Quoted Barlow, *Autobiography*, p. 153.

[4] Quoted Barlow, *The Autobiography*, p. 224 (from an unpublished letter).

[5] Erasmus Darwin (1800). *Phytologia; or the Philosophy of Agriculture and Gardening*. London: J. Johnson, p. 7.

[6] Quoted Janet Browne (1995). *Charles Darwin: Voyaging*. London: Pimlico, p. 198.

[7] Erasmus Darwin (1791). *The Loves of the Plants (The Botanic Garden, Part II)*. J. Johnson, Canto 3, pp. 411–18.

[8] Charles Darwin (1845), *Journal of Researches* (2nd edition), London: John Murray, p. 500.

The inheritance of religious attitudes is harder to ascertain, although both men were prudently discreet about their lack of adherence to orthodox Anglicanism. Charles refused to be pinned down, but one of his critics' major objections to the *Origin of Species* was the absence of a divine plan and the importance of chance events. Erasmus was less reticent about articulating his deist view that God does not directly intervene in terrestrial affairs: "That there exists a superior ENS ENTIUM, which formed these wonderful creatures, is a mathematical demonstration. That HE influences things by a particular providence, is not so evident." Proffering only vague explanations, he insisted that natural laws governed evolution, and clung adamantly to the concept of progress, "the improving excellence observable in every part of the creation."[9] Their views were so much in accord that Charles came to disagree with William Paley's *Natural Theology* (1802), which had influenced him as a young man but included several pages condemning an unnamed opponent easily identifiable as Erasmus Darwin.[10] Paley was particularly horrified by Erasmus's insistence that the processes of change had no final purpose, no overall plan imposed by God. A staple of Charles's Cambridge curriculum, *Natural Theology* is now most famous for its watchmaker analogy vindicating the belief that a Great Designer had created a teleological universe, a position that Charles implicitly refuted with his theory of evolution by natural selection. Directly refuting Paley's example of the human eye, Charles maintained that natural selection explains why eyes are often defective and wear out before their owners. In a sense, the *Origin of Species* was an extended attack on Paley's theological views, and many of Charles's subsequent experiments – on orchids, for example – were intended to provide specific examples vindicating natural adaptation against divine forethought (see also Ruse, Chapter 4, this volume).[11]

3.3   Printed Legacies

In addition to this general family environment, there was also direct textual transmission. As a student at Edinburgh, Charles was so interested in his grandfather that he read Anna Seward's biography, while marginal annotations

---

[9] Erasmus Darwin (1794–6). *Zoonomia; Or, the Laws of Organic Life*. 2 vols., London: J Johnson, vol. 1, p. 509; see James Harrison (1971). "Erasmus Darwin's View of Evolution." *Journal of the History of Ideas*, vol. 32, pp. 247–64.

[10] William Paley (1802). *Natural Theology*, pp. 463–73; *Edinburgh Review* 1 (1802–3), 301–3. See: John T. Baldwin (1992). "God and the World: William Paley's Argument from Perfection Tradition – A Continuing Influence." *Harvard Theological Review*, vol. 85, pp. 109–20; David Burbridge (1998). "William Paley Confronts Erasmus Darwin: Natural Theology and Evolutionism in the Eighteenth Century." *Science and Christian Belief*, vol. 10, pp. 49–71; Paul Elliott (2003). "Erasmus Darwin, Herbert Spencer, and the Origins of the Evolutionary Worldview in British Provincial Scientific Culture, 1770–1850." *Isis*, vol. 94, pp. 1–29; Norton Garfinkle (1955). "Science and Religion in England, 1790–1800: The Critical Response to the Work of Erasmus Darwin." *Journal of the History of Ideas*, vol. 16, pp. 376–88.

[11] Browne, *Charles Darwin*, 174–5.

demonstrate that he studied Erasmus's work closely.[12] Of particular relevance, he was familiar with both the books in which Erasmus outlined evolutionary concepts – *Zoonomia* and *The Temple of Nature*. As just a couple of specific examples, Charles appropriated his grandfather's *Temple* report of a wasp tearing off a fly's wings so that it would be easier to carry, while *Zoonomia* provided medical details for his *Expression of the Emotions*.[13]

Reluctantly following family tradition, Charles started out by studying medicine at Edinburgh, where he was directly influenced by the radical evolutionary ideas of Robert Grant. A keen supporter of Jean-Baptise Lamarck, Grant also enthused about the implications of Erasmus's *Zoonomia*, which by then had been consigned to the Vatican's list of banned publications. The chapter called "Of Generation" was particularly problematic: Charles recorded in the margin, "Lamarck concisely forestalled by my grandfather."[14] He was referring to Erasmus's proposal "that in the great length of time, since the earth began to exist ... warm-blooded animals have arisen from one living filament ... possessing the faculty of continuing to improve by its own inherent activity, and of delivering down those improvements by generation to its posterity."[15] This passage included four sacrilegious claims: that the age of the earth was far greater than in the biblical account; that life was created in a single event; that the living world had changed over time; and that this development had occurred independently of God.

Several eighteenth-century naturalists supported evolutionary concepts, so Charles and his contemporaries were considering the possibility well before he embarked on his *Beagle* voyage (see also Corsi, Chapter 2, this volume). Once safely back home, he underlined the word "Zoonomia" on the first page of a new notebook – and although in later life he neglected to mention this tribute, it was here that he sketched his first version of an evolutionary tree (on this sketch, see also Love, Chapter 9, this volume). As if recalling Erasmus's suggestion that a stag's horns or a quail's spurs might be attractive to females, Charles speculated that the key to generation might lie in sexual reproduction.[16] A few pages later, he articulated a key insight of his natural selection theory, wondering if a species might adapt in response to changes in its environment. He was also struck by Erasmus's concern with redundant features such as male nipples, and the possibility that they might be remnants from long processes of change. Charles's pencil marks reveal that he paused at this suggestion by

[12] Browne, *Charles Darwin*, 83–5; Mario A. Di Gregorio and N. W. Gill (1990). *Charles Darwin's Marginalia Vol.1*. New York: Garland.
[13] William Montgomery (1985). "Charles Darwin's Thought on Expressive Mechanisms in Evolution." *The Development of Expressive Behavior: Biology-Environment Interactions*, edited by Gail Zivin, Orlando: Academic Press, pp. 27–50, pp. 39–41.
[14] Di Gregorio and Gill, *Charles Darwin's Marginalia*, p. 185.
[15] Darwin, *Zoonomia; Or, the Laws of Organic Life*, vol. 1, p. 505.
[16] Darwin, *Zoonomia; Or, the Laws of Organic Life*, vol. 1, p. 503.

Erasmus: "Perhaps all the productions of nature are in their progress to greater perfection!"[17]

However much Charles may have tried to obscure this ancestry, his critics found it advantageous to reinforce his links to Erasmus. In the notorious Oxford debate of 1860, Samuel Wilberforce supposedly asked whether Darwin's propagandist Thomas Huxley was descended from an ape on the side of his grandfather or his grandmother (see Brooke, Chapter 13, this volume). In a savage printed review, Wilberforce elaborated this jibe at length, accusing Darwin of having inherited his bizarre evolutionary views from his "ingenious grandsire." He even reproduced a long and irrelevant extract from a vindictive poem called "The Loves of the Triangles" (1798) that superficially mocked Erasmus's *Loves of the Plants* but greatly exaggerated his political engagement by stressing the radical implications of his ideas about evolution and progress.[18] Throughout Charles's life, this satire on his grandfather continued to be acclaimed as a prime example of British wit.[19] He must have winced at every mention.

3.4   Creation and Evolution

In *The Temple of Nature*, Erasmus provided a far more detailed account than in *Zoonomia*. As if writing a poetic predecessor to the *Origin of Species*, he argued emotively by relentlessly listing examples, the same rhetorical device deployed by Charles. He did, however, supplement his verse account with substantial, detailed scholarly notes. Both Darwins were also similar in relying on close observations of the animal world. For example, Erasmus sought insights into human behavior by describing how elephants pick up coins, squirrels run around their cages and birds line their nests with moths. Positing that living organisms first appeared in the ocean, he maintained that through successive generations these minute beings gradually grew larger, acquiring new forms and functions until whales governed the seas, lions the land, and eagles the air. Apparently more daring than his grandson, he included human beings in this depiction of progressive evolution, presenting them as the culmination of continuous development.

To strengthen his own arguments, Charles effectively suppressed ideas about the creation of life that had been proposed by Erasmus. In *The Temple of Nature*,

---

[17] Erasmus Darwin (1803). *The Temple of Nature; or, The Origin of Society.* J. Johnson, 1803, p. 54n of the copy CCA.24.64 in Cambridge University Library; Darwin, *The Loves of the Plants*, note to canto 1, line 65.

[18] Samuel Wilberforce (1860). "Review of *On the Origin of Species.*" *Quarterly Review* 108, pp. 225–64.

[19] Patricia Fara (2012). *Erasmus Darwin: Sex, Science and Serendipity.* Oxford: Oxford University Press, pp. 30–42, 259–80, and *passim*.

his grandfather had elaborated his belief that the development of life on earth was governed by natural laws without direct divine intervention. Just as controversially, he maintained that life originated in some unspecified way during a single event that involved matter in motion, without being initiated by any divine spark. As both Darwins were aware, autogenesis (or spontaneous generation) was an emotive concept with great political and religious ramifications – and one from which Charles was determined to distance himself. In the *Origin of Species*, Charles deliberately avoided the topic, although in the third edition, published in 1861, he did respond to criticisms by maintaining that ignorance of life's origins provided no valid objection to his theory. Yet although he continued to insist publicly that understanding the creation of life lay beyond the capabilities of current science, privately he was considering the possibility as early as 1837, when he wrote in a notebook that "spontaneous generation [was] not improbable."[20]

Many naturalists condemned suggestions that life could appear as if by its own volition, but they did not always differentiate between two distinct positions, both circulating during Erasmus' lifetime. The more extreme of these was especially prevalent in Germany, where the eminent naturalist Johann Blumenbach promoted the idea that materialist processes of creation might have been recurring ever since the beginning of the world (see Rupke, Chapter 14, this volume). His model was based on the existence of an inborn force or *Bildungstrieb* that enabled organisms to create and repair themselves throughout their lives. Instead of adopting that scheme, Erasmus favored the less contentious view that there was one single act of creation for the first living germ, which subsequently gave rise to every other organism, including (eventually) humans. Although this opinion was shared by a substantial group of naturalists, Charles sought to wipe it from the record by presenting his own theory as the only viable alternative to repeated creative acts. This act of erasure made evolution by natural selection more appealing than might otherwise have been the case.[21]

Both Darwins expressed Malthusian ideas. In *The Temple of Nature*, Erasmus envisioned a constant battle between opposing forces of good and evil. In couplet after violent couplet, wolves tear apart innocent lambs, eagles swoop on helpless doves, and sharks gobble up shoals of fish. When celebrating fecundity, he overwhelmed his readers with illustrations of nature's teeming over-abundance – pregnant poppies, prolific aphids, sex-crazed snails, tadpoles

[20] Juli Peretó, Jeffrey L. Barrado and Antonio Lacano (2009). "Charles Darwin and the Origin of Life," *Origins of Life and Evolution of Biospheres*, 39, pp. 395–406, quotation p. 404.

[21] Nicolaas Rupke (2010). "Darwin's Choice." *Biology and Ideology from Descartes to Dawkins*, edited by Denis R. Alexander and Ronald L. Numbers. Chicago: University of Chicago Press, pp. 139–64.

and herring spawn. Death, warfare and disaster, he argued, are essential for preventing a population explosion that would outrun the world's resources.

Erasmus's argument closely parallels that of Thomas Malthus' *An Essay on the Principle of Population*, which was first published in 1798, while he was working on his poem. Although there is no documentary evidence that he read Malthus, he must at the very least have come across some of the many reviews that appeared, which – in the fashion of the time – often summarized rather than appraised the book. In his autobiography, Charles reported that reading a later edition of Malthus had proved a major impetus for considering the battle for survival in situations of scarce resources.[22] Yet two months earlier, he had already marked these Malthusian lines in his copy of *The Temple of Nature:*

> —Air, earth, and ocean, to astonish'd day
> One scene of blood, one mighty tomb display!
> From Hunger's arm the shafts of Death are hurl'd,
> And one great Slaughter-House the warring world![23]

3.5   A Victorian Hero

During his lifetime, Charles was often mocked, and by the early twentieth century his theory was so strongly contested that one hostile book was called *At the Deathbed of Darwinism*. It was not until 1942 that the term "modern synthesis" was coined to describe an evolutionary model that differed substantially from Charles's own. Although what is called "Darwinism" continues to be modified and challenged, today's formulations still carry his name as if he were the sole inventor.

Yet Charles has been elevated into a quintessentially English hero of science, an Isaac Newton of the biological world. Immediately recognizable in pictures, he features as a reclusive bearded genius who braved opposition to formulate a theory that overturned religious tradition by demoting human supremacy as dramatically as when the sun was placed at the center of the universe. In hagiographic renditions of "Darwin the lone genius," there is no room for precursors. Alfred Lord Tennyson's "Nature, red in tooth and claw" is often used to evoke Darwinian struggle, yet it appeared nine years before Charles ventured into print. Standard accounts still omit not only Erasmus Darwin and the French evolutionists of the eighteenth century (see Corsi, Chapter 2, this volume), but also Robert Chambers, author of *Vestiges of Creation*, which, published in 1844, was the book that smoothed Charles's path by defusing

---

[22] Janet Browne (1995). *Charles Darwin: Voyaging*. London: Pimlico, pp. 384–90.
[23] Erasmus Darwin. *The Temple of Nature; or, The Origin of Society*. J. Johnson, 1803, canto 4, lines 63–6 (Darwin, *Temple*, IV, 63–6).

many controversies before the *Origin of Species* was belatedly published in 1859.

Charles was also given a starring role as intrepid warrior in the struggle for survival between science and religion. According to those scenarios, just as Newton is falsely credited with banishing God from the universe, so Darwin is praised for overturning scriptural calculations that the earth was created during a single week in 4004 BCE. As historians have increasingly stressed, the accepted age of the earth had been extended decades earlier – indeed, he took advantage of that antiquity to justify the eons of time needed for his evolutionary model to work.[24]

The eighteenth century used to be less fashionable among historians of science than the Victorian era, as if it were a dormant period bereft of great advances and heroic discoverers. Erasmus's assumed irrelevance was further compounded by the absence of a biographer to champion his cause. Consigned to historical oblivion at the end of the eighteenth century, he remained an outcast until the 1960s, when he was rescued by the distinguished physicist Desmond King-Hele. For over fifty years, King-Hele campaigned to resuscitate Erasmus's reputation by publishing books, correspondence and archival manuscripts. Inspired by his research, several scholars have demonstrated Erasmus's significance as an industrial innovator, scientific popularizer, educational reformer and evolutionary pioneer.[25] His increasing importance in eighteenth-century historiography has made it clear that Erasmus had a greater effect on later events than previously appreciated, including on his grandson.

Erasmus was also side-lined by historians convinced that scientific theories are rooted in their cultural context and that biological ideas are intrinsically ideological constructs. Victorian specialists argued that because the scientists who hypothesized about evolution by natural selection were predominantly British, the theory could *only* have emerged in Britain, where the fittest reigned supreme in a cut-throat marketplace.[26] Or as Adrian Desmond and James Moore expressed it with characteristic eloquence, "Both Darwinian science and poor law society were now reformed along competitive Malthusian lines. Ruthless competition was the norm; it guaranteed the progress of life and a low-wage, high-profit capitalist society."[27] If Darwinian evolution was a Victorian product, then Erasmus had no place in the story.

---

[24] Peter Bowler (1984). *Evolution: The History of an Idea.* Berkeley: University of California Press, p. 4.

[25] Summarized in Patricia Fara (2012). *Erasmus Darwin: Sex, Science and Serendipity.* Oxford: Oxford University Press, pp. 253–4.

[26] Nicolaas Rupke (2010). "Darwin's Choice." *Biology and Ideology from Descartes to Dawkins,* edited by Denis R. Alexander and Ronald L. Numbers. Chicago: University of Chicago Press, pp. 139–64, pp. 149–53.

[27] Adrian Desmond and James Moore (1992). *Darwin.* London: Penguin, p. 265.

3.6   Conclusion

The extent of Erasmus Darwin's influence on his grandson must remain uncertain, but it cannot be dismissed. Both deliberately and unconsciously, everybody reinterprets their memories, and Charles may well have relied on his grandfather more than he acknowledged, even to himself. He had been surrounded by Erasmus's books and beliefs since childhood, and in later editions of the *Origin of Species* simultaneously acknowledged and denied his grandfather's importance by commenting in a footnote: "It is curious how largely my grandfather, Dr Erasmus Darwin, anticipated the views and erroneous grounds of Lamarck in his *Zoonomia.*"[28]

After being publicly humiliated by "The Loves of the Triangles," Erasmus destroyed much of the poem he had been composing. A substantially different version was published a year after his death as *The Temple of Nature; or, The Origin of Society*. More than fifty years later, another book on evolution appeared with a related title: *On the Origin of Species*. Inside it, Charles glossed over the contentious problem of life's origins – but he could never escape his own family origins.

[28] Quoted Ernst Krause. (1879). *Erasmus Darwin*. London: John Murray, p. 131.

Myth 4   That Darwin Always Rejected the Argument from Design in Nature and Developed His Own Theory to Replace It

Michael Ruse

4.1   The Early Years

It seems to be a shared idea among antievolutionists that Charles Darwin rejected the idea of design. For instance, we read that "Thus, Darwin developed a sophisticated anti-design theory to explain how so many creature features that clearly function for a purpose could originate apart from the agency of a Designer."[1] Or that "For the Christian theologians surveyed here, a basic consequence of Christian belief in God is the belief that the human species was designed by God. Darwin denies the latter, and thus logically denies the former as well."[2] But did he really? Here is what Darwin himself wrote in his autobiography around 1875: "The old argument of design in nature, as given by Paley, which formerly seemed to me so conclusive, fails, now that the law of natural selection has been discovered."[3] One thing we can say with certainty is that these are not the words of a man who *always* rejected the argument from design. Darwin was an undergraduate at the University of Cambridge from 1828 to 1831.[4] It was there that he read Archdeacon William Paley's (Figure 4.1) major works. Also in the *Autobiography*, Darwin tells us that Paley's "*Natural Theology* gave me as much delight as did Euclid."[5] And that: "I did not at that time trouble myself about Paley's premises; and taking these on trust I was charmed and convinced by the long line of argumentation." And yet, in the passage quoted above, Darwin tells us that the mechanism he

---

[1] Randy Guliuzza (2018). *Engineered Adaptability: Darwin's Anti-Design Doctrine*, Institute for Creation Research (www.icr.org/article/darwins-anti-design-doctrine).

[2] Jonathan Wells (1991). *Darwinism and the Argument to Design*, Discovery Institute (www.discovery.org/a/102).

[3] C. Darwin (1958). *The Autobiography of Charles Darwin 1809–1882. With the Original Omissions Restored. Edited and with Appendix and Notes by his Grand-Daughter Nora Barlow*. London: Collins, 87.

[4] J. Browne (1995). *Charles Darwin: Voyaging. Volume 1 of a Biography*. London: Jonathan Cape.

[5] Darwin, *The Autobiography*, 59.

Figure 4.1 William Paley. Source: Hulton Archive/Stringer/Archive Photos/ Getty Images

discovered – natural selection – had destroyed the argument. So, what's the story?

After Cambridge, Darwin spent five years – 1831–6 – as de facto ship's naturalist on HMS *Beagle*, as she mapped the coastlines of South America, going on from there through the Pacific, and on, via Australasia and South Africa, around the world. At the beginning of the voyage, he was a conventional Christian, intended indeed for the clergy of the Church of England. "Whilst on board the Beagle I was quite orthodox, and I remember being heartily laughed at by several of the officers (though themselves orthodox) for quoting the Bible as an unanswerable authority on some point of morality."[6] However, on the *Beagle* voyage, Darwin's religious beliefs changed from those of a theist, believing in a God who interferes in His world (most particularly by descending as Jesus Christ), to a deist, believing in a God who designed and set the apparatus in motion, and then sat back and let everything work according to unbroken law. Increasingly committed to the rule of law – the influence of Charles Lyell's *Principles of Geology*[7] was significant here – Darwin found he could no longer believe in miracles.

[6] Darwin, *The Autobiography*, 85.
[7] C. Lyell (1830–33). *Principles of Geology: Being an Attempt to Explain the Former Changes in the Earth's Surface by Reference to Causes now in Operation*. London: John Murray.

By further reflecting that the clearest evidence would be requisite to make any sane man believe in the miracles by which Christianity is supported, – that the more we know of the fixed laws of nature the more incredible do miracles become, – that the men at that time were ignorant and credulous to a degree almost incomprehensible by us, – that the Gospels cannot be proved to have been written simultaneously with the events, – that they differ in many important details, far too important as it seemed to me to be admitted as the usual inaccuracies of eye-witnesses; – by such reflections as these, which I give not as having the least novelty or value, but as they influenced me, I gradually came to disbelieve in Christianity as a divine revelation.[8]

Does this mean that Darwin gave up on a Designer? Not necessarily. It could just be that the God of deism planned it all before and then did it through law without intervening miracles. This seems to have been Darwin's position in the *Origin of Species*, reprinted in all editions including the sixth of 1872.

Authors of the highest eminence seem to be fully satisfied with the view that each species has been independently created. To my mind it accords better with what we know of the laws impressed on matter by the Creator, that the production and extinction of the past and present inhabitants of the world should have been due to secondary causes, like those determining the birth and death of the individual. When I view all beings not as special creations, but as the lineal descendants of some few beings which lived long before the first bed of the Silurian system was deposited, they seem to me to become ennobled.[9]

Without intending to put words into Darwin's mouth, I read this as saying God could have been the cause – was the cause – of design-like features in the organic world; but he got these through the action of unbroken law rather than miraculously.

4.2   Evolution and Its Cause

On returning to England, welcomed by his mentors (at Cambridge) and others, Darwin entered full-time into British scientific society, distributing and cataloging his collections from the trip, writing up his account of the *Beagle* voyage – which were published in 1839 as the *Journal of Researches into the Geology and Natural History of the Various countries visited by H.M.S. Beagle*[10] – and, very privately, opening notebooks where he speculated on the evolutionary origins of all organisms.[11] "Very privately," because Darwin knew that his mentors, most especially Cambridge professor (and future

---

[8] Darwin, *The Autobiography*, 86.
[9] C. Darwin (1859). *On the Origin of Species by Means of Natural Selection, or the Preservation of Favoured Races in the Struggle for Life*. London: John Murray, 488–9.
[10] London: Henry Colburn.
[11] C. Darwin (1987). *Charles Darwin's Notebooks, 1836–1844*. Editors P. H. Barrett, P. J. Gautrey, S. Herbert, D. Kohn, and S. Smith. Ithaca, NY: Cornell University Press.

historian and philosopher of science) William Whewell, thought that the argument from design, which they interpreted in a theistic fashion of a miracle-working God, made law-bound explanations inadequate, to the point of heresy. "Geology and astronomy are, of themselves, incapable of giving us any distinct and satisfactory account of the origin of the universe, or of its parts." Continuing, "the mystery of creation is not within the range of her legitimate territory; she says nothing, but she points upwards."[12]

Darwin had long known about evolutionary speculations. As a teenager, he had read evolutionary writings of his grandfather Erasmus Darwin (see Fara, Chapter 3, this volume). Then, before Cambridge, during a two-year stay in Edinburgh where he was a somewhat unhappy medical student, Darwin encountered evolution again, thanks to the enthusiasm of one of the older naturalists in the city, Robert Grant. Darwin's own conversion, as one might call it, came in the spring of 1837, when he learnt that the very similar looking birds he had collected on the Galapagos – an archipelago in the Pacific – were of different species, and different but similar to birds on the South American mainland (see Sulloway, Chapters 5 and 6, this volume). Evolution was the answer! Not the whole answer. As one entering the domain of professional science, Darwin knew full well that the leaders of this community – notably Whewell and the Cambridge-connected, scientist-philosopher John F. W. Herschel – venerated above all the seventeenth-century scientist, also a Cambridge man, Isaac Newton.[13] What was Newton's great achievement? Copernicus had put the sun at the center. Kepler had found that the planets go in ellipses, not circles. Galileo had discovered the laws of motion down here on Earth. Newton gave the causal explanation of it all. There is a force – gravity – between all bodies, proportional to their size and inversely proportional to the square of the distance between them. This was the physical world. What about the biological world? At the end of the eighteenth century, the great German philosopher Immanuel Kant said there will never be a Newton of the blade of grass.[14] The hugely ambitious young Darwin was determined to be that Newton! He had to find the force, the cause, behind evolutionary change.

Toward the end of the eighteenth century, Britain had had an industrial revolution.[15] People left the rural areas and moved to the cities, urban areas, and with this came a population explosion. More people had to be fed by fewer people. It was necessary to improve the quality of crops and livestock so they

---

[12] W. Whewell (1837). *The History of the Inductive Sciences (3 vols)*. London: Parker, 3, 587–8.

[13] M. Ruse (1975). Darwin's Debt to Philosophy: An Examination of the Influence of the Philosophical Ideas of John F.W. Herschel and William Whewell on the Development of Charles Darwin's Theory of Evolution. *Studies in History and Philosophy of Science*, 6, pp. 159–81.

[14] I. Kant [1790] (2000). *Critique of the Power of Judgment*. Editor P. Guyer. Cambridge: Cambridge University Press.

[15] M. Ruse (1979). *The Darwinian Revolution: Science Red in Tooth and Claw*. Chicago: University of Chicago Press.

could do better with less. Farmers – and animal fanciers, like dog and bird enthusiasts – soon realized that the key to change was artificial selection. Choose your better stock or crops and breed from them only. Improvement will follow. Fairly quickly, Darwin twigged on. Coming from an agricultural part of England, hence knowing about farming and cattle and sheep and the like, Darwin realized that selection is the key to important change in nature. If he had any doubts that there could be an analogous force in nature, causing such change, Darwin actually read a pamphlet that told him so.

A severe winter, or a scarcity of food, by destroying the weak or unhealthy, has all the good effects of the most skilful selection. In cold and barren countries no animal can live to the age of maturity, but those who have strong constitutions; the weak and the unhealthy do not live to propagate their infirmities, as is too often the case with our domestic animals. To this I attribute the peculiar hardiness of the horses, cattle, and sheep, bred in mountainous countries, more than their having been inured to the severity of climate.[16]

This was written in 1809 by Sir John Sebright, to whom Darwin refers in the first chapter of the *Origin of Species*. Darwin read the passage and, in the margin, marked its importance. "Sir J. Sebright – pamphlet most important showing effects of peculiarities being long in blood. ++ thinks difficulty in crossing race – bad effects of incestuous intercourse. – excellent observations of sickly offspring being cut off so that not propagated by nature. – Whole art of making varieties may be inferred from facts stated."[17] There is no implication here of plagiarism. In the pamphlet, Sebright gives no indication that he sees any profound or wide-ranging evolutionary consequences. Rather, Sebright's off-hand comment was the kind of clue that Darwin was seeking in his search for a cause. "Chance favors the prepared mind."

The question for Darwin now was how such a "natural" form of selection could come about. Toward the end of September 1838, Darwin read a work by the political economist (also an Anglican clergyman) Thomas Robert Malthus who, worried about how God had arranged that people would get up and work rather than spending their lives as indolent undergraduates, realized that the now-ongoing population explosion was the key.[18] Population increase outstrips the available food and space, so there will be a "struggle for existence." Only the hard working will survive, thus reproducing and passing on their good qualities to the next generations. Darwin took this natural theological argument, turned it on its head, and argued that the struggle is the force behind an ongoing natural form of selection.

[16] J. Sebright (1809). *The Art of Improving the Breeds of Domestic Animals in a Letter Addressed to the Right Hon. Sir Joseph Banks, K.B.* London: Privately published, 15–16.
[17] Darwin, *Charles Darwin's Notebooks*, C133.
[18] T. R. Malthus [1826] 1914. *An Essay on the Principle of Population (Sixth Edition)*. London: Everyman.

Population is increase at geometrical ratio in far shorter time than 25 years – yet until the one sentence of Malthus no one clearly perceived the great check amongst men. – there is spring, like food used for other purposes as wheat for making brandy. – Even a few years plenty, makes population in Men increase & an ordinary crop causes a dearth. take Europe on an average every species must have same number killed year with year by hawks, by cold &c. – even one species of hawk decreasing in number must affect instantaneously all the rest. – The final cause of all this wedging, must be to sort out proper structure, & adapt it to changes. – to do that for form, which Malthus shows is the final effect (by means however of volition) of this populousness on the energy of man. One may say there is a force like a hundred thousand wedges trying force into every kind of adapted structure into the gaps of in the oeconomy of nature, or rather forming gaps by thrusting out weaker ones.[19]

Darwin had got his cause. The biological equivalent of gravity. Darwin was going to be the Newton of biology.

4.3   Paley's Argument

What about the role, or non-role, of the argument from design in all of this? Well, what about the argument? Darwin would have encountered the argument countless times, especially at university. He was a keen beetle collector and so would have relied heavily on the classic text by Kirby and Spence: *An Introduction to Entomology: or Elements of the Natural History of Insects* (1815–28).[20] Through nature one gets to God: "no study affords a fairer opportunity of leading the young mind by a natural and pleasing path to the great truths of Religion, and of impressing it with the most lively ideas of the power, wisdom, and goodness of the Creator."[21] But turn to the most significant influence – basically for everyone and not just Darwin – Paley's exposition in his *Natural Theology*. Paley starts off with the famous lines:

In crossing a heath, suppose I pitched my foot against a *stone* and were asked how the stone came to be there, I might possibly answer, that, for any thing I knew to the contrary, it had lain there forever: nor would it perhaps be very easy to show the absurdity of this answer. But suppose I had found a *watch* upon the ground, and it should be enquired how the watch happened to be in that place; I should hardly think of the answer which I had before given, –that, for any thing I knew, the watch might have always been there. Yet why should not this answer serve for the watch, as well as for the stone?[22]

Note what is being claimed. The watch is complex in a way that the stone is not. But more than just complex. The watch serves a *purpose*. In the language of

[19] Darwin, *Charles Darwin's Notebooks*, D135.
[20] London: Longman, Hurst, Reece, Orme, and Brown.
[21] Kirby and Spencer, *An Introduction*, 35.
[22] W. Paley [1802] (1819). *Natural Theology (Collected Works: IV)*. London: Rivington, 1.

Aristotle, it exhibits a "final cause," an end, a reason for its existence. And here comes the punchline. It could not have been chance. There had to have been a watch/maker designer: "there must have existed, at some time and at some place or other, an artificer or artificers who formed it for the purpose which we find it actually to answer; who comprehended its construction, and designed its use."[23]

Now we have the analogy. The eye shows the same kind of complexity as the watch. Hence it must have a purpose, a final cause – which we know that there is, namely seeing. As the watch had a maker/designer, the eye must have had a maker designer. The Great Optician in the Sky, aka God. Compare the eye with a telescope. "As far as the examination of the instrument goes, there is precisely the same proof that the eye was made for vision, as there is that the telescope was made for assisting it."[24]

4.4   Adaptation

We have two claims here. First: the works of nature, of organisms, are design-like, they are as if made for ends, they exhibit final causes. Second: the only reasonable explanation is that a Designer God in some way miraculously intervened in nature to put them in functioning place. And it was this second part that Darwin, on the *Beagle* voyage, found he could no longer accept. Note that this occurred before the acceptance of evolution and before the discovery of natural selection. What of the first claim, that the organic world is design-like? Features, characteristics, organs are there for a purpose, they serve ends, they exhibit final causes. They are "adapted" to their circumstances to serve the needs of the organism. Adaptation! The eye is like a telescope because, just as the telescope seems designed for seeing, the eye seems designed for seeing. The final cause of both telescope and eye is seeing (see Lennox, Chapter 16, this volume). Darwin, never, ever doubted the truth of this claim. Not on the *Beagle*, not in the notebooks, not in the *Origin of Species*, not in the *Descent of Man*, not ever. He was certainly not setting out to refute Paley on this. From the *Origin of Species*:

How have all those exquisite adaptations of one part of the organisation to another part, and to the conditions of life, and of one distinct organic being to another being, been perfected? We see these beautiful co-adaptations most plainly in the woodpecker and missletoe; and only a little less plainly in the humblest parasite which clings to the hairs of a quadruped or feathers of a bird; in the structure of the beetle which dives through the water; in the plumed seed which is wafted by the gentlest breeze; in short, we see beautiful adaptations everywhere and in every part of the organic world.[25]

[23] Paley, *Natural Theology*, 3.   [24] Paley, *Natural Theology*, 14.
[25] Darwin, *On the Origin of Species*, 60–1.

I want to stress that for Darwin, this was genuine, end-directed thinking (see Lennox, Chapter 16, this volume). Organisms with the right characteristics had done well in the past. Hence, it was reasonable to think that, with these right characteristics, they would continue to do so in the future.

Thus, Darwin on the argument from design. He has dropped the Designer intervening miraculously in His creation. But he seems to accept a Designer working at a distance, putting all in motion and then standing back. He always accepted the fundamental truth that organisms seem designed-like. They exhibit final causes.

4.5   The Argument Fails

And yet Darwin argued that Paley's argument from design failed, because of natural selection. Independently, for all that he kept untouched the deistic passage in the *Origin of Species*, we know that by about 1870 he had moved to a form of agnosticism. This he held for the rest of his life. Why the change of mind? We know that there were revealed theological reasons why Darwin lost all belief.

I can indeed hardly see how anyone ought to wish Christianity to be true; for if so the plain language of the text seems to show that the men who do not believe, and this would include my Father, Brother and almost all my best friends, will be everlastingly punished. And this is a damnable doctrine.[26]

We know that there were natural theological reasons why Darwin gave up on God. The argument from evil. This is a letter he wrote to his close American friend, Asa Gray.

With respect to the theological view of the question; this is always painful to me. – I am bewildered. – I had no intention to write atheistically. But I own that I cannot see, as plainly as others do, & as I shd wish to do, evidence of design & beneficence on all sides of us. There seems to me too much misery in the world. I cannot persuade myself that a beneficent & omnipotent God would have designedly created the Ichneumonidæ with the express intention of their feeding within the living bodies of caterpillars, or that a cat should play with mice. Not believing this, I see no necessity in the belief that the eye was expressly designed.[27]

Even here Darwin seems not to have given up on God entirely. The argument from design interpreted in a deistic manner seems still to have legs. What brought on such cramps that, later in the decade, the legs gave way entirely? Darwin grew to agree with people like Whewell that it was the theist's God or no God at all. At the end of the large, two-volume work he published later in the decade – *Animals and Plants under Domestication* – he argued that he could not believe in a God who created all sorts of variations and, knowing the

---

[26] Darwin, *The Autobiography*, 87.
[27] Letter to Asa Gray, May 22, 1860; Darwin Correspondence Project-LETT-2814

winner, stepped back and let nature run its course. Suppose "an architect were to rear a noble and commodious edifice, without the use of cut stone, by selecting from the fragments at the base of a precipice wedge-formed stones for his arches, elongated stones for his lintels, and flat stones for his roof."[28] Would we truly want to say that truly the stones were not random, because God had been behind making them as they are?

The shape of the fragments of stone at the base of our precipice may be called accidental, but this is not strictly correct; for the shape of each depends on a long sequence of events, all obeying natural laws; on the nature of the rock, on the lines of deposition or cleavage, on the form of the mountain which depends on its upheaval and subsequent denudation, and lastly on the storm or earthquake which threw down the fragments. But in regard to the use to which the fragments may be put, their shape may be strictly said to be accidental.[29]

And here's the rub.

We are led to face a great difficulty, in alluding to which I am aware that I am travelling beyond my proper province. An omniscient Creator must have foreseen every consequence which results from the laws imposed by Him. But can it be reasonably maintained that the Creator intentionally ordered, if we use the words in any ordinary sense, that certain fragments of rock should assume certain shapes so that the builder might erect his edifice?[30]

Darwin is arguing against Asa Gray, who claimed that the variations used by selection are guided by God. Darwin thought (truly) that this makes selection redundant. He came to see that the variations produced by a deistic God are no less directed than the variations produced by a theistic God. The deistic God makes the variations needed, doing this deliberately. He then just covers this up by creating a slew of variations with no use.

In the end, there is no real difference between the successful variations of theist and deist. Either way, natural selection is redundant. The argument from design, as claimed by Whewell, needed guided variations. Natural selection makes guided variations unnecessary. No need to assume a Designer. Hence, even as it keeps the major premise, that organisms are as-if designed, natural selection destroys the argument from design. Let the great Catholic theologian, John Henry Newman, have the last word: "I believe in design because I believe in God; not in a God because I see design."[31]

---

[28] C. Darwin (1868). *The Variation of Animals and Plants Under Domestication*. London: John Murray, Vol. 2, 430.

[29] Darwin, *The Variation*, 2, 431.   [30] Darwin, *The Variation*, 2, 431.

[31] J. H. Newman (1973). *The Letters and Diaries of John Henry Newman, XXV.* Editors C. S. Dessain and T. Gornall. Oxford: Clarendon Press.

Myth 5   That Darwin Converted to Evolutionary Theory During His Historic Galápagos Islands Visit

Frank J. Sulloway

5.1   Introduction

In September 1835, Charles Darwin landed on San Cristóbal, the easternmost island in the Galápagos Archipelago (Figure 5.1). He was then in the fourth of his five years serving as the ship's naturalist during H.M.S. *Beagle*'s circumnavigation of the globe. Darwin scholars generally agree that Darwin was still an adherent of creationist theory at this point in the *Beagle* voyage.[1] This creationist belief, including Darwin's acceptance of the immutability of species, prompted him to remark shortly after his arrival in the Galápagos that "it will be very interesting to find from future comparison to what district or 'centre of creation' the organised beings of this archipelago must be attached."[2]

Darwin's five-week visit to these islands fundamentally altered his scientific allegiance, initiating one of the most far-reaching revolutions in the history of science. Nevertheless, the precise nature and timing of this critical influence on Darwin's thinking has been obscured by a series of myths regarding the insights he reached during his visit. At least four closely integrated myths have come to define the "Darwin-Galápagos legend": (1) that Darwin's conversion to evolution occurred in the Galápagos;[3] (2) that his iconic Galápagos finches played a key role

---

I am grateful to Kostas Kampourakis, Malcolm J. Kottler, Smriti Mehta, Gregory Radick, Michael Ruse and Daniel Sherman for helpful comments on earlier versions of this chapter.

[1] Frank J. Sulloway, "Tantalizing Tortoises and the Darwin-Galápagos Legend," *Journal of the History of Biology* 42 (2009), 3–31.

[2] Charles Robert Darwin, *Charles Darwin's* Beagle *Diary*, Richard Keynes, ed. (Cambridge: Cambridge University Press, 1988), 356.

[3] "The Galapagos facts ... must have struck him at once, – without waiting for accurate determinations, – as a microcosm of evolution": Francis Darwin, "Introduction," *Foundations of the Origin of Species: Two Essays Written in 1842 and 1844 by Charles Darwin*, ed. Francis Darwin (Cambridge: Cambridge University Press, 1909), xi–xxiv, at xiv. With regard to the myth that Darwin was converted to evolution during his famous Galápagos visit, there is still some ongoing scholarly debate. A minority of Darwin scholars continue to argue that Darwin experimented with evolutionary ideas either before, during or shortly after his Galápagos visit. See, for example, Paul D. Brinkman, "Charles Darwin's *Beagle* Voyage, Fossil Vertebrate Succession, and 'the Gradual Birth & Death of Species,'" *Journal of the History of Biology*, 43 (2010), 363–99; Niles Eldredge,

in this conversion;[4] (3) that these same finches inspired Darwin's theory of natural selection;[5] and finally, (4) that the Galápagos finches were integral to Darwin's argument for evolution in the *Origin of Species* (1859).[6] This chapter focuses

"Experimenting with Transmutation: Darwin, the *Beagle*, and Evolution," *Evolution: Education and Outreach*, 2 (2009), 35–54; Jonathan Hodge, "Darwin, the Galápagos and His Changing Thoughts about Species Origins: 1835–1837," *Proceedings of the California Academy of Sciences*, 61 (2009), Supplement II, No. 7, 89–106. The evidence presented here is consistent with the majority consensus as well as with Darwin's own testimony about the timing of his conversion. See the sources cited in note 3 as well as Charles Robert Darwin, *Darwin's Journal*, ed. Gavin de Beer, *Bulletin of the British Museum (Natural History) Historical Series*, 2, no. 1 (1959), 7; and Darwin's 1877 letter to Otto Zacharias in which he asserted: "When I was on board the Beagle I believed in the permanence of Species, but, as far as I can remember, vague doubts occasionally flitted across my mind" (Charles Robert Darwin, *The Correspondence of Charles Darwin*, vol. 25: 1877, ed., Frederick Burkhardt et al. (Cambridge: Cambridge University Press, 2018), 105. The minority position on this historical question tends to make more of Darwin's "vague doubts" than he himself apparently did, usually by interpreting in an evolutionary light evidence that was also perfectly consistent with creationism (for example, the continuity of fossil succession). As a case in point, when Charles Lyell highlighted the importance of Darwin's *Beagle* voyage fossil discoveries in early 1837, he had no trouble interpreting continuity in "the succession of living beings" within a creationist perspective (Charles Lyell, "Presidential Address to the Geological Society," *Proceedings of the Geological Society of London*, 2, no. 29 (1837), 479–523, at 520).

[4] "It all began in the Galapagos, with these finches": Sharon Weiner Green and Ira K. Wolf, *Barron's How to Prepare for the SAT 2007*, 23rd ed. (New York: Barron's Educational Series, 2006), at 582.

[5] "Darwin's finches (a unique cluster of 13 new species) provided the final evidence needed for his explanation of natural selection and the transmutation of species": Brian G. Gardiner, "Darwin and South American Fossils," *The Linnean*, 20, no 4 (2004), 16–22, at 22.

[6] "The Galapagos finches helped Darwin to solidify his idea of natural selection . . . [He] fully explored the information gained from the Galapagos Finches in his most famous book *On the Origin of Species*": Heather Scoville, "Charles Darwin's Finches," ThoughtCo, June 26, 2019, www.thoughtco.com/charles-darwins-finches-1224472. Although all four myths were almost universally endorsed in biology textbooks prior to 1980, most textbooks published since then have acknowledged their falsehood, generally citing some of my publications on the subject, including Frank J. Sulloway, "Geographic Isolation in Darwin's Thinking: The Vicissitudes of a Crucial Idea," *Studies in the History of Biology*, 3 (1979), 23–65; "Darwin's Conversion: The *Beagle* Voyage and Its Aftermath," *Journal of the History of Biology*, 15 (1982), 325–96; "Darwin and His Finches: The Evolution of a Legend," *Journal of the History of Biology*, 15 (1982), 1–53; "The Legend of Darwin's Finches," *Nature*, 303 (1983), 372; "Darwin and the Galápagos," *Biological Journal of the Linnean Society*, 21 (1984), 29–59; and "Darwin and the Galápagos: Three Myths," *Oceanus*, 30 (1987), 79–85. These publications drew on earlier research by other Darwin scholars, including Michael T. Ghiselin, *The Triumph of the Darwinian Method* (Berkeley: University of California Press, 1969); Sandra Herbert, "The Place of Man in the Development of Darwin's Theory of Transmutation. Part 1. To July 1837," *Journal of the History of Biology*, 7 (1974), 217–58; and Malcolm J. Kottler, "Darwin's Biological Species Concept and Theory of Geographic Speciation: The Transmutation Notebooks," *Annals of Science*, 35 (1978), 275–8. Occasional textbook exceptions nevertheless continue to espouse one or more of these four myths, e.g., Jonathan Bard, *Principles of Evolution: Systems, Species, and the History of Life* (New York and London: Garland Science, 2017), 206–07; Brian K. Hall and Benedikt Halgrimsson, *Strickberger's Evolution*. London and Sudbury, MA: Jones and Bartlett, 2008), 24; Kenneth A. Mason and George B. Johnson, *Biology*, 10th ed. (New York: McGraw-Hill, 2014), 418; and Peter W. Skelton, *Evolution* (Milton Keynes: Open University, 1993), 15. These same myths are now most commonly found in popular science publications and news stories, e.g., Eva Botkin-

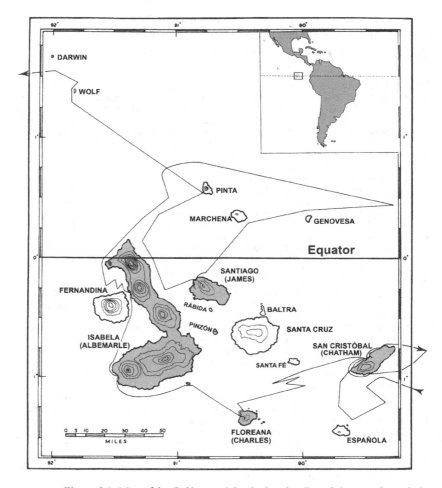

Figure 5.1 Map of the Galápagos Islands showing Darwin's route through the archipelago aboard H.M.S. *Beagle*. Darwin visited the four islands shaded in gray and spent a total of nineteen days on land during his five-week visit.[7]

Kowacki, "Darwin's Finches Are Pecking Their Way through Evolution." *The Christian Science Monitor* (2016), 22 April; Tina Hesman Saey, "Genetic History of Darwin's Finches: DNA Analysis Pinpoints Source of Birds' Varied Beaks," *Science News*, 187 (March 7, 2015), 7; Christopher Solomon, "Life in the Balance: A Warming Planet Threatens the Galápagos Species that Inspired Darwin's Theory of Natural Selection," *National Geographic*, 231 (2017), June, 52.
[7] Map adapted from Frank J. Sulloway, "Darwin and the Galapagos Giant Tortoises," in *Galapagos Giant Tortoises*, ed., James P. Gibbs, Linda J. Cayot, and Washington Tapia Aguilera (Cambridge, MA: Academic Press, 2020), 83–95, at 84.

principally around the claim that Darwin was converted to the theory of evolution during his historic Galápagos visit, although, where relevant, I also consider other aspects of the Darwin-Galápagos legend. In the next chapter I take up in greater detail the way in which Darwin's various Galápagos observations did, and did not, influence his post-voyage thinking about evolution, as well as the role that Darwin's now-iconic Galápagos finches played in this process.

5.2   Nearly Entrapped by the Myth

I begin this historical reconstruction of Darwin's Galápagos visit and its influence on him by addressing the last of the four myths I have just mentioned – the claim that Darwin's Galápagos finches played an important part in his argument for evolution by natural selection as set forth in the *Origin of Species*. While an undergraduate at Harvard College I had organized an expedition to retrace Darwin's footsteps around the South American continent and to make a documentary film about Darwin's findings.[8] With the helpful support of evolutionary biologists Ernst Mayr and Edward O. Wilson, this three-month expedition during the summer of 1968 included a brief visit to the Galápagos Islands, where my fellow expedition members and I paid special attention to filming Darwin's famous finches.

The following year I completed a study guide for an educational film produced from our 1968 filming of Darwin's finches.[9] In this study guide I asserted that "Darwin gave a prominent place to the Galápagos finches in the *Origin of Species*." Shortly thereafter I departed for the Galápagos Islands, having been awarded a year-long postgraduate fellowship by Harvard University to return to these islands as well as to study the extensive collection of Darwin manuscripts in England.[10] In the library of the Charles Darwin research station I happened to examine a facsimile of the first edition of the *Origin of Species*, edited by Ernst Mayr.[11] To my astonishment I could find no discussion of the Galápagos finches, either in the first edition or in any of the later editions.[12]

---

[8] For further details about the Harvard-Darwin Expedition (1968), see William K. Stevens, "Darwin's Voyage on 'Beagle' is Retraced by Five for Film," *New York Times*, February 11, 1969, pp. 33–8.

[9] Mark B. Adams and Frank J. Sulloway, A Series of Six Films and Film Guides on Different Aspects of "Charles Darwin's Voyage with H.M.S. Beagle" (Santa Monica, CA: BFA Educational Media, 1970).

[10] Frank J. Sulloway, "Frank J. Sulloway in the Galapagos, 1970: The Journal of a Graduate Student at Harvard Who Takes a New Look at Darwin's Revolutionary Observations," in *On Their Own: Three New Hampshire Scholars Chronicle Their Adventures Abroad*, ed. Alvah W. Sulloway (West Kennebunk, ME: The New Hampshire Historical Society and Phoenix Publishing, 2001), 247–338.

[11] Ernst Mayr, ed., *On the Origin of Species by Charles Darwin: A Facsimile of the First Edition* (Cambridge, MA: Harvard University Press, 1964).

[12] Morse Peckham, *The Origin of Species by Charles Darwin: A Variorum Text* (Philadelphia: University of Pennsylvania Press, 1959).

The realization that Darwin did not specifically cite the finches in the *Origin of Species* initially elicited a sense of panic, as my written assertion to the contrary was then in press and might already have been published. My anxiety was compounded owing to the lack of an easy way to correct my error before my return to the United States in three months' time. Back in 1970, no telephone communication existed between the Galápagos Islands and the mainland. Mail service was largely dependent on sporadic military flights or the *Cristóbal Carrier*, which visited the islands every month or so. Fortunately, I learned of a ham radio operator, Lucio Saltos Gómez, who lived near the Darwin Research Station. With Saltos's help, I eventually succeeded in contacting another ham radio operator in the USA, who then facilitated a long-distance telephone call with the film production company in Cambridge, Massachusetts, allowing me to correct my error in time.

This anecdote provides a good example of what Albert Martínez has called the "must have" principle that underlies commonly held myths in the history of science.[13] Given the seminal role Darwin's finches have played in establishing several fundamental principles in evolutionary biology, it follows quite naturally that Darwin himself must have discussed them in his most celebrated work about evolution. Why, in fact, he did not do so provides an important key to unravelling the other three pervasive myths about Darwin's Galápagos visit.

Because I had nearly been entrapped by this particular aspect of the Darwin-Galápagos legend, I began to study in detail those portions of Darwin's voyage manuscripts that related specifically to his ornithological collections. These "Ornithological Notes," transcribed and published seven years earlier by Darwin's granddaughter Nora Barlow, provided a complete catalogue of all the bird specimens that Darwin collected during the *Beagle* voyage.[14] As I studied these notes, I was particularly puzzled by seeing Darwin's reference to a "Wren" collected during his Galápagos visit (Figure 5.2). I knew enough about the ornithology of the Galápagos to know that there are no wrens present in these islands. So what was this bird? Based on the possibilities, I realized that the specimen in question had to be the Warbler Finch (*Certhidea olivacea*), one of the seventeen Galápagos species that are currently recognized in this iconic avian group.[15] In other words, Darwin had apparently mistaken this particular species for a member of another disparate avian family – a confusion that in turn reflects

---

[13] Alberto A. Martínez, *Science Secrets: The Truth about Darwin's Finches, Einstein's Wife, and Other Myths* (Pittsburgh: University of Pittsburgh Press, 2011), 248.

[14] Charles Robert Darwin, *Darwin's Ornithological Notes*, ed. with an Introduction, Notes, and Appendix by Nora Barlow. *Bulletin of the British Museum (Natural History) Historical Series*, 2 (1963 [1836]), no. 7, 265.

[15] Peter R. Grant, B. Rosemary Grant, Erik D. Enbody, Leif Andersson, and Sangeet Lamichhaney, "Darwin's Finches, an Iconic Adaptive Radiation." *Encyclopedia of Life Sciences*, 1 (2020), 672–682. An eighteenth species of "Darwin's finches" inhabits Cocos Island, located about 450 miles northeast of the Galápagos group.

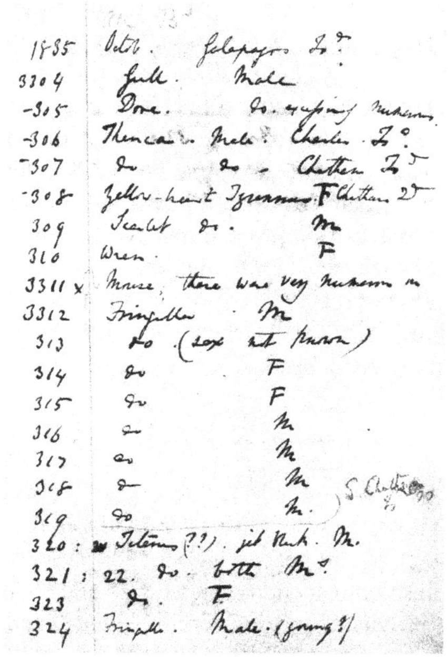

Figure 5.2 A page from Darwin's *Beagle* voyage specimen catalogue
showing some of his Galápagos birds. Specimen 3310 ("Wren") is the
Galápagos Warbler Finch. Designated below as "Icterus (??)" are Darwin's
four specimens of the Cactus Finch (3320–3). Inscribed very faintly below the
eight "Fringilla" specimens (3312–19), Darwin has written "⸮Chatham Is^d??"
This penciled comment was likely made after his return to England when he
realized the importance of having island localities for all of the finches. At this
time, he appears to have assumed that the first eight "Fringilla" specimens
might have been collected on the first island where he landed. However, the
birds listed in this catalogue are not entered in the order of the islands Darwin
visited, but rather in a topsy-turvy manner. For example, his "Wren" near the
top of the list was collected on the last of the four islands he visited, as were all
four of the cactus finches.[16] In addition, Darwin only entered specimens in
this catalogue after he had left the archipelago and was sailing to Tahiti, so by
this time all of the birds had apparently been intermingled by island. Photo:
Frank J. Sulloway, courtesy of Down House and English Heritage.

the extraordinary degree of adaptive radiation that characterizes Darwin's
Galápagos finches.

As I further perused these same "Ornithological Notes," I was also surprised
by Darwin's reference to four specimens of "Icterus" (3320–3). Once again,

---

[16] Sulloway, "Darwin and His Finches," p. 27.

this is a rather strange designation for a Galápagos bird, as the Icteridae include the blackbirds, orioles, and meadowlarks, none of which are present in the Galápagos. Given that the Icteridae are distinguished by their long, pointed bills, Darwin was clearly referring to the Cactus Finch (*Geospiza scandens*). Additionally, I was struck by the fact that Darwin's field notes distinguished between "Fringilla" (or true finches), the name he applied to the smaller ground finches, and "Grosbeaks," the term Darwin used for two subspecies of the Large Ground Finch (*G. magnirostris*). In short, the extensive variation in beak size and shape among Darwin's finches caused him to mistake these species for members of four different avian families. Thus, rather than becoming an evolutionist because of the famous finches. Darwin was initially prevented from recognizing their common ancestry because of their extensive adaptive radiation.[17]

5.3   David Lack's *Darwin's Finches* (1947)

Also missing from Darwin's field notes, as well as his later published discussions, was information about the ecological differences among the various species, including differences in resource use corresponding to the size and shape of the beaks. This information was later provided by British ornithologist David Lack in his influential 1947 book *Darwin's Finches*, which popularized the term "Darwin's finches," although this term had first been used eleven years earlier by ornithologist Percy Lowe.[18] As Lack himself fully understood, Darwin had failed to notice differences in the feeding habits of the finches because he mostly observed them as they fed together in flocks on the ground. With the exception of the Cactus Finch (*G. scandens*), Darwin believed that "the several species of Geospiza are undistinguishable from each other in habits."[19] By contrast, Lack's detailed ecological research established that the different beaks of Darwin's finches were adapted for differing diets. Such ecological diversification, Lack further concluded, reduces competition among the different finch species and promotes adaptive evolutionary radiation through continued natural selection. With these new and important findings, Lack singlehandedly elevated the Galápagos finches into the mythological role they subsequently acquired in biology textbooks, although other commentators subsequently built on the legend that grew from Lack's 1947 book.[20]

---

[17] Sulloway, "Darwin and His Finches," 8–11; "The Legend of Darwin's Finches," 372.

[18] David Lack, *Darwin's Finches: An Essay on the General Biological Theory of Evolution* (Cambridge: Cambridge University Press, 1947); Percy R. Lowe, "The Finches of the Galapagos in Relation to Darwin's Conception of Species," *Ibis*, 13th series, 6 (1936), 310–21.

[19] Charles Robert Darwin, *The Zoology of the Voyage of H.M.S. Beagle, under the Command of Captain FitzRoy, R.N., During the Years 1832–1836. Part III: Birds* (London: Smith, Elder, 1841), 99.

[20] Sulloway, "Darwin and His Finches," 45–7; "The Legend of Darwin's Finches," 372; and John van Wyhe, "Where Do Darwin's Finches Come From?" *The Evolutionary Review*, 3 (2012), 185–95.

5.4   The Galápagos Mockingbirds and Giant Land Tortoises

The next decade of my life as a Darwin scholar was devoted to gaining a fuller understanding of how the Galápagos Islands had influenced Darwin's thinking. This historical reconstruction begins not with the finches, but with the Galápagos mockingbirds. Upon landing on Floreana (the second of the four islands he visited), Darwin noticed that the resident mockingbird (*Mimus trifasciatus*) differed in its plumage from a single specimen (*M. melanotis*) he had previously collected on San Cristóbal. Having realized that there might be more than one species of mockingbirds in the Galápagos, he subsequently paid special attention to their collection, procuring an additional specimen on Isabela, and a fourth specimen on Santiago (Figure 5.3). In his "Ornithological Notes," Darwin provided island localities for these four specimens, but, interestingly, he did not do so for the finches or other birds, with the exception of four birds that he and other *Beagle* collectors had seen only on one island.

Figure 5.3 Darwin's four Galápagos mockingbird specimens, shown from top to bottom in the order he collected them: *Mimus melanotis* (Chatham Island or San Cristóbal); *M. trifasciatus* (Charles Island or Floreana); *M. parvulus* (Albemarle Island or Isabela); and *M. parvulus* (James Island or Santiago). A fourth species, *M. macdonaldi*, is confined to Española, an island Darwin did not visit. © Trustees of the Natural History Museum, London. Photo: Frank J. Sulloway.

While he was visiting Floreana, Darwin had the good fortune to dine with the Archipelago's Vice Governor, Nicholas O. Lawson. During this brief encounter, Lawson boasted that he could "tell at once from which island any one [giant tortoise] was brought,"[21] It is worth noting that Lawson's claim was considerably overstated, as even the most experienced taxonomists cannot identify the particular island of origin of every adult tortoise.[22] Darwin also tells us in his *Journal of Researches* that he did not initially appreciate the full implications of the Vice Governor's assertion.[23]

The reason for Darwin's delayed understanding is straightforward. At the time of this visit, both Darwin and Captain Robert FitzRoy were under the impression that giant land tortoises had been introduced to these islands, most likely by buccaneers who had transported the tortoises from islands in the Indian Ocean.[24] This incorrect assumption accounts for the inappropriate name – *Testudo indicus* – by which Darwin described these tortoises in his *Journal of Researches*. In fact, Galápagos tortoises are currently recognized as an endemic taxon (*Chelonoidis*) including fifteen distinct species, although some herpetologists have long considered them to be members of a single species.[25] Because Darwin and FitzRoy thought the Galápagos tortoises had been introduced to these islands, they initially concluded, following the opinion of buccaneer William Dampier, that all of the giant land tortoises on these islands were simply "varieties caused by climate, soil, food, and habits."[26]

It was for this reason that Darwin later failed to collect a specimen of adult tortoise from the island of Santiago, which would have allowed him to compare this specimen with two others that FitzRoy had previously procured on Española. In addition, and shortly before leaving the archipelago, the *Beagle* had taken on

---

[21] Charles Robert Darwin, *Journal of Researches into the Geology and Natural History of the Various Countries visited by H.M.S. Beagle, under the Command of Captain FitzRoy, R.N. from 1832 to 1836* (London: Henry Colburn, 1839), 465.

[22] Peter C. H. Pritchard, *The Galápagos Tortoise: Nomenclature and Survival Status* (Luneburg, MA: Chelonian Research Foundation, 1996), 38.

[23] Darwin, *Journal of Researches*, 474.

[24] Sulloway, "Darwin's Conversion," 338–45; "Darwin and the Galapagos Giant Tortoises," 88–9; and "Tantalizing Tortoises and the Darwin-Galápagos Legend," 18–21.

[25] Adalgisa Caccone, "Evolution and Phylogenetics," in *Galapagos Giant Tortoises*, ed. James P. Gibbs, Linda J. Cayot, and Washington Tapia Aguilera (Cambridge, MA: Academic Press, 2020), 117–38; Jack Frazier, "The Galapagos: Island Home of Giant Tortoises," in *Galapagos Giant Tortoises*, ed. James P. Gibbs, Linda J. Cayot, and Washington Tapia (Cambridge, MA: Academic Press, 2020), 3–21.

[26] William Dampier, *A New Voyage Round the World, in A Collection of Voyages*, vol. 1, 7th ed. (London: James and John Knapton, 1729), 202; these same words by Dampier are quoted by Captain Robert FitzRoy, *Narrative of the Surveying Voyages of His Majesty's Ships Adventure and Beagle, between the Years 1826 and 1836, Describing Their Examination of the Southern Shores of South America, and the Beagle's Circumnavigation of the Globe, Vol.2: Proceedings of the Second Expedition, 1831–1836, under the Command of Captain Robert FitzRoy, R.N.* (London: Henry Colburn, 1839), 505.

board thirty adult San Cristóbal tortoises as a source of fresh meat for the voyage across the Pacific.[27] Still not sufficiently aware of the tortoises' scientific importance as the *Beagle* sailed to Tahiti, Darwin stood by as all the San Cristóbal tortoises were consumed by the *Beagle* crew. The inedible remains, including the diagnostically valuable carapaces, were thrown overboard.

Darwin and his servant Syms Covington did bring back to England two baby tortoises – one from Floreana (Covington's tortoise) and one from Santiago (Darwin's tortoise). During the remainder of the voyage, these two tortoises were treated as pets.[28] Following his return to England, Darwin finally realized that these two young tortoises might provide evidence to support the Vice Governor's claims about the differences by island. Unfortunately, the two tortoises were both too immature to provide any meaningful comparison.[29]

During the *Beagle* voyage, Darwin seems to have reached the same conclusion about his four specimens of Galápagos mockingbird that he initially did about the tortoises. In a famous passage written in his "Ornithological Notes" almost a year after his Galápagos visit, Darwin reflected on these birds:

When I recollect, the fact that from the form of the body, shape of scales & general size, the Spaniards can at once pronounce, from which Island any Tortoise may have been brought. When I see these islands in sight of each other, & possessed of but a scanty stock of animals, tenanted by these birds, but slightly differing in structure & filling the same place in Nature, I must suspect they are only varieties.[30]

It is worth noting that some modern ornithologists have fully agreed with Darwin's tentative *Beagle* voyage verdict here by recognizing the Galápagos mockingbirds as only a single species, subdivided into four localized subspecies.[31] Darwin did appreciate, however, that if he were wrong about these birds being "just varieties," such novel evidence might pose a considerable empirical challenge to creationist dogma. He therefore concluded this same "Ornithological Notes" discussion about the mockingbirds by asserting, "If there is the slightest foundation for these remarks the zoology of Archipelagos – will be well worth examining; for such facts would undermine the stability of Species."[32]

---

[27] FitzRoy, *Narrative of the Surveying Voyages of His Majesty's Ships Adventure and Beagle*, 2:498.

[28] Sulloway, "Darwin's Conversion," 344; Sulloway, "Tantalizing Tortoises and the Darwin-Galápagos Legend," 21; Aaron M. Bauer and Colin J. McCarthy, "Darwin's Pet Galápagos Tortoise, *Chelonoidis darwinia*, Rediscovered," *Chelonian Conservation and Biology*, 9 (2010), 270–6.

[29] Darwin, *Journal of Researches*, 465, 628.   [30] Darwin, *Darwin's Ornithological Notes*, 262.

[31] John Davis and Alden H. Miller, "Family Mimidae," in *Check-List of Birds of the World: A Continuation of the Work of James L. Peters*, ed. Ernst Mayr and James, C. Greenway Jr. (Cambridge, MA: Museum of Comparative Zoology, 1960), Vol. 9, 447–8.

[32] Darwin, *Darwin's Ornithological Notes*, 262.

5.5   The Critical Influence of John Gould

The evidence ultimately required to "undermine the stability of Species" emerged after Darwin's return to England when ornithologist John Gould carefully examined Darwin's ornithological collections early in 1837.[33] After seeing the four Galápagos mockingbird specimens, Gould emphatically insisted that they belonged to three different species. Gould also informed Darwin that if these mockingbirds were not distinct species, then "the experience of all the best ornithologists must be given up, and whole genera must be blended into one species."[34]

Darwin first learned of Gould's taxonomic conclusions about the mocking-birds at a meeting with Gould at the London Zoological Society during the second week of March 1837. It was during this same meeting that Gould corrected Darwin's previous misclassifications of the Galápagos finches, placing all these birds in four closely related subgenera.[35] It was Gould's confident taxonomic authority that subsequently led Darwin to record the following famous entry in a private journal: "In July [1837] opened first notebook on 'Transmutation of Species' – Had been greatly struck from about Month of previous March on character of S. American fossils – & species on Galápagos Archipelago. These facts origin (especially latter) of all my views."[36]

In sum, the Galápagos mockingbirds, not the tortoises or the finches, provided the straw that finally broke the camel's back in terms of Darwin's ultimate willingness to accept the mutability of species. Only then did Darwin begin to appreciate that the finches, like the mockingbirds, might originally have evolved on different islands in the Galápagos group and become, over time, an even more remarkable example of evolutionary divergence from a common ancestor. With the Darwin-Galápagos legend set aside, we can now see that Darwin's ultimate decision to accept an evolutionary interpretation of his *Beagle* voyage evidence was heavily dependent on the seemingly mun-dane but indispensable practice of biological classification as this endeavor was performed by his scientific peers following his return to England. The Darwinian Revolution began in earnest when Darwin finally understood just how arbitrary the practice of taxonomy could be when systematists, operating

---

[33] Sulloway, "Darwin's Conversion," 359–69; John Gould, "Remarks on a Group of Ground Finches from Mr. Darwin's Collection, with Characters of the New Species," *Proceedings of the Zoological Society of London*, 5 (1837), 4–7; John Gould, "Three Species of the Genus Orpheus, from the Galapagos, in the Collection of Mr. Darwin," *Proceedings of the Zoological Society of London*, 5 (1837), 27.

[34] Darwin, *The Zoology of the Voyage of H.M.S. Beagle, Part III: Birds*, 63–4.

[35] Sulloway, "Darwin and His Finches," 21; Sulloway, *The* Beagle *Collections of Darwin's Finches (Geospizinae), Bulletin of the British Museum (Natural History) Zoology Series*, 43, no. 2 (1983), 57–8.

[36] Darwin, *Darwin's Journal*, 7.

within a creationist paradigm, were confronted with the problem of "doubtful species." This was also the crucial moment when Darwin first began to understand the novel concept of "population thinking" – that is, the realization that biological variation, properly conceived, is not just a taxonomic nuisance but rather the first step in the formation of new species.[37]

---

[37] On the revolutionary nature of Darwin's transition from typological to population thinking, see Ernst Mayr, *The Growth of Biological Thought: Diversity, Evolution, and Inheritance* (Cambridge: Harvard University Press, 1982), pp. 45–7, 487–8. In the *Origin of Species*, Darwin repeatedly uses the phrase "doubtful species," including in the running heads of chapter 2, to highlight his argument that varieties are "incipient species" (pp. 44, 47–52, 58, 409).

Myth 6   That Darwin's Galápagos Finches Inspired His Most Important Evolutionary Insights

Frank J. Sulloway

6.1   Introduction

If Darwin's iconic Galápagos finches did not play a decisive role in his conversion to the theory of evolution, as I have argued in the preceding chapter of this book, then what influence did these birds actually have in his evolutionary theorizing? Before this question can be answered, there is a notable mystery that must first be addressed. This mystery involves a crucial point of historical evidence emphasized by David Lack in his 1947 book *Darwin's Finches*.[1] According to Lack, after Darwin landed on the second of the four islands he visited in the Galápagos and learned there that the giant tortoises differed by island, he began to label the Galápagos finch specimens by island. Such attention to island localities implies that Darwin understood, at the time, that this information was necessary to establish that geographical isolation can create new species. As it turns out, Lack was wrong, but it took me almost a decade to ascertain why, and, more importantly, to understand how island localities ended up on the labels of Darwin's own type specimens, which now reside at the British Museum of Natural History (Figure 6.1).

6.2   Reconstructing the Finch Localities

The solution to this mystery concerning Darwin's Galápagos finch localities was eventually provided by two previously overlooked manuscript documents in the Darwin Archives at Cambridge University. In these two documents, which Darwin drafted sometime after his meeting with John Gould in March 1837, Darwin solicited locality information for the twenty-five finch specimens procured by three other *Beagle* collectors.[2] Unlike Darwin, these other collectors had

I am grateful to Kostas Kampourakis, Malcolm J. Kottler, Gregory Radick, Michael Ruse, Smriti Mehta and Daniel Sherman for helpful comments on earlier versions of this chapter.
[1] David Lack, *Darwin's Finches: An Essay on the General Biological Theory of Evolution.* Cambridge: Cambridge University Press, 1947), 23.
[2] Frank J. Sulloway, "Darwin and His Finches: The Evolution of a Legend," *Journal of the History of Biology*, 15 (1982), 1–53.

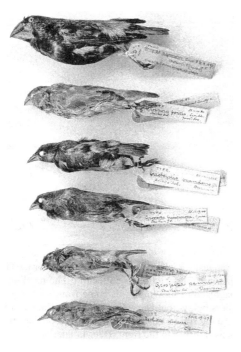

Figure 6.1 Five of the type specimens of Darwin's Galápagos finches. Shown from top to bottom are the Large Ground Finch (*Geospiza magnirostris*), the Medium Ground Finch (*G. fortis*), the Cactus Finch (*G. scandens*), the Small Ground Finch (*G. fuliginosa*), the Small Tree Finch (*Camarhynchus parvulus*), and the Warbler Finch (*Certhidea olivacea*). Confusion about the island localities can be seen on some of the tags. The first specimen is labeled as coming from "Chatham? [Charles]" but appears to have been collected on Chatham.[3] The fifth specimen, labeled as having been collected on both "Chatham" and "James?", belongs to the James Island form of this species.[4] © Trustees of the Natural History Museum, London. Photo: Frank J. Sulloway.

recorded island localities for all of their Galápagos birds (Figures 6.2, 6.3). Darwin also tried to supplement this locality information by making guesses about the localities of some of his own specimens, and he later published this presumed locality information, along with the known localities from the three other *Beagle* collections, in the *Zoology of the Voyage of H.M.S. Beagle* (1841).[5]

---

[3] Frank J. Sulloway, "The *Beagle* Collections of Darwin's Finches (Geospizinae), *Bulletin of the British Museum (Natural History)," Zoology Series*, 43, no. 2 (1983), 63, 77.

[4] Sulloway, "The *Beagle* Collections of Darwin's Finches," 65, 77.

[5] Charles Robert Darwin, *The Zoology of the Voyage of H.M.S. Beagle, under the Command of Captain FitzRoy, R.N., during the Years 1832–1836*. Part III: *Birds* (London: Smith, Elder, 1841). For evidence of Darwin's attempt to second-guess the localities for some of his own Galápagos finch specimens, see the previous chapter of this volume (Figure 5.1).

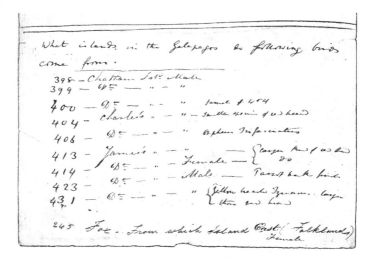

Figure 6.2 Darwin's request for island locality information for the Galápagos birds included in Captain Robert FitzRoy's collection. The heading "What islands in the "Galapagos" do following birds come from" is in Darwin's hand, as are the numbers below that Darwin copied from Captain FitzRoy's specimen catalogue. To the right of these numbers are responses made by an unidentified amanuensis. Further to the right, Darwin later added notes of his own. Written at the bottom, in the hand of a second unidentified amanuensis who is known to have worked for Darwin after the *Beagle* voyage, is an additional question about the locality of a specimen of the Falkland Islands fox, which was also answered by the first amanuensis. (DAR 29.3: 26; courtesy of the Syndics of Cambridge University library.)

It was this published locality information that curators at the British Museum of Natural History subsequently used to provide island localities for Darwin's Galápagos finch specimens, as these curators mistakenly assumed that Darwin's published localities were based on his own specimens. However, more than half of these locality assignments turned out to be erroneous, as Darwin's specimens did not necessarily come from the same islands from which other accurately labeled finch specimens had been procured. As a result of the various uncertainties introduced by these attempts to second-guess the localities for Darwin's own specimens, even some of Captain FitzRoy's valid island identifications became subject to doubt.[6] These confusions about the type specimen localities in turn created a taxonomic nightmare

[6] Sulloway, "Darwin and His Finches"; Sulloway, "The *Beagle* Collections of Darwin's Finches," 60–62; and Harry S. Swarth, "The Avifauna of the Galapagos Islands," *Occasional Papers of the California Academy of Science*, 18 (1931), 1–299.

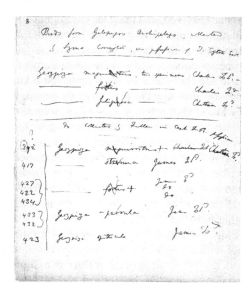

Figure 6.3 Darwin's record of responses to his request for locality information for the Galápagos finches collected by his servant, Syms Covington, whose birds had been acquired by ornithologist Thomas Eyton, as well as localities for the finches collected by Harry Fuller, which were still in the possession of Captain FitzRoy. Darwin's check marks show his collation of this information for subsequent publication in the *Zoology of the Voyage of H.M.S. Beagle* (1841). (DAR 29.3: 30; courtesy of the Syndics of Cambridge University library.)

for subsequent ornithologists who were naturally puzzled about why some of Darwin's specimens did not match the morphological profiles observed for specimens collected later on the same islands. Some ornithologists even speculated that these discrepancies were attributable to evolution having occurred since Darwin's visit.[7] Over the years, as different ornithologists added their own notes to the specimen tags, the tags became increasingly laden with contradictory information.

6.3 Darwin's Galápagos Bobolink

Even though I felt reasonably confident, based on the newly discovered manuscript evidence, that Darwin had not labeled any of his own Galápagos finches by island, I was mindful of the fact that almost none of Darwin's original

---

[7] Swarth, "The Avifauna of the Galápagos Islands," 147–9.

Beagle voyage specimen tags had survived, and that Darwin might conceivably have included locality information on his own tags, which were later discarded by museum curators and replaced with printed museum labels. As I was studying Darwin's *Beagle* type specimens at the British Museum of Natural History in 1980, it occurred to me that there might actually be a Galápagos specimen still bearing Darwin's original tag. During Darwin's nine-day stay on Santiago, his shipmate Harry Fuller shot a bobolink – a bird migrating at the time through the Galápagos Islands. As a personal favor, Fuller gave this bird to Darwin so that Darwin would have the most complete collection.[8] After Darwin's return to England, John Gould recognized this migrant specimen as belonging to an already described North American species (*Dolichonyx oryzivorus*). Because the specimen was not new, it was placed in a drawer at the British Museum of Natural History where it remained for the next century in the company of dozens of other bobolink specimens.

When I succeeded in locating this original *Beagle* specimen, I found that it contained a crude paper label with a single number inscribed on it – "3374." Written in Darwin's own hand, this is the same number Darwin assigned to the specimen in his specimen catalogue and "Ornithological Notes."[9] What is of particular importance about this original *Beagle* label is the absence of any island locality. This one surviving Galápagos specimen tag thus establishes that Darwin, after his visit to Floreana (and two weeks before acquiring this Santiago bobolink specimen), did not begin to systematically record island localities on his own specimen tags.

When I finally understood how Darwin's type specimens had acquired their conflicting island localities, I realized the need for a monographic treatment of all four *Beagle* collections, three of which were at the British Museum of Natural History and the fourth of which I had recently identified at Cambridge University.[10] Made by Captain FitzRoy's steward Harry Fuller, this fourth collection included eight Galápagos finch specimens of previously unknown provenance. These birds were clearly identifiable as *Beagle* specimens by the numbers on the tags, which were part of the same numerical sequence in Captain FitzRoy's Galápagos collection. Based on morphological evidence for beak size and wing length, as well as on the newly discovered manuscript evidence, I was able to supply what I thought were correct localities for almost all of the fifty-six *Beagle* specimens. Three decades later, when Ken

---

[8] Charles Robert Darwin, *Darwin's Ornithological Notes*, ed. with an Introduction, Notes and Appendix by Nora Barlow. *Bulletin of the British Museum (Natural History) Historical Series*, 2, no. 7 (1963 [1836]). On the *Beagle* officers giving Darwin choice of the best specimens, see Captain Robert FitzRoy's November 16, 1837 letter to Darwin, in Charles Robert Darwin, *The Correspondence of Charles Darwin*, vol. 2: 1837–1843, ed., Frederick Burkhardt and Sydney Smith (Cambridge: Cambridge University Press, 1986), 58.

[9] Darwin, *Darwin's Ornithological Notes*, 265; Sulloway, "Darwin and His Finches," 16.

[10] Sulloway, "The *Beagle* Collections of Darwin's Finches."

Petren and colleagues undertook a DNA analysis of these same *Beagle* specimens, I was relieved to find that none of the new DNA evidence contradicted
my own previous island reassignments.[11]

6.4   Darwin's Finches and the *Origin of Species* (1859)

I have recounted this story about the mixed-up localities for Darwin's
Galápagos finches because it explains, in large part, why these finches were
never mentioned by Darwin in the *Origin of Species*. Although Darwin suspected that some of the finches might be confined to different islands, he and the
other *Beagle* collectors had not done sufficient collecting to be entirely sure.
Moreover, Darwin's own speculative, post-voyage efforts to assign localities to
these specimens were plagued by errors, as were his associated claims that
some of the species were confined to different islands.[12] In fact, only four of the
seventeen species of Galápagos finches are restricted to single islands, and
none of these four species were collected during the *Beagle*'s Galápagos visit.
Wherever individual Galápagos finch species may first have evolved, most of
them have long since spread to other islands in the group, and a majority of the
islands are inhabited by eight or more of the currently recognized seventeen
species.[13] Of further relevance, after the *Origin of Species* was published, an
extensive new collection of birds from the Galápagos was made in 1868 by
Dr. Simeon Habel. In his taxonomic description of the new specimens, Osbert
Salvin found it necessary to assert that "Mr. Darwin's views as to the exceedingly restricted range of many of the [finch] species must be considerably
modified."[14]

What really gave Darwin confidence that species of birds and giant land
tortoises in the Galápagos Islands had evolved into new "representative species" owing to geographic isolation were the results of Joseph Hooker's
taxonomic assessment of Darwin's Galápagos plants, which Darwin received
in 1845 just in time to include in the second edition of his *Journal of
Researches*. Of the 109 endemic plant species that Darwin and several prior
collectors had obtained in these island, 85 (or a remarkable 78 percent) turned

[11] Kenneth Petren, Peter R. Grant, B. Rosemary Grant, Andrew A. Clack, and Ninnia V. Lescano,
"Multilocus Genotypes from Charles Darwin's Finches: Biodiversity Lost Since the Voyage of
the *Beagle*," *Philosophical Transactions of the Royal Society*, 365 (2010), 1009–18.

[12] Charles Robert Darwin, *Journal of Researches into the Natural History and Geology of the
Countries Visited during the Voyage of H.M.S. Beagle round the World. . . .* 2nd ed. (London:
John Murray, 1845), 395; Sulloway, "The *Beagle* Collections of Darwin's Finches," 60–2.

[13] Peter R. Grant, B. Rosemary Grant, Erik D. Enbody, Leif Andersson, and Sangeet Lamichhaney,
"Darwin's Finches, an Iconic Adaptive Radiation." *Encyclopedia of Life Sciences*, 1 (2020),
672–82.

[14] Osbert Salvin, "On the Avifauna of the Galapagos Archipelago," *Transactions of the Zoological
Society of London*, 9 (1876), 447–510, at 461.

out to be confined to just a single island.[15] In response to Hooker's extensive statistical findings, Darwin responded: "I cannot tell you how delighted and astonished I am at the results of your examination; how wonderfully they support my assertion on the differences in the animals of the different islands, *about which I have always been fearful.*"[16] What Darwin meant by being "fearful" about his previous Galápagos findings was the fact that taxonomic decisions about the mockingbirds and finches, as well as the giant land tortoises, were inevitably somewhat arbitrary, especially without knowing whether these geographically isolated and putative species might successfully interbreed (in accordance with the biological species concept that Darwin endorsed in the wake of his conversion to the theory of evolution).[17] Darwin himself acknowledged this problem in the 1845 edition of his *Journal of Researches* when he noted that "some of these representative species, at least in the case of the tortoise and some of the birds, may hereafter prove to be only well-marked races," although he added that "this would be of equally great interest to the philosophical naturalist."[18]

Darwin's recognition that some of his Galápagos evidence was subject to doubt highlights a historical irony. He was converted to evolution when John Gould, a creationist working with a narrow and predominantly morphological species concept, convinced Darwin that three of his four Galápagos mockingbirds were unquestionably distinct species.[19] Yet these and other taxonomic judgments soon persuaded Darwin that Gould's morphological approach to species distinctions was inherently arbitrary, at least in close cases like the mockingbirds and some of the finches.[20] What still inclined Darwin to accept the fact of evolution was the totality of the *Beagle* evidence, including numerous other instances of representative species on the South American continent

---

[15] Joseph D. Hooker, "Enumeration of the Plants in the Galapagos Islands, with Descriptions of the New Species," *Proceedings of the Linnean Society of London*, 1 (1846), 276–9; Joseph D. Hooker, "An Enumeration of the Plants of the Galapagos Archipelago; with Descriptions of Those which Are New," *Transactions of the Linnean Society of London*, 20 (1847), 163–233.

[16] Darwin, *Correspondence*, vol. 3: 1844–1846, 216 (emphasis added).

[17] Malcolm J. Kottler, "Darwin's Biological Species Concept and Theory of Geographic Speciation: The Transmutation Notebooks," *Annals of Science*, 35 (1978), 275–8; Frank J. Sulloway, "Geographic Isolation in Darwin's Thinking: The Vicissitudes of a Crucial Idea," *Studies in the History of Biology*, 3 (1979), 23–65.

[18] Darwin, *Journal of Researches*, 397.

[19] In his first notebook on the "Transmutation of Species," Darwin quoted John Gould as saying, "the beauty of species is their exactness" while responding in rebuttal that "varieties do the same" (Charles Robert Darwin, *Charles Darwin's Notebooks, 1836–1844*, edited by Paul H. Barrett, Peter J. Gautrey, Sandra Herbert, David Kohn, and Sydney Smith [Ithaca, NY: British Museum (Natural History) and Cornell University Press]), 213. Later, in the manuscript of "Natural Selection," Darwin wrote that "no one will accuse [Gould] of running varieties together" (Charles Robert Darwin, *Charles Darwin's Natural Selection: Being the Second Part of His Big Species Book Written from 1856 to 1858*, ed. R. C. Stauffer [Cambridge: Cambridge University Press, 1975], 282; see also p. 144).

[20] Sulloway, "Geographic Isolation in Darwin's Thinking," 27–8.

as well as the compelling evidence for fossil succession among edentates and other South American quadrupeds.[21]

In the *Origin of Species* Darwin again alluded to the problem of the "doubtful" taxonomic status of his various Galápagos species when he asserted: "Many years ago, when comparing, and seeing others compare, the birds from the separate islands of the Galapagos Archipelago, both with one another, and with those from the American mainland, I was much struck with how entirely vague and arbitrary is the distinction between species and varieties."[22] In this same work Darwin gave this problem a positive spin by asserting that varieties are simply "incipient species" and that many "doubtful species" inevitably exist in nature owing to the process of evolution.[23] However, for this argument to be convincing, Darwin had to show that some incipient species really have gone beyond a "doubtful" status by becoming clearly delineated "representative species." What was therefore so important to Darwin about Hooker's taxonomic enumeration of the Galápagos plants were the sheer numbers involved. Although John Gould might conceivably have been wrong about the four Galápagos mockingbird specimens belonging to three distinct species, or in how he subdivided the finches, surely Joseph Hooker was not wrong that some of the 109 endemic Galápagos plants were true representative species, endemic to individual islands within that group. Although Darwin did not single out the Galápagos finches in the *Origin of Species*, he did call attention to Joseph Hooker's impressive Galápagos botanical findings in this work.[24] In short, although the Galápagos plants have not acquired the same iconic status in modern evolutionary biology as have Darwin's famous finches, these plants provided considerably better evidence for evolution via geographic isolation than did either the mockingbirds, tortoises or finches.

Although Darwin included a brief discussion of the Galápagos finches in his unpublished manuscript "Natural Selection," his intent there was simply to provide an example of evolution from a single ancestral species into multiple descendant species, just as he had done, more guardedly, in the second edition of his *Journal of Researches*. In this 1845 edition he also included the now-iconic illustration of beak diversity among the finches (Figure 6.4). But without information concerning the dietary differences between the finch species, and also without more extensive and reliable locality information, Darwin was limited in what he could actually say about these birds in the *Origin of Species*. He nevertheless did mention the finches indirectly when he noted that "nearly every land-bird, but only two out of the eleven marine birds, are

[21] Sulloway, "Darwin's Conversion," 369, 371–9.
[22] Charles Robert Darwin, *On the Origin of Species by Means of Natural Selection, or, The Preservation of Favoured Races in the Struggle for Life* (London: John Murray, 1859), 48.
[23] Darwin, *Origin of Species*, 52, 404.   [24] Darwin, *Origin of Species*, 391, 398.

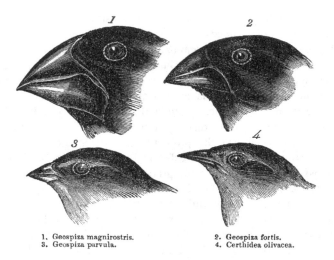

1. Geospiza magnirostris.
3. Geospiza parvula.

2. Geospiza fortis.
4. Certhidea olivacea.

Figure 6.4 Four of the Galápagos finch species, depicted in the second edition of Darwin's *Journal of Researches*. In one of the most frequently quoted passages from this work, which Darwin added to this 1845 edition, he wrote: "The most curious fact is the perfect gradation in the size of the beaks in the different species of Geospiza, from one as large as that of a hawfinch to that of a chaffinch, and (if Mr. Gould is right in including his sub-group, Certhidea, in the main group), even to that of a warbler. ... Seeing this gradation and diversity of structure in one small, intimately related group of birds, one might really fancy that from an original paucity of birds in this archipelago, one species had been taken and modified for different ends."[25]

peculiar" to the Galápagos Archipelago.[26] This, however, was the closest he ever came to mentioning these birds in this book.

As for the myth that Darwin derived his theory of natural selection from his study of the Galápagos finches and their diverse beaks, Darwin tells us clearly in his *Autobiography* how he first grasped this theory in September 1838 after reading the sixth edition of Thomas Robert Malthus's *Essay on the Principle of Population*, published in 1826.[27] There he recounts:

Being well prepared to appreciate the struggle for existence which everywhere goes on from long-continued observation of the habits of animals and plants, it at once struck me

---

[25] Darwin, *Journal of Researches*, 379–380.   [26] Darwin, *Origin of Species*, 390.

[27] Thomas Robert Malthus, *An Essay on the Principle of Population; or, a View of Its Past and Present Effects on Human Happiness; with an Enquiry into Our Prospects Respecting the Future Removal or Mitigation of the Evils which It Occasions*, 6th ed., 2 vols. (London: John Murray, 1826).

that under these circumstances favourable variations would tend to be preserved, and unfavourable ones to be destroyed. The result would be the formation of new species. Here, then, I had at last got a theory by which to work.[28]

Neither the Galápagos finches, nor any of the other Galápagos species, played a role in this revolutionary insight.

6.5   Functions and Consequences of the Darwin-Galápagos Legend

Collectively, the four myths I have reviewed in this and the preceding chapter of this book represent a relatively typical instance of how, in the retelling, history tends to become telescoped into a more compelling and memorable story. In the case of important scientific discoveries, this retelling often involves a "heroic" element that sometimes includes a sudden moment of brilliant insight befitting the mind of a genius.[29] With its various engaging components, the Darwin-Galápagos legend conveys the story of a man who recognized something quite extraordinary during a long and dangerous voyage of exploration, and who then returned to face the intense controversy that his new scientific discoveries unleashed upon an unprepared and hostile world. In the end, so the story also goes, Darwin triumphed not only because the scientific evidence he presented in the *Origin of Species* was remarkably compelling but also because the primary source of this evidence, Darwin's iconic Galápagos finches, has continued to reaffirm the brilliance and validity of his theories. The myth was so powerful that it even led to a falsification of the actual scientific evidence so that Darwin might appear, contrary to historical fact, to have carefully recorded island localities for all of his own specimens of the now-famous Galápagos finches.

But while the myths are compelling, they nevertheless do us a considerable disservice if we really wish to understand the nature of science as well as the often complex processes entailed in scientific discovery. In particular, the myths underestimate how a prior belief in creationist theory, which offered plausible and often persuasive explanations for the same evidence pointing toward evolution, made it difficult for Darwin, as well as most of his scientific contemporaries, to see their way clear to the scientific revolution that now bears his name. The myths also obscure the manner in which scientific insight is typically a process involving many intermediate steps – including errors, doubts, and revisions of previous conclusions – before a more coherent

[28] Charles Robert Darwin, *The Autobiography of Charles Darwin*, ed. Nora Barlow (London: Collins, 1958 [1876]), 120.

[29] Frank J. Sulloway, *Freud, Biologist of the Mind: Beyond the Psycho-analytic Legend* (New York: Basic Books, 1979), 445–95; Alberto A. Martínez, *Science Secrets: The Truth about Darwin's Finches, Einstein's Wife, and Other Myths* (Pittsburgh: University of Pittsburgh Press, 2011).

understanding eventually emerges and finds its way into print. Instead, the myths portray a certainty of insight that only came much later.

Additionally, by presenting a "heroic" version of history, the myths about Darwin's famous Galápagos visit overlook the collective nature of most scientific research. In this particular instance, the myths fail to recognize the crucial assistance provided by Darwin's *Beagle* shipmates in helping to rectify his failure to label his own Galápagos birds by island, as well as the insightful taxonomic analyses that John Gould and other British naturalists supplied after Darwin's return to England. As it turns out, it was this considerable assistance that, for Darwin, finally tipped the scales in favor of evolutionary theory by facilitating a much fuller and more informed understanding of his *Beagle* voyage findings.

The Darwin-Galápagos legend also overlooks another of Darwin's most important biological insights. Prior to publication of the *Origin of Species*, most naturalists considered variation as having only secondary importance relative to the primary objective of identifying new Linnaean species and providing them with formal taxonomic descriptions. Once a new species was suitably named and described, a carefully selected "type" specimen was then placed in an accessible museum collection to provide an exemplar of the species' essential characteristics. Throughout the *Beagle* voyage, Darwin's own collecting procedures reflected this typological approach to natural history, as he usually collected only one or two specimens of each species (including a male and female whenever sexual dimorphism was present). In the Galápagos Islands, Darwin procured only thirty-one specimens of the now-famous finches, leaving John Gould with the almost impossible task of properly ascertaining the limits of within-species variability in this diverse avian group. By contrast, when the California Academy of Science returned to the Galápagos Islands in 1905 – informed by a Darwinian perspective on the importance of individual variation – they collected more than 5,000 specimens of Darwin's finches. It was David Lack's later study and detailed statistical analyses of this massive collection that finally allowed these birds to assume the iconic status they have today in evolutionary theory.

When Darwin, in early 1837, witnessed the inability of experienced ornithologists to agree on what were, and what were not, new species among his *Beagle* collections, he began to understand that species can never be adequately represented by a single specimen, no matter how carefully any individual specimen has been chosen to exemplify the "type." Rather, he realized that species are the sum of all their individual variants, and that this evidence of within-population variability is not only required to resolve taxonomic disputes but is also crucial for appreciating that evolution sometimes precludes clear-cut Linnaean subdivisions in nature.[30] Once Darwin fully understood this new way

---

[30] On the importance of taxonomy in Darwin's post-voyage evolutionary thinking, see Mary P. Winsor, "Taxonomy Was the Foundation of Darwin's Evolution," *Taxon*, 58 (2009), 43–9.

of thinking about species, he also recognized the folly of some of his own voyage collecting procedures. Shortly after his momentous insight into the principle of natural selection in September 1838, Darwin asserted in one of his "Transmutation Notebooks" that the "naming of mere single specimens in skins [is] worse than useless – I may say this, having myself aided in such sins."[31] In short, it was only after Darwin appreciated just how much variability there is in nature, and how this overlooked variability undermines a Linnaean belief in immutable essences, that he was fully prepared to appreciate how a mechanism like natural selection might give rise to new species.

[31] Darwin, *Charles Darwin's Notebooks*, "Notebook E," 52.

Myth 7   That Darwin Was a Recluse, and a Theoretician Rather Than a Practical Scientist

Alison M. Pearn

7.1   Introduction

In April 1855 the *Gardeners' Chronicle and Agricultural Gazette*, a "newspaper of rural economy and general news" published a letter from "Charles Darwin, Down, Farnborough, Kent." The letter asked readers to suggest plants Darwin could use in an experiment. He was testing whether seeds would still germinate after being soaked in salt-water, hoping to demonstrate that islands could have been colonized by plants from distant locations. Darwin wanted to prove that the existence of such apparently isolated communities of related species was not evidence for separate acts of creation, and did not undermine theories of common descent. "I have in small bottles out of doors ... Cress, Radish, Cabbages, Lettuces, Carrots, Celery, & Onion seed; 4 great Families," Darwin wrote to his botanist friend, Joseph Hooker: "These after immersion for exactly one week, have all germinated, which I did not in the least expect ... for the water of nearly all & of the cress especially, smelt very badly, & the cress-seed emitted a wonderful quantity of mucus."[1]

If you find it hard to picture Darwin up to his elbows in smelly, slimy water, you are not alone. The image of Darwin as a lone thinker, a "supreme theorizer"[2] rather than a hands-on scientist, and a "recluse"[3] who worked

---

Any discussion of Darwin's life necessarily relies on the work of my colleagues past and present in the Darwin Correspondence Project, but I especially acknowledge in this context the invaluable insights and expertise of my fellow editor, Shelley Innes.

[1] *Correspondence* no. 1667, letter to J. D. Hooker, 13 April [1855]. All references to Darwin's *Correspondence* refer to Frederick Burkhardt, James A. Secord, Samantha Evans, et al. eds. *The Correspondence of Charles Darwin*. 30 vols. Cambridge: Cambridge University Press 1985–2023. All of Darwin's letters may also be accessed online by letter number at darwinproject.ac.uk.

[2] "When he landed in October 1836, the vicarage had faded, the gun had given way to the notebook, and the supreme theorizer – who would always move from small causes to big outcomes – had the courage to look beyond the conventions of his own Victorian culture for new answers." www.britannica.com/biography/Charles-Darwin/Evolution-by-natural-selection-the-London-years-1836-42.

[3] "In September, 1842, he went into retirement at Down, an out-of-the-way village in Kent. There, partly compelled by ill-health, he dwelt as a recluse for forty years, serenely contemplating nature

largely in isolation, is stubbornly persistent. Accounts that do feature him as a practical researcher tend to emphasize the domestic setting of his work, focusing on experiments that can be replicated in a modern house, garden or school.[4] But Darwin was an ingenious and innovative experimenter, keenly aware of advances in science, and often at the cutting edge both in the nature of his investigations and in the technologies he employed.

His letter to the *Gardeners' Chronicle*, four years before the publication of the *Origin of Species* illustrates Darwin's highly practical approach to science.[5] He was constantly devising ways to test his theories: at the same time as soaking seeds in salt water he was also freezing them in vats of snow and feeding them to fish, stocking an aviary with fancy pigeons for experiments on cross-breeding, and commissioning schoolchildren to gather and count the seeds of campion and parsnip plants.

7.2   An Empirical Science Based on Collaboration

The "gathering and grinding of facts," both through close observation and through experiment, were fundamental to Darwin's approach. "True science" for Darwin involved rigorous testing and a willingness to recognize when new facts challenged theory. "I have long thought," he wrote, "that naturalists make far too few experiments."[6] Asked to give advice to young biologists, he replied that few good or original discoveries had ever been made without speculation. "He who speculates much ought, however, to learn," he continued, "and must learn to a certain extent, if his work is to be worth anything, to give up repeatedly and manfully his most cherished views, if facts run counter, as they generally do, against every first-formed theory."[7] He stressed the importance of thorough documentation and the value of negative results, writing to one correspondent, "I hope you will record all facts, such as the number of plants you experimentise on, their names &c &c. . . . Negative facts (ie failures) are as important to know as successes."[8]

Darwin's way of working was iterative: he told Alfred Russel Wallace that he was "a firm believer, that without speculation there is no good & original

and diligently gathering information, but seldom emerging into the world from which his richly-stored and phenomenally creative intellect had little to gain but to which it never ceased to give, during the remainder of his life." Charles F. Cox, "The Individuality of Charles Darwin," in Hovey, Edmund Otis ed. 1909. Darwin memorial celebration. *Annals of the New York Academy of Sciences* 19, No. 1, Part 1 (31 July): 1–40.

[4] For a recent example see James T. Costa. 2017. *Darwin's Backyard: How Small Experiments Led to a Big Theory.* New York: W. W. Norton.

[5] C. R. Darwin. 1859. *On the Origin of Species by Means of Natural Selection, or the Preservation of Favoured Races in the Struggle for Life.* London: John Murray.

[6] *Correspondence* no. 2939, Charles Darwin to Daniel Oliver, 5 October [1860].

[7] *Correspondence* no. 7778F, Charles Darwin to C. L. Balch, 29 May 1871.

[8] *Correspondence* no. 1137, to Abraham Clapham [29 October 1847?].

observation,"[9] but speculation needed facts to guide it.[10] Having a "theory by which to work"[11] provided focus to his research, but he was wary of too much untested theorizing, joking about the "sin of speculation." Keenly aware that he had been unable to test his proposed mechanism of inheritance, "Pangenesis," through experiment, he characterized it as "abominably wildly, horridly speculative."[12] Although he published it, and never gave up the idea entirely, it was carefully quarantined in an appendix to his book *Variation under domestication*, where it is cautiously labelled a "provisional hypothesis" (Burkhardt, Chapter 8, this volume).

Far from working alone on gathering facts and grinding out his theories, Darwin actively sought collaboration with all sorts of people around the world, both asking for their help and encouraging their own investigations. He was conscious of how much he owed to the help of others, flattering one obliging correspondent by declaring that if any proof of the goodness of his fellows was wanted, one had only to do as Darwin was doing and "pester them with letters."[13]

So where does Darwin's "lone thinker" reputation come from, and why is it so hard to shift?

7.3   Origins of the Myth

There is no single explanation for the dominance of the popular image, but it is perhaps not surprising. Darwin worked largely at home, never held down a paying job, never gave a public lecture, and never performed any formal scientific function other than a stint as secretary to the Geological Society. He never traveled abroad again after the *Beagle* voyage, and he rarely appeared at scientific meetings.

Surviving images of Darwin do nothing to dispel the myth. There are no photographs or portraits of him with any of the tools of the trade – no photographs in front of a blackboard like Thomas Henry Huxley, or holding a geological hammer like Hugh Falconer, or on a field expedition like Joseph Hooker – and there is only one image of him with another human being, a photograph taken with his baby son, William, in 1840. This is also one of only a very few photographs of Darwin without the beard he grew in his fifties. No wonder we think of him as a solitary old man.

Darwin's scientific reputation was first established through his theoretical work on the role of large-scale subsidence in the formation of coral reefs,

---

[9] *Correspondence* no. 2192, to A. R. Wallace, 22 December 1857.
[10] *Correspondence* no. 2703, to Charles Lyell, 18 [and 19 February 1860].
[11] Darwin's "Autobiography." Cambridge University Library DAR 26.
[12] *Correspondence* no. 5051, to J. D. Hooker [9 April 1866].
[13] *Correspondence* no. 5986, Charles Darwin to J. J. Weir [6 March 1868].

a beautifully constructed argument, but not demonstrated experimentally until the US Atomic Energy Commission drilled on Eniwetak atoll, Northern Marshall Islands, in 1952.[14] For most of his life Darwin's main vehicle for communicating his ideas to a wider public was the *Origin of Species* and that did little to redress the balance. He was forced for the sake of speed and brevity to leave out most of his supporting evidence, and most of the experimental work he referred to was done by others. Of those he had carried out himself, besides the seed-salting experiment, he described in any detail only one other – on the origins of honeycomb-making behavior in bees – but credited its design to the poultry breeder, William Bernhard Tegetmeier. It was only in the last ten years of his life that Darwin published detailed accounts of his experiments in popularly accessible books, as opposed to scientific papers. The exception is his 1862 work on orchid fertilization, which he thought might interest a wider readership.[15]

Darwin's own language about his practical work underplays both his skill and its role in his achievements. When he talked about "work" he generally meant time spent writing, which he found hard. By contrast, Darwin really enjoyed experiments and thought of these "hobbies" as guilty pleasures. Of his pioneering study of the reaction of plants to external stimuli, he wrote, "My hobby-horse at present is Tendrils; they are more sensitive to a touch than your finger; & wonderfully crafty & sagacious."[16] He was fond of making "fools' experiments" and asking "foolish questions," but found foolish questions often led to good results, and that fools' experiments "tho' rarely successful in a direct manner, have often led to interesting side-results."[17]

7.4   Francis Darwin as Mythmaker

One of the chief culprits in crafting Darwin's posthumous image is his son Francis. The first widely available account detailing Darwin's working life was in the edition of letters that Francis published five years after Darwin's death.[18] Writing with nostalgic affection, and perhaps keen to make his father human and relatable, Francis opened his account by introducing a gangling, shambling, physically awkward man of limited artistic ability (that last is demonstrably true). Although he goes on to describe the meticulous way his father

---

[14] Brian Rosen. 1982. "Darwin, Coral Reefs, and Global Geology." *BioScience* 32 (6): 519–25.

[15] C. R. Darwin. 1862. *On the Various Contrivances by Which British and Foreign Orchids Are Fertilised by Insects, and on the Good Effects of Intercrossing. By Charles Darwin*. London: John Murray.

[16] *Correspondence* no. 4199, to W. E. Darwin [25 July 1863].

[17] *Correspondence* nos. 3898, to J. D. Hooker, [3 January 1863], and 13214, to J. B. Hannay, [22 June 1881].

[18] Francis Darwin, ed. 1887–8. *The Life and Letters of Charles Darwin, Including an Autobiographical Chapter*. 3 vols. London: John Murray.

conducted and recorded experiments, Francis emphasized Darwin's preference for "simple methods and few instruments," favoring an old-fashioned single lens rather than compound microscope, and dissecting specimens on a piece of board on the windowsill. His tools had an air of "simpleness, make-shift, and oddness": measuring devices were old and inaccurate, or made by the local carpenter, or doubled as household utensils, and although Francis denied that this invalidated the results of the experiments, the cumulative impression is of a rather charming naivety. Although he referred to the importance of his father's discoveries, reinforcing Darwin's scientific reputation was not Francis's aim. His is a deliberately intimate account of a much-missed parent and is rich in anecdote. And it is of course written from the perspective of a younger generation. Francis was only thirty-four when his father died, and although he was formally employed as his father's assistant for the last eight years of Darwin's life, his account, while closely observed, is at first-hand only for the very end of a long career.[19]

7.5   An Extensive Network of Correspondents

A quite different picture emerges however from the contemporary evidence of Darwin's letters, including those exchanged with Francis himself.

The very fact of the letters' existence – more than 15,000 survive, exchanged with many hundreds of correspondents around the world – is evidence of Darwin's active engagement with a wide range of collaborators. The letters he received were an integral part of his working papers, interleaved with notes, cut up and stuck in his experiment book, scribbled on, and filed in research portfolios. He cultivated useful contacts. His correspondence networks were populated with a changing cast of characters as his research needs changed: plant and animal breeders were joined by medics and anthropologists as Darwin turned to writing on human evolution, by physiologists and chemists as he studied movement and digestion in plants, and archaeologists as he explored the behavior of earthworms.

Darwin was expert at cultivating and exploiting contacts. His forty-year friendship with Joseph Hooker, son of the director of Kew Gardens and later director himself, gave Darwin access to the resources of the foremost botanical research facility of the time, the hub of an international network of botanic gardens. Hooker and his colleagues provided Darwin with exotic specimens, identified plants, made observations, and performed experiments on his behalf.

Darwin's correspondence networks resembled distributed research groups. He often asked correspondents to perform experiments on his behalf or arranged to have his own experiments repeated by others: Tegetmeier not

---

[19] Darwin, *The Life and Letters*, 1: 145–50.

only undertook to repeat one of Darwin's experiments but arranged for a contact in Ireland to do so too.[20] When Francis Darwin, a botanist in his own right, spent time working in university laboratories in Germany, Darwin tapped into the expertise and resources there, suggesting experiments on plant physiology, and getting information on the latest techniques. He corresponded with the brothers Fritz and Hermann Müller, one in Brazil and the other in Germany, and the three of them exchanged seeds to observe the effects of different habitats, repeated one another's experiments, and suggested new lines of enquiry. Asa Gray, professor of botany at Harvard, and Darwin's chief North American correspondent, recruited others such as the independent researcher, Mary Treat, whose work Darwin encouraged and who in return sent Darwin observations and results of her own experiments on insectivorous plants.

7.6   Family and Friends at Work

Darwin's version of working from home was far from solitary. He was surrounded not only by a large family who actively engaged with his work, but by governesses, gardeners, and other servants. The German governess, Camilla Ludwig, translated books and letters, the gardeners helped with plant-breeding experiments, and his extended family made observations, wrote letters, edited and illustrated his books, and took part in experiments. Although even today his home at Downe in Kent feels quite rural, it was deliberately chosen to be within easy reach of London. He may not often have gone to scientific meetings, but he enjoyed visiting fellow scientists, and they visited him. Asa Gray's wife, Jane, wrote a detailed description of one stay with the Darwins: there were seven house guests in all, and a succession of neighbors dropped in for meals. Although Darwin never breakfasted with his guests he always dropped in "with some merry remark," and dinners were pleasant and sociable. Discussions were wide-ranging and lively, and at one point the guests found themselves actively contributing to Darwin's research – they became experimental subjects. Jane Gray's account of meeting Darwin underlines what a warm and engaging host he was with an ever-present sense of humor. He had, she said, "the sweetest smile, the sweetest voice, the merriest laugh": "he is full of his great theories & sees the smallest things that bear upon them, & laughs more merrily than anyone at any flaw detected, or fun made – Full of warmest feelings & quick sympathy."[21]

The experiment involved no fewer than twenty-five visitors, both at Down and in London, between March and November 1868. Darwin asked his guests

---

[20] *Correspondence* no. 4979, from W. B. Tegetmeier [after 24 January 1866].
[21] Jane Gray to Susan Loring, 28 October–2 November 1868. Archives of the Gray Herbarium, Box G AG-B10: 8. For a transcription, see www.darwinproject.ac.uk/people/about-darwin/family-life/visiting-darwins.

to look at a selection of photographic portraits supposedly capturing particular emotions. The photographs had been taken by a French physiologist with some expressions having been produced by artificial stimulation of the facial muscles. Darwin wanted to test whether the photographs were convincing. He recorded the results in a series of tables and refined his technique as he went along: he began by asking whether his visitors thought the photographs were accurate portrayals, recording only "yes/no" answers, but switched to asking instead what emotion they thought the image conveyed.[22]

7.7   A Diligent Experimenter

Many of Darwin's experiments were unglamorous. He spent nearly a decade carefully crossing generations of plants, excluding insects, pollinating them by hand, and painstakingly counting the seeds they produced to establish the greater vitality of crossed varieties. As he reported in a follow-up letter to the *Gardeners' Chronicle*, his seed-salting experiments were successful in showing the resilience of seeds, but they threw up another problem: the seeds sank. So he started again on a new tack. Might they survive by being eaten by migrating birds? Could the seed-bearing plants float? Or float long enough to be eaten by fish, eaten in turn by birds? Might birds transport seeds on their feet? He tested all these possibilities over the following years, soliciting muddy birds' feet from several correspondents and successfully raising eighty-two plants from seeds stuck to a diseased partridge foot.[23]

When he needed technical solutions, however, Darwin was ingenious in designing and refining them. This is particularly true of his investigation into animal-like behavior in plants, beginning with the kinds of movement that allow them to twine and climb, and moving on to the ability of some plants not only to attract and trap insect prey, but to digest it. He devised methods for recording the circular movements of tendrils, leaf-tips, and roots by attaching the finest possible glass filaments to them in order to amplify their movement (Figure 7.1).

From the beginning of his career Darwin took a keen interest in the equipment he used, selecting and, if necessary, adapting instruments to get the best tools for the job. He commissioned a particular type of iron drag net for collecting shells in advance of the *Beagle* voyage.[24] He took at least two, and

---

[22] For details of the experiment see www.darwinproject.ac.uk/commentary/human-nature/expres sion-emotions/emotion-experiment.

[23] *Correspondence* nos. 4436, to J. D. Hooker, 26[–7] March [1864] and 5300, to J. D. Hooker 10 December [1866].

[24] *Correspondence* nos. 123, to J. S. Henslow, 9 [September 1831], and 124, from John Coldstream, 13 September 1831.

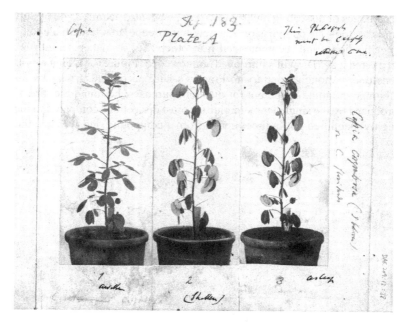

Figure 7.1 Darwin's experimental specimen of *Cassia corymbosa* "awake," "shaken," and "asleep." The photographs formed the basis for an illustration in his book *The Power of Movement in Plants*. Credit: Cambridge University Library, DAR 209.12: 38. Reproduced by permission of the Syndics.

possibly three, microscopes with him for use both on board and in the field, and later published detailed advice for other travelers.[25]

Working from home for Darwin did not mean a second-rate working environment. For his early work on adaptation in orchids Darwin relied on a steady stream of exotic plants from Kew Gardens, but in order to continue to experiment more widely he built, at considerable expense, a series of temperature-controlled hothouses in the grounds of Down House so that he could cultivate them himself. One of his first acquisitions was the insectivorous pitcher plant, *Nepenthes*. He was wealthy enough to buy the best equipment and to replace it when it became outdated, and he kept up to date with trends. To undertake the eight-year detailed taxonomic study of barnacles that followed the *Beagle* voyage, work that still shapes the field, Darwin not only became expert in the

[25] C. R. Darwin. 1849. "On the Use of the Microscope on Board Ship." C. R. Darwin, "On the use of the microscope on board ship," within R. Owen, "Zoology," in J. F. W. Herschel (ed.), *A Manual of Scientific Enquiry; prepared for the Use of Her Majesty's Navy: and adapted for Travellers in General* (London, 1849), pp. 389–95.

techniques of dissecting under a microscope but also acquired and modified the tools to do it. He commissioned a new microscope, this time made to his own specifications, and recommended it to others.[26] The makers, Smith & Beck, marketed it as "Darwin's improved" microscope (Figure 7.2). This was a single lens one, although, contrary to Francis's later account, Darwin did also use a compound microscope but for different purposes: he found the direct image produced by the single lens superior for most dissections compared to the more highly magnified but indirect image from a compound one.[27] When the same

Figure 7.2 Compound microscope bought from Smith & Beck in 1847 and used by Darwin for his research on barnacles. Credit: Image (c) Whipple Museum, Cambridge (Wh.3788).

[26] *Correspondence* no. 1167, to John Stevens Henslow [1 April 1848]. For more on Darwin's use of the microscope see Boris Jardine. 2009. "Between the *Beagle* and the Barnacle: Darwin's Microscopy, 1837–1854." *Studies in History and Philosophy of Science*, 40: 382–95.
[27] *Correspondence* no. 4136, to Isaac Anderson-Henry, 2 May [1863].

company later developed a stereoscopic microscope Darwin knew about it in advance and advised his son William to trade up.[28]

Francis Darwin spent two summers working in German laboratories and described the equipment being used there to his father. They commissioned Francis's brother Horace, an engineer, to make improved versions of both a klinostat, designed to rotate plants and exclude the effects of geotropism, and an auxanometer, which was used for measuring plant movement.[29] It was also Horace who designed and made the "worm stone" that Darwin installed in the garden at Down to measure the effect of worm action on soil level.

When he needed equipment that he couldn't access himself Darwin tapped into the facilities and expertise of a number of institutions. As part of his work on insectivorous plants, he exposed them to smaller and smaller doses of various substances, acquiring some, such as cobra poison – not readily available at home even then – from contacts working in laboratories. When Darwin could no longer achieve sufficiently controlled doses or the levels of purity required, the physiologist John Burdon Sanderson continued Darwin's experiments for him and expanded them to use the kind of electrical equipment employed in animal experimentation.

Although he never worked in one himself, Darwin was an active advocate for the importance of laboratories. He helped fund the establishment of an international marine biological research station in Naples and wrote in strong support of a similar one in France.[30] When it seemed as if the practice of vivisection might be banned, he lobbied the government to allow it to continue as a regulated practice, orchestrating the drafting of an alternative bill to send to parliament, and even appearing before a Royal Commission. In 1878 Darwin was consulted on equipment for a new laboratory at Kew Gardens. His response encapsulates his practical, well-informed approach to science:

I have a very strong opinion that it would be the greatest possible pity if the Phys: Lab., now that it has been built, were not supplied with as many good instruments as your funds can possibly afford. It is quite possible that some of them may become antiquated before they are much or even at all used. But this does not seem to me any argument at all against getting them ... I should think the German laboratories would be very good guides as to what to get; but Timiriazeff of Moscow who travelled over Europe to see all Bot. Labs, & who seemed so good a fellow, would I should think give the best list of the most indispensable instruments. Lately I thought of getting Frank or Horace to go to Cambridge for the use of the Heliostat there ... it would make a great difference if a man knew that he could use a particular instrument without great loss of time.[31]

---

[28] *Correspondence* no. 3447, to William Erasmus Darwin, 14 February [1862].

[29] *Correspondence* nos. 11613, from Francis Darwin [before 17 July 1878] and 11541, to Francis Darwin, 3 June [1879].

[30] *Correspondence* no. 13719, to Jules Barrois [after 6 March 1882].

[31] *Correspondence* no. 11425, to W. T. Thiselton-Dyer, 14 March 1878.

Of course, things did not always go well: "I am rather low today about all my experiments," he complained to his cousin, "everything has been going wrong – the fan-tails have picked the feathers out of the Pouters in their Journey home – the fish at the Zoological Gardens after eating seeds would spit them all out again – Seeds will sink in salt-water – all nature is perverse & will not do as I wish it."[32] But by the next day he'd picked himself back up again: "I am tremendously busy with all sorts of experiments."[33]

7.8   Conclusions

The term "lone genius" has a disputed history of its own quite separate from its use for Darwin, and there is no single origin for its application to him, but it is useful shorthand for a number of entwined popular misconceptions: that he was a theoretician – a thinker rather than a practical scientist, that his ideas were wholly new and developed in isolation from other developments in science, and that he was reclusive to the point of pathology. Contemporary evidence demonstrates by contrast that Darwin was not just a hands-on, innovative experimenter, but at the cutting-edge of technology, and that he was socially skilled – a gifted networker, who engaged with a varied cast of collaborators at a local, national, and international level.

---

[32] *Correspondence* no. 1678, to W. D. Fox 7 May [1855].
[33] *Correspondence* no. 1679, to Charles Lyell, 8 May [1855].

Myth 8   That Darwin Rejected Lamarck's Ideas of Use and Disuse and of the Inheritance of Acquired Traits

Richard W. Burkhardt, Jr.

8.1   Introduction

For students who are taught that Darwin's theory of evolution by natural selection succeeded where Jean Lamarck's (Figure 8.1) theory of evolution by the inheritance of acquired characters failed,[1] it is easy to believe (as their teachers usually believe as well), that Darwin rejected the idea of the inheritance of acquired characters. Here is how a widely used textbook put it:

A rival theory [to Darwin's theory of natural selection], championed by the prominent biologist Jean-Baptiste Lamarck, was that evolution occurred by the inheritance of acquired characters. According to Lamarck, individuals passed on to offspring body and behavior changes acquired during their lives. . . . In Darwin's theory, by contrast, the variation is not created by experience, but is the result of preexisting genetic differences among individuals.[2]

Darwin himself was keen to distance his theorizing from Lamarck's. For instance, in a letter to J. D. Hooker of January 11, 1844, he wrote: "Heaven forfend me from Lamarck nonsense of a 'tendency to progression' 'adaptations from the slow willing of animals' &c,— but the conclusions I am led to are not widely different from his – though the means of change are wholly so."[3] However, he strongly supported the idea of the inheritance of acquired characters, treating it as an important supplement to natural selection, the primary

---

[1] I considered this myth previously in Richard W. Burkhardt Jr., "That Lamarckian evolution relied largely on use and disuse and that Darwin rejected Lamarckian mechanisms," in Ronald L. Numbers and Kostas Kampourakis (eds.), *Newton's Apple and Other Myths about Science* (Cambridge, MA: Harvard University Press, 2015), pp. 80–7. There I gave equal attention to Lamarck and Darwin's respective ideas on the subject. Here my focus is on Darwin and afterward. For more on Lamarck's side of the equation, see the above chapter and Richard W. Burkhardt, *The Spirit of System: Lamarck and Evolutionary Biology* (Cambridge, MA: Harvard University Press, 1st ed. 1977, 2nd ed. 1995) and "Lamarck, evolution, and the inheritance of acquired characters," *Genetics*, 194 (2013), pp. 793–805.

[2] Peter H. Raven and George B. Johnson, *Biology*, 6th ed. (New York: McGraw Hill, 2002), p. 422.

[3] Darwin Correspondence Project, "Letter no. 729," accessed on October 20, 2022, www.darwinproject.ac.uk/letter/?docId=letters/DCP-LETT-729.xml.

Figure 8.1 Jean Baptiste Pierre Antoine de Monet, chevalier de Lamarck.
Photogravure by Schutzenberger after C. Thévenin, 1801. Credit: Wellcome
Collection. Public Domain Mark

process of his theory. But this aspect of Darwin's theorizing has come to be
forgotten over time. Instead, the acquired belief that Darwin rejected the idea of
the inherited effects of use and disuse has been transmitted from one generation
of students to the next. It is a particularly ironic case of cultural evolution
mimicking the theory of biological evolution now linked with Lamarck's name.

The belief that Darwin rejected Lamarck's idea of the inheritance of acquired
characters may not rise to the level of a myth, but it is at the very least a common
misunderstanding. One has only to read any of Darwin's major writings to recognize its falsity. From the first edition of the *Origin of Species* in 1859 to the sixth and
last edition of 1872, Darwin represented the hereditary effects of use and disuse as
a significant supplement to natural selection in bringing about organic change (see
also Love, Chapter 9, this volume). Likewise, in his *Variation of Animals and
Plants Under Domestication* (first edition 1868, second edition 1875), he cited
numerous instances of the inheritance of acquired characters and included inheritance of this sort among the "grand classes of facts" that his theory of variation,
heredity, and development (his "provisional hypothesis of pangenesis") was
designed to explain. Finally, in both editions of *The Descent of Man* (1871 and
1874), he stated explicitly that in the crucial period when the ancestors of
humans were changing into bipeds, "natural selection would probably have been
greatly aided by the inherited effects of the increased or diminished use of

the different parts of the body."[4] Darwin, in sum, did not reject the idea of the inheritance of acquired characters.

8.2   Lamarck and Species Change

The centerpiece of Lamarck's theory of species change (though not his explanation of organic change writ large), is commonly and properly seen to have been constituted by the two famous "laws" he offered in his *Philosophie Zoologique* (Zoological Philosophy) of 1809.[5]

First Law: In every animal that has not gone beyond the period of its development, the more frequent and sustained use of any organ whatever strengthens this organ little by little, develops it, enlarges it, and gives to it a power proportionate to the duration of this use; while the constant disuse of such an organ insensibly weakens it, deteriorates it, progressively diminishes its faculties, and finally causes it to disappear.

Second Law: All that nature has caused individuals to acquire or lose by the influence of the circumstances to which their race has been exposed for a long time, and, consequently, by the influence of a predominant use of such an organ or constant disuse of such a part, she conserves through generation in the new individuals that descend from them, provided that these acquired changes are common to the two sexes or to those which have produced these new individuals.[6]

The way these laws fit into Lamarck's view of species change was straightforward. By his account, animals develop new habits in response to changes in the environmental circumstances surrounding them, thereby using some organs more and some organs less than previously. The resulting changes in instincts, faculties, and structures are passed on by generation, and little by little, over immense periods of time, species change occurs.

The hereditary effects of use and disuse constituted the second, more subsidiary part of Lamarck's evolutionary theory. The first and more fundamental factor was what he called "the power of life," which he considered responsible for the increasing organic complexity displayed in the animal scale. However, when it came to thinking about change at the species level, the "two laws" that he announced in the *Philosophie Zoologique* sum up his thinking nicely.

Lamarck offered a variety of examples of animal structures shaped by habits, including the webbed feet of swimming birds, the long necks and legs of

---

[4] C. Darwin, *The Descent of Man, and Selection in Relation to Sex* (London: John Murray, 1871), vol. 1, p. 121.

[5] J.-B.-P.-A. Lamarck, *Philosophie zoologique, ou exposition des considérations relatives à l'histoire naturelle des animaux; à la diversité de leur organisation et des facultés qu'ils en obtiennent; aux causes physique qui maintiennent en eux la vie et donnent lieu aux mouvemens qu'ils exécutent; enfin à celles qui produisent, les unes le sentiment, et les autres l'intelligence de ceux qui en sont doués*, 2 vols. (Paris: Dentu, 1809) 235.

[6] Lamarck, *Philosophie zoologique*, vol. 1, p. 235.

shorebirds, the reduced wings of flightless birds, the reduced eyes of animals that burrow underground or live in caves, and more. The giraffe example was not the first that occurred to him, but it later became the example with which he is most frequently associated.[7]

8.3   Darwin's Examples of the Inheritance of Acquired Characters

In the *Origin of Species*, in his chapter on "Laws of Variation," Darwin explicitly addressed the topic of "effects of use and disuse," writing, "I think that there can be little doubt that use in our domestic animals strengthens and enlarges certain parts, and disuse diminishes them; and that such modifications are inherited."[8] Arguing that comparable effects could be seen in the wild, he offered "the nearly wingless condition of several birds," such as the ostrich, as an example of a change in structure caused by disuse.[9] When it came to wingless beetles, however, such as those found on the island of Madeira, he was inclined to think that their wingless state was primarily the result of natural selection, "but combined probably with disuse."[10] (His argument was that beetles that flew more than others would have been more likely than the others to blow out to sea and perish.) As for the rudimentary eyes of the moles and the blindness of many cave animals, he decided they were due either to disuse exclusively or to some combination of disuse and selection.[11] To explain pointing in hunting dogs, he called on a combination of artificial selection and the inherited effects of habit, but when it came to the tameness of domestic rabbits, he thought this was a case where "habit alone ... has sufficed."[12] He did not address the question of how the giraffe got its long neck until St. George Mivart pressed him on the subject, whereupon Darwin in 1872, in the sixth edition of the *Origin of Species*, explained that natural selection could account for the feature but then added, "combined no doubt in a most important manner with the inherited effects of the increased use of parts."[13]

We thus find Darwin in the *Origin of Species* assessing the relative importance of natural selection and the inheritance of acquired characters on a case-by-case basis, sometimes giving one factor precedence over the other and sometimes giving them roughly equal weight. He was unequivocal in his overall assessment, however, that natural selection was the primary process of evolutionary change

---

[7] Lamarck, *Philosophie zoologique*, vol. 1, pp. 256–7.

[8] C. Darwin, *On the origin of species by means of natural selection, or the preservation of favoured races in the struggle for life* (London: John Murray, 1859, 1st ed.), p. 134.

[9] Darwin, *Origin* (1859), p. 134.   [10] Darwin, *Origin* (1859), p. 136.

[11] Darwin, *Origin* (1859), pp. 137–9.   [12] Darwin, *Origin* (1859), pp. 213–15.

[13] Darwin, *Origin* (London: John Murray, 1872, 6th ed.), p. 178. Later in the same volume, pp. 198–9, he said the same thing: natural selection was sufficient for producing the giraffe's long legs and neck "but the prolonged use of all the parts together with inheritance will have aided in an important manner in their co-ordination."

and that the inheritance of acquired characters was secondary.[14] Beginning with the 1866 (fourth edition) of the *Origin of Species*, he entitled his discussion of use and disuse, "The Effects of the increased Use and Disuse of Parts, *as controlled by Natural Selection*."[15]

Darwin also took special pride, from the first edition of the *Origin of Species* onward, in pointing to specific adaptations that natural selection could explain but the inheritance of acquired characters could not, most notably, the structures and instincts of neuter castes of insects. "Habit," he wrote, no doubt sometimes comes into play in modifying instincts; but it certainly is not indispensable, as we see, in the case of neuter insects, which leave no progeny to inherit the effects of long-continued habit."[16] Interestingly enough, this is the only place in Darwin's published writings that he explicitly associated the idea of use-inheritance with Lamarck's name, stating, "I am surprised that no one has advanced this demonstrative case of neuter insects, against the well-known doctrine of Lamarck."[17]

8.4   The Inheritance of Acquired Characters and Pangenesis

Additional instances where Darwin thought that use-inheritance *had* operated to produce organic change are abundant in his 1868 work, *The Variation of Animals and Plants Under Domestication*. He allowed that it was "notorious" (a word he used for something very well known) "that increased use or action strengthens muscles, glands, sense-organs, &c; and that disuse, on the other hand, weakens them."[18] (This is the equivalent of Lamarck's "First Law," though Darwin did not identify it as such.) With respect to the effects of use or disuse being *inherited*, Darwin offered evidence for that too, citing, for example, the drooping ears found in many different domestic animals as compared to the erect ears of their wild counterparts. "The incapacity to erect the ears," he wrote, "is certainly in some manner the result of domestication." The cause, he decided, was "probably" the "effects of disuse" (as some writers had already suggested). He explained, wild animals "constantly use their ears like funnels to catch every passing sound," but "animals protected by man" did not have the same need or same habit.[19]

Darwin conducted careful studies on anatomical differences between wild and domesticated rabbits, concluding with respect to the brain of the domestic rabbit, that its smaller size, proportional to body weight, was the result of disuse.[20] He likewise compared the skeletons of the wild mallard ducks and

---

[14] The other such causes he identified he were "correlated growth" and "the direct action of the conditions of life."

[15] Darwin, *Origin* (London: John Murray, 1866, 4th ed.), p. 160 (emphasis added).

[16] Darwin, *Origin* (1859), p. 474.   [17] Darwin, *Origin* (1859), p. 242.

[18] C. Darwin, *Variation* (London: John Murray, 1868, 1st ed.), Vol. II, 295.

[19] Darwin, *Variation*, II, 301.   [20] Darwin, *Variation*, Vol. I, 129.

different domesticated ducks, showing that in the latter the wing bones were smaller and the leg bones larger. He attributed this to domesticated ducks flying less and walking more than wild ducks (an example to which Lamarck had pointed more than half a century earlier).[21] He attributed not only structural changes but also behavioral and mental changes to the inherited effects of use and disuse.[22]

On the other hand, he cautioned that one should not be too ready to ascribe organic changes to use and disuse or the direct action of changed conditions, when the natural selection of individual variations might be the true cause. "For instance," he wrote,

it is possible that the feet of our water-dogs, and of the American dogs which have to travel much over the snow, may have become partially webbed from the stimulus of widely extending their toes; but it is far more probable that the webbing, like the membrane between the toes of certain pigeons, spontaneously appeared and was afterwards increased by the best swimmers and the best snow-travellers being preserved during many generations.[23]

Darwin's "Provisional Hypothesis of Pangenesis" appeared as chapter XVII of his *The Variation of Animals and Plants Under Domestication*. With this hypothesis he sought to account for a host of phenomena (some genuine, others now recognized as fictive) that he associated with variation, heredity, and development. "Pangenesis" rested on two postulates: (1) that the organic units of the body (essentially, the cells) "throw off minute granules or atoms, which circulate freely throughout the system," and (2) that these granules "when supplied with proper nutriment multiply by self-division, subsequently becoming developed into cells like those from which they were first derived."[24] Calling his hypothetical particles "gemmules," Darwin endowed them with such extraordinary versatility that they could account (in principle) for such diverse phenomena as variation, reversion, regeneration, normal growth, and the inheritance of acquired characters. As he put it:

It is equally or even more unintelligible on any ordinary view, how the effects of the long-continued use or disuse of any part, or of changed habits of body or mind, can be inherited. A more perplexing problem can hardly be proposed; but on our view we have

---

[21] Darwin, *Variation*, I, 284–6; II, 371. Lamarck, *Philosophie zoologique*, I, 260.

[22] Darwin, *Variation*, II, 371–2.   [23] Darwin, *Variation*, II, 419.

[24] Darwin, *Variation*, II, 374. A great many historical papers address Darwin's hypothesis of pangenesis. Among them see Gerald L. Geison, "Darwin and Heredity: The Evolution of His Hypothesis of Pangenesis," *Journal of the History of Medicine and Allied Sciences*, 24 (1969), 375–411; M. J. S. Hodge, "Darwin as a Life-long Generation Theorist," in David Kohn (ed.), *The Darwinian Heritage* (Princeton: Princeton University Press, 1985), pp. 207–43; Jim Endersby, "Darwin on Generation, Pangenesis and Sexual Selection," in M. J. S. Hodge and Gregory Radick (eds.), *The Cambridge Companion to Darwin*, 2nd ed. (Cambridge: Cambridge University Press, 2003), pp. 73–95; Kate Holterhoff, "The History and Reception of Charles Darwin's Hypothesis of Pangenesis," *Journal of the History of Biology*, 47 (2014), 661–95.

only to suppose that certain cells become at last not only functionally but structurally modified; and that these throw off similarly modified gemmules.[25]

Darwin also had an explanation for why it took generations before changes of habit produced changes in structure: "the gemmules derived from the same cells after modification, naturally go on increasing under the same favouring conditions, until at last they become sufficiently numerous to overpower and supplant the old gemmules."

8.5   Change of Habit as an Agent in Evolution

During his *Beagle* voyage, Darwin made diverse observations that stuck in his mind and that later, as he thought back on them, appeared to bear on the question of species mutability. With these in mind, Darwin hazarded in the first edition of the *Origin of Species* the following scenario. It has something of the flavor of several of Lamarck's examples of organic change, but here natural selection rather than the acquired characters is the purported cause:

In North America the black bear was seen by [the explorer Samuel] Hearne swimming for hours with widely open mouth, thus catching, like a whale, insects in the water. Even in so extreme a case as this, if the supply of insects were constant, and if better adapted competitors did not already exist in the country, I can see no difficulty in a race of bears being rendered, by natural selection, more and more aquatic in their structure and habits, with larger and larger mouths, till a creature was produced as monstrous as a whale.[26]

Darwin's contemporaries made fun of the example.[27] The bear and whale story did not disappear completely in later editions of the *Origin of Species*, but he pared it down severely, leaving it simply to read: "In North America the black bear was seen by Hearne swimming for hours with widely open mouth, thus catching, almost like a whale, insects in the water."[28] For our purposes, what is interesting about Darwin's example in his original rendering of it is that it represented a scenario where change in habit preceded change in structure, *but not in a Lamarckian fashion.* Unlike Lamarck's accounts of the origins of the long legs and neck of the wading birds or the giraffes, Darwin portrayed a situation in which a race of bears, after taking up new habits in a new setting, were transformed, in competition with other forms, via the natural selection of small, individual variations.

This scenario of animals adopting new habits and initiating evolutionary change by doing so, but without the inheritance of acquired characters and without natural selection driving the process initially, was reintroduced at the

---

[25] Darwin, *Variation*, II, 395.   [26] Darwin, *Origin* (1859), p. 184.
[27] See the Darwin Correspondence Project and various of Darwin's letters, including Darwin to Andrew Murray, 28 April [1860]: www.darwinproject.ac.uk/letter/DCP-LETT-2772.xml.
[28] Darwin, *Origin*, 2nd ed. (London: John Murray, 1860), p. 184.

end of the century and came to be called "the Baldwin effect."[29] The idea was championed again in the middle of the twentieth century by Alister Hardy, who argued eloquently for the idea of behavioral change as a driving force in evolution (at the level, at least, of birds and mammals). For animals with that level of intelligence, changes of habit could spread by imitation or learning, thereby promoting moves into new niches, which then set the stage for the selection of any mutations or genetic rearrangements that would then make the new habits more efficient. The idea of behavior initiating evolution has found increasing favor since the 1980s, for example in Patrick Bateson's idea of the "adaptability driver": "an organism's choices, its construction of a niche for itself, its adaptability, and its mobility, have all played important roles in biological evolution."[30] The mechanism is not Lamarckian, insofar as it does not involve the inheritance of acquired characters. However, it does represent a view in which the organism is an active rather than passive agent in the evolutionary process.

8.6   The Emergence of the Idea That Darwin Did Not Believe in the Inheritance of Acquired Characters

Darwin was irked by critics who maintained that he believed that evolution depends only on natural selection.[31] St. George Mivart was one of the culprits. In his 1871 book, *On the Genesis of Species*, Mivart referred variously to "the pure Darwinian position;" "pure Darwinism, which makes use *only* of indirect modifications through the survival of the fittest" (italics in the original); and "the purely Darwinian theory, which relies upon the survival of the fittest by means of minute fortuitous indefinite variations." Not surprisingly, he called those who believed in the above-named positions "pure Darwinians."[32] Whereas Mivart opposed "pure Darwinism," however, Alfred Russel Wallace promoted it. In his 1889 book entitled *Darwinism*, Wallace maintained that he was advocating "pure Darwinism," that is, the belief that natural selection was by far the preeminent factor in evolution. In the volume, he offered "new

---

[29] On Morgan, Baldwin, Osborne, and the Baldwin effect, see Robert J. Richards, *Darwin and the Emergence of Evolutionary Theories of Mind and Behavior* (Chicago: University of Chicago Press, 1987).

[30] For Hardy's views, see Alister S. Hardy, *The Living Stream* (London: Collins, 1965). For later work, see, for example, J. S. Wyles, J. G. Kunkel, and A. C. Wilson, "Birds, behavior and anatomical evolution," *Proceedings of the National Academy of Sciences of the United States of America*, 80 (1983), 4394–7; and Patrick Bateson, "New thinking about biological evolution," *Biological Journal of the Linnaean Society*, 112 (2014), 268–75.

[31] See, for example, Darwin, *Origin of Species*, 6th ed. (London: John Murray 1872), p. 176; *The Descent of Man*, 2nd ed. (London: John Murray, 1874), p. v; "Sir Wyville Thomson and natural selection," *Nature*, 23 (November 11, 1880), 32.

[32] St. George Mivart, *On the Genesis of Species* (London: Macmillan, 1871), pp. 22, 75, 76, 83, 110.

(a)   (b)

Figure 8.2 Francis Galton (left; credit: from the frontispiece of Pearson, Karl. 1914–30. The life, letters and labors of Francis Galton. Cambridge: Cambridge University Press. S405.b.92.2 Cambridge University Library) and August Weismann (right; Credit: L. I. Gardner, Molecular genetics. Wellcome Collection. Attribution 4.0 International (CC BY 4.0))

evidence" for the "non-heredity of acquired characters" and "a proof that the effects of use and disuse, even if inherited, must be overpowered by natural selection."[33]

By this point, the heredity of acquired characters had already received two significant challenges, the first from Darwin's cousin, Francis Galton, the second from the German embryologist August Weismann (Figure 8.2). Galton's challenge emerged in a stepwise fashion. It appeared first in his 1865 paper on "hereditary talent and character," where he allowed that "there are but a few instances in which habit even seems to be inherited." He proceeded to dismiss such instances, including the familiar example of pointers pointing, as susceptible of alternative explanation.[34] He followed this up six years later with

[33] Alfred Russel Wallace, *Darwinism* (London: Macmillan, 1889), pp. xi–xii. Wallace's advocacy of "pure Darwinism" included his opposition to Darwin's views on sexual selection.

[34] Francis Galton, "Hereditary talent and character," *Macmillan's Magazine*, 12 (1865), 157–66, 318–27, quote from p. 322. On Galton see Ruth Schwartz Cowan, "Nature and nurture: the

his critique of pangenesis, which came after his blood transfusion and breeding experiments designed to support Darwin's theory of "gemmules" produced only negative results.[35] From there he went on to develop his own theory of heredity, which had no place for the inheritance of acquired characters. He did not, however, launch a direct campaign against Lamarck's signature idea, but he simply continued to allow that the inheritance of acquired characters was at most a rare and inconsequential occurrence.

August Weismann's critique of the inheritance of acquired characters, in contrast, was a frontal assault. His opening salvo was his 1883 essay, "On heredity," where he fixed his sights on "the transmission of acquired characters which has been hitherto assumed to occur." Making a sharp distinction between germ cells and body cells, he posited that the former transfer their hereditary tendencies from one generation to the next, "always uninfluenced in any corresponding manner, by that which happens during the life of the individual which bears it." If his hypothesis were correct, he said, then "the whole principle of evolution by means of exercise (use and issue), as proposed by Lamarck, and accepted in some cases by Darwin, entirely collapses."[36]

Weismann's attack proceeded on two fronts. The first was his broad theoretical claim that germ cells remain separate from, and wholly uninfluenced by, the body cells. The second was his analysis and debunking of the diverse cases that had been cited in support of the inheritance of acquired characters. Weismann's experiments cutting off mouse tails for successive generations, designed to establish the noninheritance of mutilations, fit into this latter category. The English zoologist Edward B. Poulton would later call Weismann's attack on use-inheritance "the most stimulating shock that has been received by the biological world since the appearance of the *Origin of Species*."[37]

"Lamarckian" theories of various kinds existed aplenty at the turn of the century,[38] but Weismann's challenge put the proponents of the idea of

interplay of biology and politics in the work of Francis Galton," *Studies in the History of Biology*, 1 (1977), 133–208; Michael Bulmer, "The development of Francis Galton's ideas on the mechanism of heredity," *Journal of the History of Biology*, 32 (1999), 263–92.

[35] Francis Galton, "Experiments in Pangenesis, by breeding from rabbits of a pure variety, into whose circulation blood taken from other varieties had previous been largely transfused," *Proceedings of the Royal Society*, 19 (1871), 393–410. See Bulmer, "The development," pp. 272–6.

[36] August Weismann, "On Heredity, 1883," in *Essays Upon Heredity and Kindred Biological Problems*, E. B. Poulton, S. Schönland, and A. E. Shipley, eds., 2 vols. (Oxford: The Clarendon Press, 1889), 1, 67–106, quote from p. 69.

[37] Edward Bagnall Poulton, *Essays on Evolution* (Oxford, 1908), p. 139.

[38] See Vernon L. Kellog, *Darwinism To-Day* (New York: Henry Holt, 1907) and Peter J. Bowler, *The Eclipse of Darwinism: Anti-Darwinian Evolution Theories in the Decades around 1900* (Baltimore: Johns Hopkins University Press, 1983) on the diversity of evolutionary theorizing at the turn of the century.

use-inheritance on the defensive, and the decades that followed were not kind to the Lamarckians.[39] With experimentation becoming increasingly important in biology, as exemplified by the triumphs of Thomas H. Morgan's school of experimental genetics, the lack of experimental confirmation of the inheritance of acquired characters was a conspicuous liability for the Lamarckian side. As for *disconfirmation* of the idea, Weismann's famous mouse-tail-mutilation experiments, designed to show the noninheritance of mutilations, should not be overrated. Weismann himself maintained that it was not experimentation that would be the downfall of Lamarckism, but instead the realization that "the observed phenomena of transformation" could be explained without it.[40] Doubly damned, both as unsupported by experiments and as an unnecessary hypothesis, the inheritance of acquired characters was left out of the evolutionary synthesis of the 1930s and 1940s.

The architects of the evolutionary synthesis, like the neo-Darwinians and neo-Lamarckians of a generation earlier, remained aware that Darwin had endorsed the inheritance of acquired characters.[41] For textbook writers, however, the idea that Darwin denied Lamarckian use-inheritance may have had the attractive simplicity of a "winners-versus-losers" tale where Darwin got it right and Lamarck got it wrong.

8.7   The Conditions Under Which the Myth Might Cease to Be

The author's experience is that biologists and biology students do not display strong resistance – only surprise – when told that Darwin believed in the inheritance of acquired characters. It seems unlikely that the myth could disappear, however, simply by broadcasting what Darwin specifically said in support of the idea. The way the myth might erode, however, is if modern studies claiming some quasi-Lamarckian effect (recent work in epigenetics, for example) gained increasing traction among biologists generally.[42] It would not matter if the organic changes in question bore little resemblance to the kinds of changes Lamarck had in mind when he wrote about the acquired effects of new

---

[39] For an overview of the decline of Lamarckism in Britain and the United States in the first half of the twentieth century see Richard W. Burkhardt, Jr., "Lamarckism in Britain and the United States," in Ernst Mayr and William Provine, eds., *The Evolutionary Synthesis* (Cambridge, MA: Harvard University Press, 1980), pp. 82–9.

[40] See Weismann's 1888 essay, "The supposed transmission of mutilations," in *Essays Upon Heredity*, 1, 421–48. On Weismann, see Frederick B. Churchill, *Weismann: Development, Heredity, and Evolution* (Cambridge, MA: Harvard University Press, 2015).

[41] For example, George Gaylord Simpson, "The History of Life," in Sol Tax, ed., *Evolution after Darwin*, 3 vols. (Chicago: The University of Chicago Press, 1960), I, pp. 117–80 (see p. 119 fn.); Ernst Mayr, "Prologue: some thoughts on the evolutionary synthesis," in Mayr and Provine, *The Evolutionary Synthesis*, pp. 1–48 (see p. 15).

[42] Eva Jablonka and Snait Gissis, eds., *Transformations of Lamarckism: From Subtle Fluids to Molecular Biology* (Cambridge, MA: M.I.T. Press, 2011).

habits maintained over a great many generations. The more it is demonstrated that some kinds of external influences on the individual organism can be incorporated into the genome, the more likely will it be recognized (as it once was) that Darwin had a place for the inheritance of acquired characters in his own evolutionary theorizing.

Myth 9   That Darwin's Theory Was Essentially Complete Once He Came Up with the Idea of Natural Selection

Alan C. Love

9.1   Commonsense Assumptions About Scientific Reasoning

One of the most persistent myths about the process of scientific inquiry is the instantaneous moment of discovery.[1] Many recognize the common label – "Eureka moments" – derived from the myth that Archimedes exclaimed "Eureka!" (roughly "I have found it") after entering a bath and recognizing that the water level rose, leading to the insight that the volume displaced must be equal to the volume of what had been submerged. This was especially useful for measuring the volume of irregular objects with precision and Archimedes apparently leapt out of the tub naked, eager to announce his discovery, and ran through the streets of Syracuse. Although a memorable narrative (it would have gone viral on TikTok), it arose several hundred years after Archimedes lived and is, alas, not true.[2]

The discovery myth, like many others about scientific reasoning, arises from a simplification of complex cognitive processes, many of which are unobservable, that feed into new or increased understanding about the workings of the natural world. A version of this type of myth is reflected in comments on Charles Darwin's pivotal "discovery" of the concept of natural selection as an explanation for adaptive evolution: "In 1839 he shut his last major evolution notebook, his theory largely complete";[3] "In May 1839, some two and a half years after Darwin's return to England, ... he had had much time to consider what he had observed and arrived at the theory of evolution by natural selection."[4] "Discovery" is in scare

I am grateful for research assistance from Anna Sekerak in preparing the manuscript, comments and suggestions from other volume contributors at the July 2022 online meeting where my initial ideas were aired, and helpful feedback on earlier versions from Kostas Kampourakis and Anya Plutynski.

[1] R. L. Numbers and K. Kampourakis, eds. *Newton's Apple and Other Myths about Science*. 2015, Cambridge, MA: Harvard University Press.

[2] D. Biello, *Fact or fiction? Archimedes coined the term "Eureka!" in the bath*. Scientific American, 2006. www.scientificamerican.com/article/fact-or-fiction-archimede/.

[3] A. J. Desmond, *Charles Darwin*, in *Encyclopedia Britannica* (www.britannica.com/biography/Charles-Darwin/Evolution-by-natural-selection-the-London-years-1836-42).

[4] H. Plotkin, *Evolutionary Worlds without End*. 2010, New York: Oxford University Press, p. 52.

quotes because it is more accurately (though awkwardly) rendered as an "extended sequence of formulation and reformulation" that began before 1838 and continued long after.[5] The word "discovery" contributes to the mythology since its everyday meaning is finding something at a specific moment in time ("The shipwreck of the Titanic was *discovered* on 1 September 1985"). Repackaging the subtle variations and changes in a scientist's thinking over months, years, or decades into a simpler form is misleading, because it belies the time, energy, and contingency involved in generating new ideas or systematic explanations of natural phenomena.

In addition to the myth of a discovery moment, there is another commonsense assumption reflected in how this Darwinian myth is frequently portrayed. It involves a simplified conception of what counts as a scientific theory and how it is structured. What does it mean to talk about the "major parts" or "main elements" of a theory? How are scientific theories organized or structured? How do we know a theory is "essentially complete" (and what does that mean)? These questions are not so easy to answer in a general fashion, a conclusion that has emerged from sustained analyses by historians and philosophers of science.[6] Here I concentrate on the details of this Darwinian myth and explore three primary questions: What is "Darwin's theory"? What is meant by "essentially complete"? How do we understand "natural selection"? Along the way, we will observe simplifying assumptions about scientific inquiry interwoven in this fictional narrative. As with many myths, there is a kernel of truth in the claim that Darwin's theory was essentially complete once he came up with the idea of natural selection. However, it is best appreciated once we revise our expectations about the nature of science, giving up not only instantaneous moments of discovery but also cartoon representations of scientific theory structure.

9.2   Unraveling the Myth

According to a major textbook on evolution, the spark that led Darwin to his sudden insight is clear: "The idea of natural selection came to him in September 1838, as he read Thomas Robert Malthus' *Essay on Population*."[7] Why does this seem so crystal clear in retrospect? Darwin himself bears part of the blame. In his *Autobiography*, written forty years later, the Eureka moment stands out:

In October 1838, that is, fifteen months after I had begun my systematic enquiry, I happened to read for amusement Malthus on *Population*, and being well prepared to

[5] D. Ospovat, *The Development of Darwin's Theory: Natural History, Natural Theology, and Natural Selection, 1838–1859*. 1981, New York: Cambridge University Press.

[6] A. C. Love, *Theory is as theory does: Scientific practice and theory structure in biology*. Biological Theory, 2013. 7: p. 325–37. R. G. Winther, The structure of scientific theories, in *The Stanford Encyclopedia of Philosophy*, E. N. Zalta, Editor. 2021. https://plato.stanford.edu/archives/spr2021/entries/structure-scientific-theories/.

[7] N. H. Barton, D. E. G. Briggs, J. A. Eisen, et al., *Evolution*. 2007, Cold Spring Harbor, NY: Cold Spring Harbor Laboratory Press, p. 16.

appreciate the struggle for existence which everywhere goes on from long-continued observation of the habits of animals and plants, it at once struck me that under these circumstances favourable variations would tend to be preserved, and unfavourable ones to be destroyed. The result of this would be the formation of new species.[8]

Although this seems incontrovertible ("it at once struck me"), Darwin was kind to future historians and saved almost every scrap of paper he wrote on, much of it organized with his own filing system. Gems amidst this archival treasure trove are his transmutation notebooks.[9] The Eureka moment doesn't "pop" from the notebook page with quite the same clarity that Darwin nostalgically recalled.

Population is increase at geometrical ratio in FAR SHORTER time than 25 years – yet until the one sentence of Malthus no one clearly perceived the great check amongst men. – there is spring, like food used for other purposes as wheat for making brandy. – Even a *few* years plenty, makes population in Men increase & an *ordinary* crop causes a dearth. take Europe on an average every species must have same number killed year with year by hawks, by cold &c. – even one species of hawk decreasing in number must affect instantaneously all the rest. – The final cause of all this wedging, must be to sort out proper structure, & adapt it to changes. – to do that for form, which Malthus shows is the final effect (by means however of volition) of this populousness on the energy of man. One may say there is a force like a hundred thousand wedges trying force ~~into~~ every kind of adapted structure into the gaps ~~of~~ in the oeconomy of nature, or rather forming gaps by thrusting out weaker ones.[10]

There is clearly a new idea being worked out in light of what Malthus said ("The final cause of all this wedging, *must be* to sort out proper structure, & adapt it to changes"). However, no further reflections from Darwin on this supposedly momentous insight appear immediately after the entry, which is followed by notes on how human skin color changes quickly or slowly when different groups interbreed and the incorrigible curiosity of monkeys. Historian Jonathan Hodge summarizes the situation aptly: "From the middle of September 1838 Darwin's pace slows strikingly. The theory of natural selection emerges gradually, from late September 1838 to mid-March 1839, in languid, intermittent notebook theorising work; so discrediting any stereotypes – fostered by Darwin's later reminiscences and much scholarship in the same vein – of a single moment of decisive insight during intense activity."[11] Most importantly, as historian Dov Ospovat has shown, Darwin had not given up on a traditional picture of the economy or harmony of nature derived from British natural theology. "While his

---

[8] N. Barlow, ed. *The Autobiography of Charles Darwin 1809–1882. With the original omissions restored. Edited and with appendix and notes by his grand-daughter Nora Barlow.* 1958, London: Collins, p. 120.

[9] P. H. Barrett, P. J. Gautrey, S. Herbert, et al., eds. *Charles Darwin's Notebooks, 1836–1844: Geology, Transmutation of Species, Metaphysical Enquiries.* 1987, Ithaca: Cornell University Press.

[10] Barrett, et al., *Charles Darwin's Notebooks*, p. 375–6. Strikethrough in original.

[11] M. J. S. Hodge, *The notebook programmes and projects of Darwin's London years*, in *The Cambridge Companion to Darwin*, M. J. S. Hodge and G. Radick, Editors. 2003, Cambridge: Cambridge University Press, pp. 40–68 (pp. 59–60).

reading of Malthus gave Darwin a new theory of organic change and adaptation, it did not immediately alter his conception of nature, nor was his new theory free from the effects of his old assumptions."[12] This naturally prompts a question about what exactly counts as Darwin's theory.

9.3   What Is Darwin's Theory?

Contemporary textbook authors frequently offer summaries of what counts as Darwin's theory or his main theses. In the fourth edition of his graduate level introduction, evolutionary biologist Doug Futuyma states it in the following manner: "*The Origin of Species* has two major theses. The first is Darwin's theory of descent with modification. ... The second theme of *The Origin of Species* is Darwin's theory of the causal agents of evolutionary change. This was his theory of natural selection."[13] For many textbooks, in the sections devoted to discussing evolution, this is offered as a general template for how scientific theories are structured: "[S]cientific theories usually have two components. The first is either a claim about a pattern that exists in nature or a statement that summarizes a series of observations about the natural world. In short, the pattern component is about facts. The second component is a process that produces the pattern or set of observations."[14] Thus, Darwin's theory is represented as containing claims about a pattern found in the biological world (common descent) and claims about a putative process – natural selection – that accounts for this pattern.

Is this a good summary of Darwin's theory? Not really. Of course, we can recognize that this representation is formulated for pedagogical purposes and therefore highly idealized. And it does illuminate why one might talk of natural selection making Darwin's theory "essentially complete" since it seems to add the necessary causal explanation for why a particular pattern is observed in nature. However, from a close reading of Darwin's transmutation notebooks, historian David Kohn has recognized four earlier Darwinian theories, none of which concentrates on natural selection. Instead, "those passages of Darwin's notebooks ... are most cogently devoted to the search for an evolutionary mechanism."[15] Whereas contemporary readers envision Darwin's theory to be the one whose primary evolutionary process is natural selection, as laid out in the *Origin of Species*,[16] there are four reasons to think this is a problematic way to understand Darwin's reasoning related to natural selection in 1838.

---

[12] Ospovat, *The Development*, p. 37.
[13] D. J. Futuyma, *Evolution*. 4th ed. 2005, Sunderland, MA: Sinauer Associates, Inc, p. 7.
[14] S. Freeman, *Biological Science*. 2002, Upper Saddle River, NJ: Prentice Hall, p. 412.
[15] D. Kohn, *Theories to work by: Rejected theories, reproduction, and Darwin's path to natural selection*. Studies in the History of Biology, 1980. 4: pp. 67–170 (p. 67).
[16] C. Darwin, *On the Origin of Species: A Facsimile of the First Edition*. 1964 [1859], Cambridge, MA: Harvard University Press.

First, Darwin's thinking about natural selection underwent key changes between 1838 and 1859.[17] We will return to this below when exploring the meaning of "essentially complete" (Section 9.4) and scrutinizing the concept of natural selection more closely (Section 9.5). The bottom line is that Darwin's 1859 theory of evolution by natural selection is not the same as what he envisioned after reading Malthus in 1838.

Second, Darwin's 1859 theory of evolution by natural selection was not simply offering a process for evolution but an explanation for *adaptive* evolution.

In considering the Origin of Species, it is quite conceivable that a naturalist, reflecting on the mutual affinities of organic beings, on their embryological relations, their geographical distribution, geological succession, and other such facts, might come to the conclusion that each species had not been independently created, but had descended, like varieties, from other species. Nevertheless, such a conclusion, even if well founded, would be unsatisfactory, until it could be shown how the innumerable species inhabiting this world have been modified, so as to acquire that perfection of structure and coadaptation which most justly excites our admiration.[18]

The pattern in view for Darwin was not only common descent but also adaptation and therefore many textbook representations of the two-part structure of Darwin's theory are misleading about the "facts" he was trying to explain.

Third, in the context of Darwin's 1859 discussion, he appealed to other mechanisms besides natural selection to account for adaptive evolution, most notably use and disuse. This reason is intimately intertwined with another myth: that Darwin rejected Lamarck's laws of use and disuse and of the inheritance of acquired traits (see Burkhardt, Chapter 8, this volume). As one might guess, he did no such thing. Chapter 5 of the *Origin of Species* ("Laws of Variation") has a dedicated subheading entitled: "Use and disuse, combined with natural selection." In response to criticisms of his 1859 presentation,[19] Darwin *continuously* appealed to use and disuse in subsequent editions of the *Origin of Species*.[20] Darwin's own conception of his theory included use and disuse as a different causal process relevant for explaining evolutionary patterns of biological adaptation.

Finally, there is a complexity in Darwin's theory presentation of 1859 that cannot be understood well by the two-part theory structure of pattern and

---

[17] Ospovat, *The Development*, p. 86.   [18] Darwin, *On the Origin of Species*, p. 3.

[19] D. L. Hull, ed. *Darwin and His Critics: The Reception of Darwin's Theory of Evolution by the Scientific Community.* 1973, Cambridge, MA: Harvard University Press.

[20] M. Peckham, ed. *The Origin of Species by Charles Darwin: A Variorum Text.* 1959, Philadelphia: University of Pennsylvania Press; T. Hoquet, *The evolution of the Origin (1859–1872)*, in *The Cambridge Encyclopedia of Darwin and Evolutionary Thought*, M. Ruse, Editor. 2013, Cambridge University Press: Cambridge. pp. 158–64.

process. There are thirteen chapters in the *Origin of Species* and the structure of the book has been notoriously difficult to parse.[21] Although there is agreement that natural selection is laid out in the earlier chapters and common descent is primarily addressed in later chapters, this does not follow the expectation of a two-part structure: setting out the evidence showing a particular pattern (such as common descent or coadapted traits), followed by an account of the process of evolution (such as natural selection or use and disuse). Some have even argued that Darwin wrote his own book in the wrong order![22]

Clearly the organization of Darwin's theory is more complicated than the simple two-part structure implies. This two-part structure also neglects Darwin's appeal to use and disuse (process) and misses his primary focus on adaptation rather than common descent (pattern). Thus, not only are assumptions about continuity between what Darwin discovered in 1838 and presented in 1859 shaky, it is also unclear what theory components are being compared such that his theory could be designated as "essentially complete" at the earlier juncture.

9.4   What Is Meant by "Essentially Complete"?

All scientific theories are works in progress. They are formulated as attempts to understand and explain features of the natural world and then tested against empirical evidence. We expect scientific theories to change over time as scientists adjust them in various ways that become apparent through this process of gathering data and thinking through the consequences that derive from different aspects of a theory. This picture helps us make sense of why Darwin's theory has a more complicated structure than the two-part idealization found in many textbooks. It also makes sense of why biologists continue working on the theory of evolution by natural selection today, more than 150 years later. Among the possibilities of what "essentially complete" could mean for Darwin's theory, it cannot plausibly be read as "basically finished."

Another candidate for how "essentially complete" might apply to a theory is in terms of empirical adequacy. That is, the theory can handle or address the natural phenomena for which it was formulated in the first place, even if it only addresses them imperfectly or not to everyone's satisfaction. This standard is very difficult to apply in the case of Darwin's first formulation of natural selection in 1838. David Kohn recognized that "we at least need a colloquial definition of theory" to talk of Darwin having a theory and Kohn adopted one offered by Ken Schaffner: "Theories can be construed as attempts to capture the

---

[21] C. K. Waters, *The arguments in the Origin of Species*, in *The Cambridge Companion to Darwin*, M. J. S. Hodge and G. Radick, Editors. 2003, Cambridge: Cambridge University Press, pp. 116–39.

[22] E. Sober, *Did Darwin Write the Origin Backwards? Philosophical Essays on Darwin's Theory.* 2011, Amherst, NY: Prometheus Books.

essentials of the subject areas which they explain."[23] However, on this criterion, Darwin's theorizing before 1838 was "essentially complete" in the same way his later theorizing might be labeled. This makes claims about a theory being "essentially complete" a trivial matter and leaves no basis for picking out natural selection as distinctive in Darwin's thinking.

Even after his initial essay of 1844, which offered the first full version of Darwin's theoretical ideas and whose structure is recognizably similar to the later *Origin of Species*, Darwin was concerned about the evidential support for different aspects of his theory. Importantly, this is not the same as being fearful of publishing because of the possible reactions of others to his evolutionary ideas (see van Wyhe, Chapter 10, this volume).[24] This is reflected in his request to have the essay published (upon revision) if he was to die, and in reluctantly giving up on the large book he was working on, which would be called *Natural Selection*,[25] to publish the "abstract" *Origin of Species* under pressure from Wallace's 1858 manuscript (Ruse, Chapter 11, this volume). Nowhere in his publications or personal correspondence does Darwin exhibit an explicit confidence that the 1859 presentation of his theory is "essentially complete" in the sense of being empirically adequate. This is consistent with his ongoing tinkering as he made modifications in different editions of the *Origin of Species* in response to criticisms.[26]

One last possibility for the meaning of "essentially complete" could be "the basic framework is intact." This would not require a commitment of empirical adequacy but rather conceptual adequacy in that core principles were in place but simply needed testing. Unfortunately, this also is historically problematic because the basic principles of Darwin's theorizing underwent revisions after the 1844 essay had been copied out. These revisions – additions, extensions, elaborations, and modifications – were prompted by what other scientists were discovering and formulating, forcing Darwin to react and adjust in real time. One example was the "branching" conception of the history of life. Although Darwin had pictured evolution as a simple branching process before his reading of Malthus (Figure 9.1), work by embryologists, morphologists, and paleontologists in the 1840s and 1850s led to a more refined conception of branching that represented lineages as moving from generalized ancestors to more specialized descendants via common descent.[27] This understanding of "branching" was not present in Darwin's notebook theorizing. There it had only been a basic

[23] K. F. Schaffner, *The peripherality of reductionism in the development of molecular biology.* Journal of the History of Biology, 1974. 7: pp. 111–39 (p. 112).

[24] J. van Wyhe, *Mind the gap: did Darwin avoid publishing his theory for many years?* Notes and Records of the Royal Society, 2007. 61(2): pp. 177–205.

[25] R. C. Stauffer, ed. *Charles Darwin's Natural Selection, Being the Second Part of His Big Species Book Written from 1856 to 1858.* 1975, Cambridge: Cambridge University Press.

[26] Peckham, *The Origin of Species by Charles Darwin: A Variorum Text*; Hoquet, *The evolution of the Origin (1859–1872).*

[27] Ospovat, *The Development*, Ch. 6.

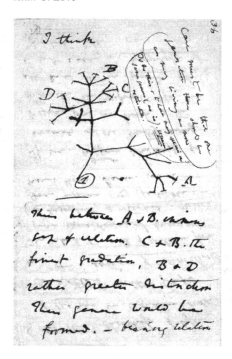

Figure 9.1 Darwin's famous branching diagram from Notebook B in 1837.
The accompanying notebook text indicates that the pattern is drawn "so as to
keep number of species constant," which presumes a set number of places in
the economy of nature (i.e., a natural theological assumption). This is not the
same branching conception that appears in the *Origin of Species* (see text for
discussion). Credit: Public domain, via Wikimedia Commons

expectation of common descent; in the *Origin of Species* it was a more detailed
set of empirical patterns that implied an evolutionary tendency of divergence.
That tendency required a special explanation, something more than natural
selection as he had previously conceptualized it (see Section 9.5). In the language
of the two-part theory structure, aspects of the facts that composed a pattern
Darwin thought he needed to account for were not yet available in 1838.

9.5   How Do We Understand "Natural Selection"?

In claiming that Darwin discovered natural selection in 1838, how do we
understand this principle or concept? Modern renditions often focus on an
abstract tripartite formulation: (1) variation in traits; (2) fitness differences

attach to this trait variation; and (3) this variation is heritable.[28] However, this leaves out the Malthusian elements of limited resources and a struggle for existence.[29] A more detailed reconstruction sensitive to Darwin's 1859 presentation is hard to compress into a slogan:

1. Species are comprised of individuals that vary ever so slightly from each other with respect to their many traits
2. Species have a tendency to increase in numbers over generations at a geometric rate
3. This tendency is checked . . . by limited resources, disease, predation, and so on, creating a struggle for survival among the members of a species
4. Some individuals will have variations that give them a slight advantage in this struggle, variations that allow more efficient or better access to resources, greater resistance to disease, greater success at avoiding predation, and so on
5. These individuals will tend to survive better and leave more offspring
6. Offspring tend to inherit the variations of their parents
7. Therefore, favorable variations will tend to be passed on more frequently than others and thus be preserved.[30]

Darwin did offer a slogan – "This preservation of favourable variations and the rejection of injurious variations, I call Natural Selection"[31] – but we can now see this is a simplification. For example, it elides a commitment to the kind of trait variation Darwin envisaged ("that vary ever so slightly from other"). Even this more detailed formulation is abstracted from a broader context of theorizing. In 1838 (and for some years after), Darwin thought of natural selection as occurring in a larger framework of natural theology with laws set up to accomplish preordained goals. It is this changing context that historians have documented between 1838 and 1859, which serves as a warning: "We know so well what natural selection is that it requires some care not simply to read that knowledge into Darwin's notes."[32]

Perhaps the most distinctive place where this change is evident emerges in his formulation of a new idea in 1856. "One other principle, which may be called the principle of divergence plays, I believe, an important part in the origin of species. The same spot will support more life if occupied by very diverse forms: we see this in the many generic forms in a square yard of turf."[33]

[28] R. Lewontin, *The units of selection*. Annual Review of Ecology & Systematics, 1970. 1: pp. 1–14.
[29] J. G. Lennox and B. E. Wilson, *Natural selection and the struggle for existence*. Studies in the History and Philosophy of Science, 1994. 25: pp. 65–80.
[30] J. G. Lennox, *Darwinism*, in *The Stanford Encyclopedia of Philosophy*, E. N. Zalta, Editor. 2019. https://plato.stanford.edu/archives/fall2019/entries/darwinism/.
[31] Darwin, *On the Origin of Species*, p. 81.   [32] Ospovat, *The Development*, p. 62.
[33] Charles Darwin to Asa Gray, September 5, 1857, *Darwin Correspondence Project, Letter #2136:* www.darwinproject.ac.uk/letter/?docId=letters/DCP-LETT-2136.xml.

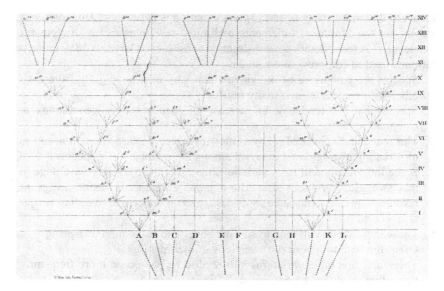

Figure 9.2 Darwin's only figure in the *Origin of Species*. It illustrates the principle of divergence, something absent from his thinking until the mid-1850s. In the accompanying text he says, "Now let us see how this principle of great benefit being derived from divergence of character, combined with the principles of natural selection and of extinction, will tend to act. The accompanying diagram will aid us in understanding this rather perplexing subject" (p. 116). Credit: Wellcome Collection. Public Domain Mark

In earlier formulations of natural selection, Darwin had taken divergence in traits between species arising from a common ancestor to result primarily from the struggle for existence and the passage of time. The principle of divergence was formulated because he recognized (from the work of his contemporaries) that it was a distinct tendency that influenced the process of natural selection.[34] This revised formulation of natural selection or new theoretical structure was definitively not what Darwin was thinking in 1838, let alone 1844. In Darwin's estimation, his theory was not complete without it – the earlier conception of natural selection was conceptually inadequate – and it is not surprising that the only figure in the *Origin of Species* is devoted to illustrating how the principle of divergence works (Figure 9.2).

---

[34] Ospovat, *The Development,* Ch. 7. R. J. Richards, *Darwin's principles of divergence and natural selection: Why Fodor was almost right.* Studies in History and Philosophy of Science Part C: Studies in History and Philosophy of Biological and Biomedical Sciences, 2012. 43(1): pp. 256–68.

Other changes occurred in conjunction with Darwin's development of this new principle. For example, his conception of organismal "place" in 1844 was framed in terms of geology and climate, concerns of traditional natural theology: an organism was located physically. By 1859, organisms were located in relation to other organisms where the principle of divergence in combination with the struggle for existence operated.[35] This shift – from thinking about organisms as situated primarily in geographical and climatic space to emphasizing interspecific and intraspecific interactions among organisms – also tracks Darwin's conceptual evolution from thinking that organisms are perfectly adapted to their physical environments to a full-blown relative adaptation of organisms to their biotic environment in the mid-1850s.[36] Overall, the concept of natural selection and several associated ideas evolved in substantial ways between 1838 and 1859.

9.6   Concluding Reflections

Famously, after his nostalgic but false recall of what happened on September 28, 1838, Darwin claimed: "Here then I had at last got a theory by which to work."[37] As we have seen, this is not true because he had already explored several working theories.[38] Yet it can lead us to recognize the importance of Darwin's extended formulation of natural selection as an explanation of adaptive change. What Darwin had when the idea of natural selection was first formulated was not "essentially complete." Rather, he had a *new* "theory by which to work" in which natural selection was a candidate process to explain adaptive evolution. However, this process needed to be elaborated both conceptually and empirically. There were several distinct questions pending an answer:

• Did natural selection exist as a cause? An analogy with artificial selection helped to establish this (at least for some).
• Was natural selection competent to produce evolutionary change? The same analogy along with Darwin's thought experiments suggested it was,[39] though controversially.
• Was natural selection responsible for evolutionary change? Darwin thought so, but no consensus emerged among his contemporaries.

However, establishing the existence, competence, and responsibility of natural selection as a genuine cause corresponded to what many of Darwin's contemporaries expected of a scientific investigation.[40] The formulation of the

---

[35] Ospovat, *The Development*, Ch. 7.   [36] Ospovat, *The Development*.
[37] Barlow, *The Autobiography*, p. 120.   [38] Kohn, *Theories to work by.*
[39] J. G. Lennox, *Darwinian thought experiments: a function for just so stories*, in *Thought Experiments in Science and Philosophy*, T. Horowitz and G.J. Masset, Editors. 1991, Savage, MD: Rowman and Littlefield, pp. 223–45.
[40] M. J. S. Hodge, *The Structure and Strategy of Darwin's 'Long Argument'*. British Journal for the History of Science, 1977. 10: pp. 237–46.

concept of natural selection was a major advance in biological theorizing and central to his own scientific reasoning.[41]

None of this requires a commitment to the myth of instantaneous discovery and essential completeness. While there is value in finding pithy formulations of Darwin's accomplishments, and for the theoretical advances of other scientists also, the myth ultimately obscures the time, energy, and contingency involved in generating new ideas and systematic explanations of natural phenomena. Scientific theories are historical creatures with a high degree of complexity that take on different forms, changing through time in the hands of individual scientists or research communities. To think of Darwin's theory as "essentially complete" once "natural selection" was first formulated in 1838 is to reify what counts as "Darwin's theory" and presume an omniscient point of view whereby a concept that was worked out slowly and painfully is conceived of as the explanatory skeleton key arrived at via a stroke of genius. The temptation of this myth is perennial, and even those most primed to recognize the complexity and heterogeneity of intellectual development fall into this framing: "Without giving it a name, Darwin had hit upon the concept of natural selection. The words in his notebook practically leap from the page"; "Darwin's intellectual development did not stop on 18 September 1838. However, on that day he articulated the core elements of natural selection."[42] The myth is cognitively sticky and difficult to resist.

The "two-part" view of Darwin's theory associated with the myth encourages a flattened image of scientific theory structure that maps poorly onto Darwin's actual reasoning practices. Yet a richer understanding of the way scientific inquiry operates has great potential for yielding a more robust justification of its methods and outcomes. We can have more confidence in the importance of natural selection for explaining adaptive evolution because the concept was worked out over many years in a milieu of new empirical findings from the community of biologists around Darwin. It would be quite strange to think that the most important and lasting scientific ideas emerged fully formed in a flash of insight, like Athena from the head of Zeus. There is good reason to separate myth from reality.

---

[41] Lennox, *Darwinism*.

[42] E. J. Browne, *Charles Darwin Voyaging: A Biography*. Vol. 1. 1995, Princeton: Princeton University Press, p. 388. Kohn, *Theories to work by*, p. 148.

Myth 10   That Darwin Delayed the Publication of His Theory for Twenty Years, Being Afraid of the Reactions It Would Cause

John van Wyhe

10.1   Introduction

Richard Dawkins wrote in a 2003 introduction to an edition of Darwin's *Origin of Species* and *Voyage of the Beagle*: "Why he delayed so long before publishing is one of the great mysteries of the history of science."[1] And in 2009, David Masci from Pew Research Center's Religion & Public Life Project wrote: "Darwin had expected no less – fear of a backlash from Britain's religious and even scientific establishment had been the primary reason he had delayed publicizing his ideas."[2] For many years one of the central beliefs about Charles Darwin has been that he delayed or held back his theory of evolution for many years and kept his belief in evolution a secret because he was afraid. Accounts of such a twenty-year delay in publishing appear in countless thousands of publications, documentaries, films, the internet and even artworks. But it is a myth whose advent has been traced to the mid-twentieth century. Postponed publication was not discovered in the historical evidence. Instead, the story evolved along with the changing attitudes of modern writers.

The longevity of stories of Darwin's delay may be due to the many boxes it ticks for different types of people. For historians of science it fits one of the central principles in the field, namely to emphasize that science does not exist in a vacuum but is embedded in a specific historical context and that social forces inform and shape scientific knowledge. Darwin's delay is also a historiographical theme that is not hagiographic, shielding serious historians from the suspicion of writing about a scientific hero – an ever-present danger when writing about famous scientific figures rather than the vast

This chapter is based partly on John van Wyhe, "Mind the gap: Did Darwin avoid publishing his theory for many years?" *Notes and Records of the Royal Society* 61 (2007): 177–205 and John van Wyhe, *Dispelling the Darkness*. (Singapore: WSP, 2013) chapter 10.

[1] C. Darwin, *The Origin of Species and The Voyage of the 'Beagle'* (London: Everyman's Library, 2003).

[2] www.pewresearch.org/religion/2009/02/04/darwin-and-his-theory-of-evolution/.

majority of the lesser known. The delay fulfilled these roles so well that it came to be accepted almost entirely without question and became a sign of being a properly informed and accredited historian of science. Although a few writers did so, to criticize it could be construed as contrary to much of what the field had achieved during the decades of its professionalization. Or worse, it could trigger a campaign of persecution and character assassination against purported apostates.

For popular audiences, however, Darwin's delay *can* serve a hagiographic role. It is irresistibly dramatic – the famous Darwin's theory of evolution was too explosive to reveal to a deeply prejudiced and hostile world and a reactionary religious orthodoxy. But the story could then culminate with the triumph of Darwin and evolution over powerful odds.

Additional elements continue to arise such as that he instructed his wife to publish his theory *only* after his death (he asked her to publish an early draft if he died prematurely) to the claim that he worried so much about the backlash that it made him ill (he complained only of the strain and overwork of the task), to the most recent – that Darwin hid an early draft under the stairs (it was kept among his working papers until superseded and only found there after his widow's death in 1896). Down House museum today even has a packet wrapped with string under the stairs labelled "Only to be opened in the event of my death." Many have opined that without the surprise interruption by A. R. Wallace in 1858, Darwin would never have published his theory. In fact Darwin had been steadily writing his "big book" chapter by chapter since May 1856 and would probably have finished it by 1860.

Figure 10.1 is one of the most popular images of Darwin today. It appears on the covers of books, the *Times Literary Supplement*, postage stamps, postcards, in paintings, drawings, murals, merchandise such as mobile phone cases, the internet and has even been tattooed onto many Darwin fans. The image was created for the 2009 Natural History Museum Darwin exhibition.[3] This is a reversed version of an original c.1880 photograph by Elliott & Fry with a hand "photoshopped" onto it.[4] The gesture is an obvious reference to Darwin keeping a secret. The myth of Darwin's delay is so ubiquitous, that the original website and poster needed no text to tell viewers what it meant. The fact that a faked photograph modified to reflect a twentieth/twenty-first-century version of history should be preferred by modern audiences to a more austere Victorian original is perhaps not surprising.

---

[3] The American Museum of Natural History website for the 2005–6 Darwin exhibition proclaimed in the largest font except for Darwin's name "FOR 21 YEARS HE KEPT HIS THEORY SECRET."

[4] Paul van Helvert and John van Wyhe, *Darwin: A Companion* (Singapore: WSP, 2021), 178 which lists all known Darwin photographs (among 1,000 unique portraits) with details, including seven newly rediscovered photographs.

Figure 10.1 Natural History Museum Darwin Exhibition poster, 2009

10.2   Delay, Where?

First things first. Nowhere in the millions of surviving words of Charles Darwin does he ever indicate, even obliquely, that he held back his theory or kept his belief in evolution a secret. In fact, he repeatedly said the opposite. On many occasions, he said that he worked on the *unfinished* theory during the years after 1837.[5] Consider this letter to Asa Gray from 1857: "It is not a little egotistical, but I sh^d. like to tell you, (& I do not *think* I have) how I view my work. Nineteen years (!) ago it occurred to me that whilst otherwise employed on Nat. Hist, I might perhaps do good if I noted any sort of facts bearing on the question of the origin of species; & this I have since been doing."[6] This is an explicit contradiction of the delay thesis in Darwin's own words and precisely the argument of my article "Mind the Gap."[7]

---

[5] Two of Darwin's letters in which there is a reference on his long work on the theory are: To ? 23 Oct. 1880 Burkhardt et al eds., *The Correspondence of Charles Darwin* (hereafter CCD), 28:350, Darwin Correspondence Project, "Letter no. 12771," accessed on 3 January 2024, www.darwin project.ac.uk/letter/?docId=letters/DCP-LETT-12771.xml; and to F. Powell [after 3 Dec. 1881] CCD29:582, Darwin Correspondence Project, "Letter no. 13529," accessed on 3 January 2024, www.darwinproject.ac.uk/letter/?docId=letters/DCP-LETT-13529.xml.

[6] Darwin to Gray 20 July [1857] CCD6:432.

[7] J. van Wyhe, "Mind the gap: Did Darwin avoid publishing his theory for many years?" *Notes and Records of the Royal Society* 61 (2007): 177–205. At the time of writing that article, I had overlooked this apt quotation.

If Darwin spent many years postponing or keeping secret his famous theory, why was this astonishingly dramatic fact never mentioned by any of his family, friends or other contemporaries? Instead, they referred to Darwin working on the theory for twenty years. Charles Lyell, in his *Antiquity of Man* (1863), wrote that Darwin had been "patiently" working on the theory for twenty years.[8] When the *Origin of Species* appeared in 1859 the *Saturday Review* referred to it as "the work upon which [Darwin] was known to have been long engaged."[9] Where are the surprised reactions of the hundreds of those who provided assistance with his "secret" project over so many years?

This absence of "Darwin's delay" continued throughout the vast literature on him for more than half a century. Writers during this period did not give much attention to, or explanation for, the widely dispersed dates for conceiving and publishing (see also Love, Chapter 9, this volume). Most simply followed Darwin's rendition on the first page of the *Origin of Species*. He told his readers that he began to work on the subject in 1837, then in 1842 "drew up some short notes; these I enlarged in 1844 into a sketch of the conclusions, which *then* seemed to me probable: *from that period to the present day I have steadily pursued the same object.*"[10]

How, then, did such a radically different version of events (Darwin's delay) replace the earlier version? Only in the 1940s–1950s did ideas of a delay/postponement begin tentatively and gradually to appear.[11] After the 1960s, most writers took it for granted that Darwin had, in fact, held back his theory for many years.

Postponement because of fear became the central theme of the Darwin story in a 1974 book by the psychologist Howard Gruber. He wrote: "we need some explanation for Darwin's long delay in publishing his views, and we need some understanding of the way in which this delay affected his inner life."[12] Gruber, assuming Darwin had held back, argued that fear was the main reason. Fear has remained the most popular explanation. Far from a revision of old-fashioned accounts by serious modern scholarship in the history of science, the delay/secret idea began as the speculation of historically untrained writers in the 1940s–1970s.

Many today find it sufficiently convincing that twenty years elapsed between Darwin becoming a transmutationist and the publication of his matured theory in 1858/9. Hence Darwin must have held back. Since the twenty-year gestation of the *Origin of Species* seems so extraordinary, surely there must have been an extraordinary reason? Not at all when we put the

[8] C. Lyell, *Geological Evidences of the Antiquity of Man* (London: Murray, 1863), 408.
[9] Anon., "On the origin of species," *Saturday Review* (Dec. 24, 1859): 775–6.
[10] C. Darwin, *On the Origin of Species* (London: Murray, 1859), 1. Emphasis added.
[11] This gradual emergence is traced in van Wyhe, "Mind the gap."
[12] H. Gruber, *Darwin on Man* (London: Wildwood, 1974).

book in the context of Darwin's manner of work. His book *Insectivorous Plants* was published sixteen years after his first observations on the subject. His theory of pangenesis was published twenty-seven years after conceiving of it. His book *Orchids* (1862) was published thirty years after Darwin began observing orchids. *Cross and Self Fertilisation* (1876) was published thirty-seven years after he began working on that subject. His notes on the development of his infant son were published thirty-seven years later. He first noticed the actions of worms for modifying the landscape in 1837 and published his book on the subject forty-two years later. Darwin was explicit that long-term research projects were his normal and even preferred type of work. When answering Francis Galton's questionnaire in 1873, did he have "Energy of mind?" Darwin replied that he did: "Shown by rigorous and long-continued work on same subject, as 20 years on the 'Origin of Species' and 9 years on Cirripedia."[13] In 1881 Darwin wrote to J. D. Hooker (Figure 10.2) "I have not the heart or strength to begin any investigation lasting years, which is the only thing which I enjoy."[14]

Figure 10.2 Photograph of Joseph Dalton Hooker by Lock & Whitfield 1878. Credit: Author's collection

[13] F. Darwin, ed. *Life and letters of Charles Darwin* (London: Murray, 1887), vol. 3, p. 179. This and all other contemporary works cited here are from John van Wyhe ed. 2002–. *The Complete Work of Charles Darwin Online*. (http://darwin-online.org.uk/).

[14] Darwin, *Life*, vol. 3. p. 356.

10.3　What Secret?

Probably the most insistent on the idea that Darwin must have kept it all secret are biographers Desmond and Moore. They believed Darwin and his contemporaries must have seen "evolution as a social crime." They even described Darwin's personal transmutation notebooks as "secret," "clandestine" and "covert." Darwin preferred, in their view, "living a lie."[15]

Nevertheless, Darwin's letters, notebooks, notes and the writings of others demonstrate that he told very many people during the years before publication. The editors of Darwin's correspondence observed: "Darwin is usually depicted as having been very careful to keep secret his heretical views on species, but the correspondence does not bear out this view, if what is meant is that Darwin was afraid to divulge his conviction that species had evolved."[16] In my article "Mind the Gap" a list of around forty individuals who knew is given. This was enlarged, thanks largely to information helpfully communicated by other historians, in my book *Dispelling the Darkness* (2013) to around fifty-six. There were no doubt many more. From a previously unpublished diary, it seems likely that Darwin discussed evolution with his neighbor and fellow magistrate G. W. Norman.[17] It is therefore indisputable that Darwin told family, friends, neighbors, colleagues (both known and unknown to him), paid copyists and even casual acquaintances at health spas and fellow guests at luncheons and dinners.

Darwin's first known recorded doubts about the stability of species are in the famous passage in his ornithological notes from 1836: "such facts ^{would} undermine the stability of Species." What is seldom realized is that these notes were not private. They were written in a collection catalogue prepared to give to a specialist along with the specimens when the *Beagle* returned.[18]

Darwin came to accept that evolution was true in mid-March 1837.[19] And his first public statement of interest in the origin of species came only about two months later in a paper read on May 31, 1837. Darwin noted that studying coral islands was important as "some degree of light might thus be thrown on the question, whether certain groups of living beings peculiar to small spots are the remnants of a former large population, or a new one springing into existence."[20]

---

[15] A. Desmond and J. R. Moore, *Darwin* (Harmondsworth: Penguin, 1992): xv, xvi, 236, 273, 228, 231, 232, 239, 292, 657.

[16] CCD2:xvi.

[17] John van Wyhe and C. Chua, *Charles Darwin: Justice of the peace. The Complete Records (1857–1882).* 2021. http://darwin-online.org.uk/converted/pdf/2021_John_van_Wyhe_&_Christine_Chua,_Charles_Darwin._Justice_of_the_Peace_A2115.pdf.

[18] N. Barlow, "Darwin's ornithological notes," *Bulletin of the British Museum (Natural History), Historical Series* 2(7) (1963): 201–78, 262.

[19] F. J. Sulloway, "Darwin's conversion," *Journal of the History of Biology.* 15 (1982): 327–98.

[20] C. Darwin, "On certain areas of elevation and subsidence in the Pacific and Indian oceans," *Proceedings of the Geological Society of London* 2 (1837): 552–4. This was widely reprinted.

He modified this in *Journal of Researches* to: "that most mysterious question, – whether the series of organized beings peculiar to some isolated points, are the last remnants of a former population, or the first creatures of a new one springing into existence."[21] And elsewhere he wrote: "the admirable laws, first laid down by Mr. Lyell, on the geographical distribution of animals, as influenced by geological changes. The whole reasoning, of course, is founded on the assumption of the immutability of species; otherwise the difference in the species in the two regions, might be considered as superinduced during a length of time."[22]

By the time the 1845 edition of *Journal of Researches* was written, John Gould had examined the birds. Darwin added the now famous passage on the Galapagos finches: "Seeing this gradation and diversity of structure in one small, intimately related group of birds, one might really fancy that from an original paucity of birds in this archipelago, one species had been taken and modified for different ends"[23] (see also Sulloway, Chapters 5 and 6, this volume). And even more revealing: "Hence, both in space and time, we seem to be brought somewhat near to that great fact – that mystery of mysteries – the first appearance of new beings on this earth."[24] In his 1851 monograph on fossil barnacles, Darwin wrote: "This, the most ancient genus of the Lepadidæ, seems also to be the stem of the genealogical tree."[25] Given these and many other published remarks, it is little wonder then, that when the anonymous evolutionary *Vestiges* appeared in 1844, Darwin heard it "has been by some attributed to me."[26] And, naturally enough, he expressed no surprise.

A Darwin letter discovered by the present author in 2022 is yet another source. Darwin wrote in 1844 to the editor of the *Gardeners' Chronicle* (personally unknown to him), agreeing with a recent editorial on transmutation in plants: "a subject [which] is well worth investigation." The editor mentioned Darwin's name and quoted from the letter in the *Gardeners' Chronicle*.[27]

Near the start of his theorizing in 1838, Darwin wrote in his *Notebook C*: "State broadly scarcely any novelty in my theory, only slight differences, 'the opinion of many people in conversation'."[28] In the sixth edition of the *Origin of Species*, published in 1872, Darwin responded to criticism that he exaggerated his originality: "I formerly spoke to very many naturalists on the subject of evolution, and never once met with any sympathetic agreement."[29] In his

---

[21] C. Darwin, *Journal and Remarks* (London: Colburn, 1839), 569.   [22] *Ibid.*, 400.

[23] C. Darwin, *Journal of Researches*. 2nd ed. (London: Murray, 1845), 380.   [24] Ibid., p. 378.

[25] C. Darwin, *A Monograph on the Fossil Lepadidæ* (London: Palæontographical Society, 1851), 48.

[26] To W. D. Fox [24 April 1845] CCD3: 180.

[27] *Gardeners' Chronicle* (November 23, 1844): 779. http://darwin-online.org.uk/content/frame set?pageseq=1&itemID=F3455&viewtype=text

[28] Paul H. Barrett, Peter J. Gautrey, Sandra Herbert, David Kohn, and Sydney Smith, et al., eds. *Charles Darwin's notebooks* (London: British Museum, 1987), 294.

[29] Darwin, *Origin*. 6th ed., 424.

autobiography he recalled: "I occasionally sounded not a few naturalists, and never happened to come across a single one who seemed to doubt about the permanence of species. ... I tried once or twice to explain to able men what I meant by Natural Selection, but signally failed."[30] Notice the distinction Darwin made between discussing evolution with "not a few" but natural selection only "once or twice." It must be stressed that the present discussion is about Darwin's belief in and work on evolution, not sharing the details of natural selection.

The diary entries of the paleobotanist and Lyell's brother-in-law Charles Bunbury provide fascinating glimpses as to how Darwin's views were sometimes communicated. At a luncheon at the home of mutual friend in 1845:

We had a good deal of pleasant talk on scientific matters, especially on the geographical distribution of plants and animals. He [Darwin] spoke of the extraordinary local peculiarity of the productions of the Galapagos islands ... He avowed himself to some extent a believer in the transmutation of species, though not, he said, exactly according to the doctrine either of Lamarck or of the "Vestiges." But he admitted that all the leading botanists and zoologists, of this country at least, are on the other side.[31]

Eleven years later in 1856, another entry shows how those who knew Darwin regarded the pre-publication years as time spent working toward the theory: "Darwin came in at breakfast time, and I had an interesting talk with him about *species*, and the various questions connected with their origin, distribution and diffusion. ... it is that to which Darwin has long devoted himself."[32] Bunbury mentioned this point again (after the reading of the Darwin-Wallace papers) on July 14, 1858: "Darwin has been engaged for nearly twenty years in a work on the general question of species."[33]

10.4   Like Confessing a Murder?

Probably the most widely known quotation purportedly supporting the delay/secret thesis is the famous line "it is like confessing a murder." It is now one of the most oft quoted passages by Darwin and is the title of numerous books, articles and other productions. Many writers starting in the 1950s have felt this passage means that Darwin was very afraid. Perhaps scholars in the twentieth century were so imbibed with Freudian expectations that they were convinced there must be some buried tension and anguish. Freud's biographer Ernest Jones wrote about "the psychology of discoverers" in 1959: "[Discovering] the relation of Natural Selection to Evolution, ... meant displacing God from His position ... Darwin, the one who stood in such awe of his own father, said it

---

[30] N. Barlow ed., *Autobiography of Charles Darwin* (London: Collins, 1958), 123.
[31] H. Lyell ed. *Life of Sir Charles J. F. Bunbury*. vol. 1 (London: Murray, 1906), 213.
[32] *Ibid.*, vol. 2, p. 98.   [33] *Ibid.*, vol. 2, p. 129.

was 'like committing [sic] murder' – as, indeed, it was unconsciously; in fact, parricide. He paid the penalty in a crippling and lifelong neurosis."[34]

The line comes from an 1844 letter to J. D. Hooker.

I am almost convinced (quite contrary to opinion I started with) that species are not (it is like confessing a murder) immutable. Heaven forfend me from Lamarck nonsense of a "tendency to progression" "adaptations from the slow willing of animals" &c, – but the conclusions I am led to are not widely different from his – though the means of change are wholly so – I think I have found out (here's presumption!) the simple way by which species become exquisitely adapted to various ends.[35]

Desmond and Moore wrote of this: "When Darwin did come out of his closet and bare his soul to a friend, he used a telling expression. He said it was 'like confessing a murder.' Nothing captures better the idea of evolution as a social crime in early Victorian Britain."[36] Many find it hard to imagine that there can be any other interpretation. Yet originally it *was* read differently. This quotation has been in print since *Life and Letters* in 1887. I have been unable to find any writer before the 1950s who interpreted this passage as evidence of fear.

Darwin barely knew Hooker at the time and was, as so often, humorously melodramatic in telling his new correspondent, probably somewhat embarrassed, that he held an unorthodox view. He never asked Hooker to keep the matter confidential. Hence, "confessing a murder" is quoted out of context. It is in fact entirely typical humorous language for Darwin. Considered next to the language Darwin used all the time, the correct interpretation becomes clear.

When work on his books felt overwhelming he would write "the descent half kills me" or "I am ready to commit suicide."[37] When Hooker was preparing to travel overseas, Darwin wrote "I will have you tried by a court martial of Botanists & have you shot." Darwin once playfully remarked to A. R. Wallace "may all your theories succeed, except that on oceanic islands, on which subject I will do battle to the death."[38] Even the word murder was typical Darwin hyperbole: "if [the plant] dies, I shall feel like a murderer." "You ought to have seen your mother she looked as if she had committed a murder & told a fib about Sara going back to America with the most innocent face." "I fear that I shall kill the splendid specimen of Sarracenia, which Hooker sent: it is downright murder, but I cannot help it." "I had long considered the Scotch Deer Hound a mongrel, par Excellence. I tell any Scotch so, or I shall be murdered." "When I saw your bundle of observations, I felt as if I had

---

[34] E. Jones, *Free Associations: Memoirs of a Psychoanalyst* (New York: Basic Books, 1959), 203–4.
[35] Darwin to Hooker [11 Jan. 1844] CCD3:2.   [36] Desmond and Moore, *Darwin*, pp. xviii; 313.
[37] Darwin to Hooker 10 Feb. [1875] CCD23:61.
[38] Darwin to Wallace 22 Dec. 1857 CCD6:514.

committed theft, arson or murder." This last remark, over embarrassment that a correspondent had taken too much trouble for Darwin, is very much stronger than the famous "confessing a murder."[39]

10.5   Popular Explanations for the "Delay"

Darwin delayed because of his wife's religious feelings? No. There is not a shred of evidence for this. No one seems to have ever suggested this before 1967. Darwin did, however, several times state in later years that he refrained from discussing religion in his publications because of the feelings of some members of his family.

Darwin delayed after witnessing the hostile reception of *Vestiges* in 1844? In fact he completed his 1844 essay and moved back to his *Beagle* publishing projects *before* that popular book appeared. *Vestiges* had nothing to do with his own research and publishing trajectory.

His eight years of barnacle work are frequently cited as undertaken in order to avoid the species theory or because he felt he could not publish it without bolstering his credentials. A single quote is often used to justify this. In 1845 Hooker wrote about the speculative work on species by Frédéric Gérard. "I am not inclined to take much for granted from anyone treats the subject in his way and who does not know what it is to be a specific Naturalist himself."[40] Darwin responded that this remark could also include himself. "How painfully (to me) true is your remark that no one has hardly a right to examine the question of species who has not minutely described many." But the following two sentences (often omitted) refute the interpretation that Darwin felt he needed to bolster his credentials:

I was, however, pleased to hear from Owen (who is vehemently opposed to any mutability in species) that he thought it was a very fair subject & that there was a mass of facts to be brought to bear on the question, not hitherto collected. My only comfort is, (as I mean to attempt the subject) that I have dabbled in several branches of Nat. Hist: & seen good specific men work out my species & know something of geology; (an indispensable union) & though I shall get more kicks than half-pennies, I will, life serving, attempt my work.[41]

Here Darwin twice says he will do his species theory anyway and lists why he feels qualified already. There are several other statements of such resolve in the face of anticipated backlash.

---

[39] Darwin to Hooker 24 Nov. 1873, CCD21:519; to H.E. Litchfield 4 Oct. [1877] CCD25:400; to W. T. Thiselton-Dyer 16 Nov. 1881, CCD29:546; to W.D. Fox 8 Mar. [1856] CCD6:50; to O. Salvin 12 Oct. [1871] CCD19:634.

[40] Darwin Correspondence Project, "Letter no. 914," accessed on January 3, 2024, www.darwin project.ac.uk/letter/?docId=letters/DCP-LETT-914.xml.

[41] Darwin to J. D. Hooker [10 Sept. 1845] CCD3:252–3.

The barnacle studies were the culmination of decades of work, not a diversion or prop for the species theory.[42] Marine invertebrates had been Darwin's central area of biological interest and expertise since 1825. Richard Keynes noted that "well over half of the pages of the Zoology Notes [on the *Beagle*] were concerned with marine invertebrates."[43] Hooker wrote to Darwin's son Francis in 1885 on why his father had spent eight years on barnacles: "Your father had Barnacles on the brain, from Chili onwards! [1833] He talked to me incessantly of beginning to work at his 'beloved Barnacles' (his favorite expression) long before he did so methodically."[44]

10.6   Not Finished

Darwin did not regard his species theory as finished in 1844 when, between other in-progress projects, he wrote out the 230-page essay. It was a consolidation of thoughts to date, "never intended for publication."[45] His letters show that at the time he foresaw that a work on species would ensue after years of work and research were completed. There were many problems to be solved and hundreds of books and journals to read, and many practical experiments to conduct before the five or so years of composition he envisaged.

Darwin later recalled that in 1844 "I overlooked one problem ... the tendency in organic beings descended from the same stock to diverge in character as they become modified."[46] Janet Browne, in reviewing Dov Ospovat's book *The Development of Darwin's Theory: Natural History, Natural Theology, and Natural Selection, 1838–1859*, praised how he showed "that Darwin became dissatisfied with the 'Essay,' pursued lines of thought not found there or in the species notebooks, and eventually recast his theory not once, but twice, before settling down in 1858 to write the Origin of Species." And Ospovat "believed that intense work from 1854 to 1858 led Darwin to a new view of the evolutionary process, and thence to a theory that Darwin considered complete. This theory hung upon his own 'principle of divergence'."[47]

Another problem which Darwin called in 1848 "the greatest *special* difficulty I have met with" was neuter insects.[48] Only in 1854 did he feel he had

[42] See Marsh Richmond's important Appendix II in CCD2:388–409.

[43] R. Keynes, *Charles Darwin's Zoology Notes & Specimen Lists from H.M.S. Beagle* (Cambridge: Cambridge University Press, 2000), x.

[44] Hooker to F. Darwin 31 Dec. 1885, in L. Huxley ed., *Life and Letters of Sir Joseph Dalton Hooker.* vol. 2 (London: Murray, 1918), 299.

[45] van Wyhe, "Mind the gap," 188.   [46] Barlow, *Autobiography*, 120.

[47] J. Browne, "New developments in Darwin studies?" *Journal of the History of Biology* 15 (1982): 276.

[48] Darwin, C. R. 6.1848. CUL-DAR73.21–22. Edited by John van Wyhe http://darwin-online.org.uk/content/frameset?pageseq=1&itemID=CUL-DAR73.21-22&viewtype=text.

solved this. He experimented extensively in the 1850s with natural means of dispersal and individual variations and crossings of pigeons and other domesticated varieties. He also experimented with hive bees, attempting to explain their geometrical constructions with natural selection. This work made up a major part of the *Origin of Species*.

10.7   Conclusion

As the years roll by and Darwin continues to be a subject of interest, new myths continue to emerge and proliferate more widely than the more accurate accounts that preceded them. There is the old belief that seeing the beaks of the finches on the Galapagos converted Darwin to evolution or that he realized their ecological significance (see Sulloway, Chapters 5 and 6, this volume). There is the now quite orthodox belief that he was not the official naturalist on the *Beagle*, the ship's surgeon actually was.[49] There is the story that the death of his daughter Annie killed off his religious faith.[50] Creationists repeat a story called "Darwin's bodysnatchers" in which Darwin sought the skulls of aboriginal peoples of Tasmania.[51] And there is the belief that he expressed concern that because he married his first cousin, his children would be congenitally prone to ill health. Instead, he every time expressed fear that they had inherited *his* bad constitution. Even the purported contemporary nickname of one of his greatest defenders, T. H. Huxley, is a myth. Far from being widely known as "Darwin's bulldog," this famous epithet is posthumous (see also Bowler, Chapter 12, this volume).[52] The only remedy to all these myths, if there is one, is continued exposure.

---

[49] F. J. Sulloway, "Darwin and his finches," *J. Hist. Biol.* 15(1) (1982): 1–53; John van Wyhe, "My appointment received the sanction of the Admiralty," *Studies in History and Philosophy of Biological and Biomedical Sciences* 44(3) (Sept. 2013): 316–26.

[50] John van Wyhe and Mark Pallen, "The Annie Darwin hypothesis," *Centaurus* 54 (2012): 1–19.

[51] John van Wyhe, "Darwin's body-snatchers?" *Endeavour* 41(1) (Dec. 2016): 29–31.

[52] John van Wyhe, "Why there was no 'Darwin's bulldog'," *The Linnean*, 35(1) (Apr. 2019): 26–30.

Myth 11 That Wallace's and Darwin's Theories Were the Same, and That Darwin Did Not Reveal Wallace's 1858 Letter and Theory Until He Ensured His Own Priority

Michael Ruse

11.1 Introduction

On June 30, 1858, two contributions – one by the already-established naturalist Charles Robert Darwin, and the other by the collector Alfred Russel Wallace (Figure 11.1) – were submitted to the Linnaean Society of London, by the geologist Charles Lyell and the botanist Joseph Hooker. Read on July 1, 1858 – neither Darwin (sick) nor Wallace (abroad) was present – these contributions supported the (already well-known but highly controversial) idea of organic evolution by proposing a cause – what has come to be known (although it was exclusively Darwin's term) as "natural selection."[1] This much is certain. Then the controversies begin. Who got to the theory first? In a letter to Charles Lyell of June 18, 1858, Charles Darwin wrote "You said this when I explained to you here very briefly my views of 'Natural Selection' depending on the Struggle for existence.— I never saw a more striking coincidence. If Wallace had my M.S. sketch written out in 1842 he could not have made a better short abstract. Even his terms now stand as Heads of my Chapters."[2] Hence the conspiracy theory against Wallace:

The explanation for Wallace's disappearance from history is threefold. In part, he was a victim of a conspiracy by the scientific aristocracy of the day and was robbed in 1858 of his priority to the proclaiming of the theory. "This delicate arrangement", Huxley' son Leonard termed the incident. "This delicate situation" Darwin characterized it in a letter to Sir Joseph Dalton Hooker, the great botanist and intimate friend of Darwin.[3]

---

[1] C. Darwin and A. Wallace. 1858. On the Tendency of Species to Form Varieties; and on the Perpetuation of Varieties and Species by Means of Selection. *Proceedings of the Linnaean Society, Zoological Journal* 3: 46–62.

[2] Darwin Correspondence Project, "Letter no. 2285," accessed on October 20, 2022, www.darwin project.ac.uk/letter/?docId=letters/DCP-LETT-2285.xml.

[3] A. C. Brackman. 1980. *A Delicate Arrangement: The Strange Story of Charles Darwin and Alfred Russel Wallace*. New York: Times Books, p. xi.

Figure 11.1 Alfred Russel Wallace (Popular Science Monthly Volume 1, 1877, Public domain, via Wikimedia Commons)

This short discussion will be divided into five parts. First, it will be shown that Darwin and Wallace did indeed, separately, hit on the same cause or process of change. Second, it will be argued that, claims to the contrary notwithstanding, Darwin behaved in an exemplary moral and professional manner in the events leading up to the reading of the two papers. Also, that what happened afterwards, that people took for granted that it was Darwin by far who was the major player in the evolution-through-natural-selection story, was perfectly justified. Wallace was not unfairly deprived of credit. Third, fourth, and fifth, focusing now on the overall notion of "theory" as opposed to "cause," it will be shown that there were very significant differences between the thinking of Darwin and Wallace.

11.2   The Cause of Evolution

Darwin and Wallace came up with the idea of natural selection independently. Darwin came to the key insight late in 1838 (see also Love, Chapter 9, this volume), and by 1842 he felt sufficiently confident to write out a 35-page "Sketch" of his theory of evolution, one where the prime cause was natural selection, and he extended this to a 230-page "Essay" in 1844.[4] It was an extract from the latter that was the first part of the contribution of Darwin to the

---

[4] J. Browne. 1995. *Charles Darwin: Voyaging. Volume 1 of a Biography*. London: Jonathan Cape.

Linnaean Society. In the case of Wallace, he had been an evolutionist for over ten years, after reading Robert Chambers' (anonymously published) *Vestiges of the Natural History of Creation.*[5] It was in February 1858 that Wallace, recovering from sickness, likewise came up with the idea of natural selection, an idea that he wrote up quickly as an essay – "On the Tendency of Varieties to Depart Indefinitely from the Original Type" – that was his contribution to the Linnaean Society.

Darwin wrote:

Lighten any check in the least degree, and the geometrical powers of increase in every organism will almost instantly increase the average number of the favoured species ... Reflect on the enormous multiplying power inherent and annually in action in all animals; reflect on the countless seeds scattered by a hundred ingenious contrivances, year after year, over the whole face of the land; and yet we have every reason to suppose that the average percentage of each of the inhabitants of a country usually remains constant ...

But let the external conditions of a country alter. If in a small degree, the relative proportions of the inhabitants will in most cases simply be slightly changed; but let the number of inhabitants be small, as on an island, and free access to it from other countries be circumscribed, and let the change of conditions continue progressing (forming new stations), in such a case the original inhabitants must cease to be as perfectly adapted to the changed conditions as they were originally ... Now, can it be doubted, from the struggle each individual has to obtain subsistence, that any minute variation in structure, habits, or instincts, adapting that individual better to the new conditions, would tell upon its vigour and health? In the struggle it would have a better chance of surviving; and those of its offspring which inherited the variation, be it ever so slight, would also have a better chance. Yearly more are bred than can survive; the smallest grain in the balance, in the long run, must tell on which death shall fall, and which shall survive. Let this work of selection on the one hand, and death on the other, go on for a thousand generations, who will pretend to affirm that it would produce no effect, when we remember what, in a few years, Bakewell effected in cattle, and Western in sheep, by this identical principle of selection?[6]

Wallace wrote:

Now let some alteration of physical conditions occur in the district–a long period of drought, a destruction of vegetation by locusts, the irruption of some new carnivorous animal seeking "pastures new"–any change in fact tending to render existence more difficult to the species in question, and taking its utmost powers to avoid complete extermination; it is evident that, of all the individuals composing the species, those forming the least numerous and most feebly organized variety would suffer first, and, were the pressure severe, must soon become extinct.

The same causes continuing in action, the parent species would next suffer, would gradually diminish in numbers, and with a recurrence of similar unfavourable conditions

[5] R. Chambers. 1844. *Vestiges of the Natural History of Creation.* London: Churchill.
[6] Darwin and Wallace, On the Tendency, 48–9.

might also become extinct. The superior variety would then alone remain, and on a return to favourable circumstances would rapidly increase in numbers and occupy the place of the extinct species and variety.

The *variety* would now have replaced the *species*, of which it would be a more perfectly developed and more highly organized form. It would be in all respects better adapted to secure its safety, and to prolong its individual existence and that of the race. Such a variety *could not* return to the original form; for that form is an inferior one, and could never compete with it for existence.[7]

For both Darwin and Wallace, you start with the Malthusian struggle for existence (or more accurately, struggle for reproduction). Because of population pressures, not all organisms can survive and leave offspring. Those that do will, on average, do so because they have superior – more helpful – characteristics. Over time, this natural form of selection – in the sense of differential survival and reproduction – will bring on major, permanent change. Not random change. Rather, change in the sense that the new organisms will be better adapted than those which went before, that is, have features that will enable them to survive and reproduce. Not better in some absolute sense, but in the sense that they can do better in the present surroundings than those successful in older surroundings.

11.3   Darwin's Behavior

It is almost certain that Wallace's paper was sent to Darwin in March 1858.[8] Wallace had had some previous correspondence on evolution with Darwin, so there was no surprise that it was to Darwin of all people he sent it. Darwin received it on June 18, 1858, and at once wrote to Lyell telling him of the essay and suggesting that he (Darwin) should stand aside and let Wallace take full credit. Lyell, long a friend and mentor of Darwin, would have none of this and so, bringing in Hooker as support, the ideas of both men were presented together at the Linnaean society. This was obviously done without Wallace's permission, but to get that permission would have taken six months and clearly Wallace did want something done with his paper, for he had asked himself Darwin to pass it on to Lyell.

In short, the behaviors of Darwin, and of his friends Lyell and Hooker, were honest and generous. They could easily have satisfied their consciences and published Darwin first, on the undeniable grounds that he was by far the first with the discovery. It is true that, later, it was always Darwin who got the major credit; but, after all, it was he not Wallace who wrote and published the *Origin of Species*, where selection is shown to explain through the whole realm of the

---

[7] Darwin and Wallace, On the Tendency, 58–9.

[8] D. Kohn. 1981. On the Origin of the Principle of Diversity: Review of Brackman, *A Delicate Arrangement. Science* 213: 1105–8.

life sciences – behavior, paleontology, biogeography, and more.[9] Wallace always felt that he had been treated most honorably, as he was.

What is not true is that Darwin suppressed the arrival of Wallace's paper until he had read it and cribbed from it. It has been claimed that this happened and that specifically Darwin took credit for (an idea hitherto unknown to him) the "Principle of Divergence" – namely the idea of and reason for the splitting of groups into separate units, the branching one sees in the tree of life.[10] However, countering this, in a letter written in 1857 to the American botanist Asa Gray – the second piece by Darwin that was presented at the Linnaean Society – there is an unambiguous statement of the Principle.

Another principle, which may be called the principle of divergence, plays, I believe, an important part in the origin of species. The same spot will support more life if occupied by very diverse forms ... [It follows] that the varying offspring of each species will try (only few will succeed) to seize on as many and as diverse places in the economy of nature as possible. Each new variety or species, when formed, will generally take the place of, and thus exterminate its less well-fitted parent. This I believe to be the origin of the classification and affinities of organic beings at all times; for organic beings always seem to branch and sub-branch like the limbs of a tree from a common trunk, the flourishing and diverging twigs destroying the less vigorous – the dead and lost branches rudely representing extinct genera and families.[11]

Wallace also had the Principle, but if more evidence is needed that Darwin did not crib it from Wallace, there is the fact that Wallace was really more interested in one form succeeding another. He rather took divergence for granted. He wrote about "the many lines of divergence from a central type,"[12] and left things at that. Indeed, he was mainly interested in showing that the arrival of a new species means the death of the old ancestral species, giving no indication that there might be two new species.

11.4 The Analogy from Artificial Selection

Darwin and Wallace really did hit upon the same process of change. And they both thought it sufficient to lead to evolution, the tree of life. This is but part of the story. As noted, if we are thinking in terms of "theory" rather than just "cause," major differences appear. First, Darwin and Wallace differed over the importance of the analogy from artificial selection. For Darwin, it was central. We have seen this, even in the early essay extracted for the Linnaean Society.

[9] C. Darwin. 1859. *On the Origin of Species by Means of Natural Selection, or the Preservation of Favoured Races in the Struggle for Life*. London: John Murray.

[10] A. C. Brackman. 1980. *A Delicate Arrangement: The Strange Case of Charles Darwin and Alfred Russel Wallace*. New York: Times Books.

[11] Darwin and Wallace, On the Tendency, 52–3.

[12] Darwin and Wallace, On the Tendency, 62.

"Let this work of selection on the one hand, and death on the other, go on for a thousand generations, who will pretend to affirm that it would produce no effect, when we remember what, in a few years, Bakewell effected in cattle, and Western in sheep, by this identical principle of selection?"[13] For Wallace, however, from the beginning this was a false analogy, to be countered.

One of the strongest arguments which have been adduced to prove the original and permanent distinctness of species is, that varieties produced in a state of domesticity are more or less unstable, and often have a tendency, if left to themselves, to return to the normal form of the parent species; and this instability is considered to be a distinctive peculiarity of all varieties, even of those occurring among wild animals in a state of nature, and to constitute a provision for preserving unchanged the originally created distinct species. ... It will be observed that this argument rests entirely on the assumption, that varieties occurring in a state of nature are in all respects analogous to or even identical with those of domestic animals, and are governed by the same laws as regards their permanence of further variation. But this is the object of the present paper to show that this assumption is altogether false, that there is a general principle in nature which will cause many varieties to survive the parent species, and to give rise to successive variations departing further and further from the original type, and which also produces, in domesticated animals, the tendency of varieties to return to the parent form.[14]

When he read the *Origin of Species*, clearly Wallace changed his mind somewhat and bought into the analogy. He even called his first collection of essays, *Contributions to the Theory of Natural Selection*,[15] although, as we shall see, Darwin was not entirely appreciative of these "contributions." But Wallace never really liked the analogy, and, on July 2, 1866, he wrote at length to Darwin, about how many people seemed to think that natural selection brought in a thinking designer:

The two last cases of this misunderstanding are, 1st. The article on "Darwin & his teachings" in the last "Quarterly Journal of Science", which, though very well written & on the whole appreciative, yet concludes with a charge of something like blindness, in your not seeing that "Natural Selection" requires the constant watching of an intelligent "chooser" like man's selection to which you so often compare it; – and 2nd., in Janet's recent work on the "Materialism of the present day", reviewed in last Saturday's "Reader", by an extract from which I see that he considers your weak point to be, that you do not see that "thought & direction are essential to the action of 'Nat. Selection'." The same objection has been made a score of times by your chief opponents, & I have heard it as often stated myself in conversation.

Now I think this arises almost entirely from your choice of the term "Nat. Selection" & so constantly comparing it in its effects, to Man's selection, and also to your so frequently personifying Nature as "selecting" as "preferring" as "seeking only the good of the species" &c. &c. To the few, this is as clear as daylight, & beautifully suggestive,

[13] Darwin and Wallace, On the Tendency, 49.
[14] Darwin and Wallace, On the Tendency, 53–4.
[15] A. R. Wallace. 1870. *Contributions to the Theory of Natural Selection*. London: Macmillan.

but to many it is evidently a stumbling block. I wish therefore to suggest to you the possibility of entirely avoiding this source of misconception in your great work, (if not now too late) & also in any future editions of the "Origin", and I think it may be done without difficulty & very effectually by adopting Spencer's term (which he generally uses in preference to Nat. Selection) viz. "Survival of the fittest."[16]

Darwin was sufficiently appreciative of this letter that, for the later editions of the *Origin of Species*, he added Spencer's term as an alternative (see Depew, Chapter 15, this volume). The title of the chapter dealing with natural selection, in the 1869 fifth edition, was altered to NATURAL SELECTION, OR THE SURVIVAL OF THE FITTEST. This said, Darwin never gave up on his own term and it is clear from the *Descent of Man*, appearing in 1871, he always thought of the main mechanism of change as something akin to what a breeder does in a farmyard.

11.5   Selection: Individual or Group?

As the very title of his essay shows, Wallace always thought that selection – as we may call it – can operate between groups as well as between individuals – group selection as opposed to individual selection. He never really distinguished the two.

But this new, improved, and populous race might itself, in course of time, give rise to new varieties, exhibiting several diverging modifications of form, any of which, tending to increase the facilities for preserving existence, must, by the same general law, in their turn become predominant. Here, then, we have progression and continued divergence deduced from the general laws which regulate the existence of animals in a state of nature, and from the undisputed fact that varieties do frequently occur . . . Variations in unimportant parts might also occur, having no perceptible effect of the life-preserving powers; and the varieties so furnished might run a course parallel with the parent species, either giving rise to further variations or returning to the former type. All we argue for is, that certain varieties have a tendency to maintain their existence longer than the original species, and this tendency must make itself felt; for though the doctrine of chances or averages can never be trusted to on a limited scale, yet, if applied to high numbers, the results come nearer to what theory demands, and, as we approach to an infinity of examples, becomes strictly accurate.[17]

Darwin, however, was adamant always that selection is between individuals, where this might be the interrelated nest or hive – giving rise to what today is known as "kin selection."[18] In the *Origin of Species*, there is a careful argument showing that features (caused by what we today call "genes") can be

---

[16] Darwin Correspondence Project, "Letter no. 5140," accessed on December 29, 2023, www.darwinproject.ac.uk/letter/?docId=letters/DCP-LETT-5140.xml.
[17] Darwin and Wallace, On the Tendency, 59.
[18] J. Maynard Smith. 1964. Group Selection and Kin Selection. *Nature* 201: 1145–7.

transmitted vicariously through relatives, and hence even the apparently disinterested altruist is truly serving its own reproductive ends. Wallace approached Darwin, suggesting that selection could account for the sterility of hybrids like mules. Rather than just by chance – the different reproductive systems do not work together (as suggested in the *Origin of Species*) – when sterility occurs this is of value to the parent species. Hybrids are just not that well adapted. Darwin countered strongly. He simply could not see how, having committed oneself to producing a hybrid, it would now be of advantage to either parent that the offspring, such as a mule, be sterile. It must simply be a function of different inheritances not meshing and working well together.

Let me first say that no man could have more earnestly wished for the success of N. selection in regard to sterility, than I did; & when I considered a general statement, (as in your last note) I always felt sure it could be worked out, but always failed in detail. The cause being as I believe, that natural selection cannot effect what is not good for the individual, including in this term a social community.[19]

Darwin was true to his commitments. In the *Descent of Man,* talking of the evolution of morality, he made it clear that it is individual selection that is at work.

It must not be forgotten that although a high standard of morality gives but a slight or no advantage to each individual man and his children over the other men of the same tribe, yet that an advancement in the standard of morality and an increase in the number of well-endowed men will certainly give an immense advantage to one tribe over another. There can be no doubt that a tribe including many members who, from possessing in a high degree the spirit of patriotism, fidelity, obedience, courage, and sympathy, were always ready to give aid to each other and to sacrifice themselves for the common good, would be victorious over most other tribes; and this would be natural selection.[20]

"Victorious over most other tribes"? Surely this is an appeal to group selection? Not at all! Immediately after this passage, Darwin implied that (what today is known as) "reciprocal altruism" is a major causal factor. You scratch my back and I will scratch yours: "as the reasoning powers and foresight of the members [of a tribe] became improved, each man would soon learn from experience that if he aided his fellow-men, he would commonly receive aid in return."[21] This is not the disinterested altruism of group selection. The individual alone is benefiting: individual selection. (There is also a veiled appeal to kin selection. The members of a tribe are interrelated, or think they are, and so help for others is – or is thought to be – help for oneself.)

---

[19] Letter to Wallace April 6, 1868, Darwin Correspondence Project, "Letter no. 6095," accessed on December 29, 2023, www.darwinproject.ac.uk/letter/?docId=letters/DCP-LETT-6095.xml.

[20] C. Darwin. 1871. *The Descent of Man, and Selection in Relation to Sex*. London: John Murray, Vol. 1, 166.

[21] Darwin, *The Descent of Man*, Vol. 1, 163.

11.6   Spiritualism

Third, finally, and most importantly, in the 1860s, Wallace became a spiritualist and started arguing – in his *Contributions to Natural Selection* book! – that human evolution requires nonphysical forces from without.

> I have shown that the brain of the lowest savages, and, as far as we yet know, of the pre-historic races, is little inferior in size to that of the highest types of man, and immensely superior to that of the higher animals; while it is universally admitted that quantity of brain is one of the most important, and probably the most essential, of the elements which determine mental power. Yet the mental requirements of savages, and the faculties actually exercised by them, are very little above those of animals. The higher feelings of pure morality and refined emotion, and the power of abstract reasoning and ideal conception, are useless to them, are rarely if ever manifested, and have no important relations to their habits, wants, desires, or well-being. They possess a mental organ beyond their needs. Natural Selection could only have endowed savage man with a brain a little superior to that of an ape, whereas he actually possesses one very little inferior to that of a philosopher.[22]

To this, as further examples where natural selection would have been inadequate, Wallace added skin and the fact that humans are not hairy, our feet and hands – they seem "unnecessarily perfect for the needs of savage man"[23] – and the human larynx – it is "beyond the needs of savages, and from their known habits, impossible to have been acquired either by sexual selection, or by survival of the fittest."[24] Hence, no selection involved.

Darwin was appalled. He was spurred to write the *Descent of Man*, explicitly countering Wallace's claims of human nonnatural origins. We have seen just above how he took on Wallace's claim that "savages" have no sense of (selection-caused) morality. Amusingly, poking Wallace one in the eye, Darwin made great use of one facet of his mechanism of sexual selection, one which is very much individual selectionist in intent – "female choice," where groups of males within the species compete for females. (A mechanism regarded dubiously by Wallace, at least inasmuch as it implied that nonhumans have the equivalent of aesthetic appreciation, as in the peahen opting for her favorite peacock tail.) Darwin and Wallace never fell out at the personal level, but it truly was the parting of the ways. Darwin and his followers were fighting for the scientific respectability of his theory (of evolution generally) and the odor – the stench – of spiritualism was anathema. Little wonder that people were happy to put a line between the thinking of Darwin and Wallace.

---

[22] Wallace, *Contributions*, 355–6.   [23] Wallace, *Contributions*, 356.
[24] Wallace, *Contributions*, 357.

11.7   Why the Differences?

Charles Darwin and Alfred Russel Wallace both conceived of natural selection as the chief cause of evolutionary change. Yet, as we have just seen, there were major differences. Why? In a way, it is a mistake to look for just one overwhelming factor, but one significant thing does stand out. Although living on family money, Darwin was always a professional scientist and recognized as such. Wallace was a collector and was always somewhat of an outsider. He could never get a professional job, for instance.[25] This shows in their different ways of thinking. Darwin seized on artificial selection because it gave him an analogy for natural selection, the chief mark of what the scientist-philosopher John F. W. Herschel called (following Newton) a *vera causa*.[26] Wallace never had mentors to direct him that way. Darwin was knowledgeable about breeding and studied it as a professional to see the implications for levels of selection. And above all, the conservative professional Darwin could never have accepted spiritualism. Darwin's success came from his professionalism and Wallace's success came because he was always thinking out of the loop. With natural selection, the two approaches paid off handsomely.[27]

---

[25] P. Raby. 2001. *Alfred Russel Wallace: A Life*. Princeton: Princeton University Press.

[26] M. Ruse. 1975. Darwin's Debt to Philosophy: An Examination of the Influence of the Philosophical Ideas of John F. W. Herschel and William Whewell on the Development of Charles Darwin's Theory of Evolution. *Studies in History and Philosophy of Science* 6: 159–81.

[27] A comprehensive account of the subject of this chapter can be found in John van Wyhe. 2013. *Dispelling the Darkness: Voyage in the Malay Archipelago and the Discovery of Evolution by Wallace and Darwin*. Hackensack World N.J.: Scientific Publishing Company.

Myth 12　That Huxley Was Darwin's Bulldog and Accepted All Aspects of His Theory

Peter J. Bowler

12.1　Introduction

It is certainly true that Thomas Henry Huxley (Figure 12.1) was recognized as the most active of Darwin's defenders, and this has led many commentators to assume that he accepted the whole Darwinian theory. In fact, he had substantial reservations about the theory's explanatory powers, and he did not gain the sobriquet "Darwin's Bulldog" until much later. Nevertheless, when the *Origin of Species* was attacked on religious grounds, he insisted that the theory of natural selection opened the way to scientific investigation of evolution. He explored the major steps in the evolution of life on earth, including a demonstration of our close relationship to the great apes. But although Huxley saw natural selection as a valid component of the evolutionists' toolkit, he did not accept that it offered a complete explanation of how living things evolve. It was a vital inspiration to scientists seeking to explain the diversity of life, but it was not the whole story.[1]

For Darwin, the variation that supplies the raw material of selection was essentially undirected, making the species plastic in the sense that it could be molded in any direction by environmental pressure. Adaptation to the environment was the only directing agent of evolution. In contrast, Huxley did not believe that the environment was all-powerful because he assumed that the biological character of the species was also responsible for the actual production of new characters. He complained that Darwin had ignored the possibility of new characters appearing as the sudden transformations known as "sports of nature" or saltations. Selection might prevent some maladaptive variants succeeding in the external world, but it could only work with material offered to it by the potentials inherent in the organism.

---

[1] Michael Bartholomew first drew attention to the complexity of Huxley's position; see his "Huxley's Defence of Darwinism," *Annals of Science*, 32 (1975): 525–35. For details of Huxley's career see Adrian Desmond, *Huxley: The Devil's Disciple* (London: Michael Joseph, 1994) and *Huxley: Evolution's High Priest* (London: Michael Joseph, 1997).

Figure 12.1 Thomas Henry Huxley in 1877. Lock & Whitfield, *Men of mark.*
London (Wellcome Collection. Public Domain Mark)

To understand why Huxley mistrusted Darwin's reliance on adaptation
we need to explore their backgrounds and interests. Although both agreed
on the need for purely naturalistic explanations in science, they came
from very different social origins and followed different research agendas. Darwin came from a wealthy middle-class family and always thought
of himself as an old-fashioned field naturalist. He wanted to understand
how animals and plants fitted into the world around them. Huxley's
origins were in a lower social stratum and he made his way by becoming
a professional scientist, using the latest techniques to study the internal
structure of animals. He was a morphologist who dissected animals to
understand their fundamental similarities and differences. He also became
an expert in the analysis of fossil skeletons. From this perspective he
could appreciate how the superficial features of species might be shaped
by adaptation, but he was much more likely to see the defining features of
the main animal groups as the product of internal, purely biological
factors.

12.2   Huxley and the *Origin of Species*

Like Darwin, Huxley gained much of his early scientific experience on a Royal Navy surveying vessel, although he was the ship's surgeon, not the captain's guest. Aboard H.M.S. *Rattlesnake* he collected marine specimens on the Great Barrier Reef and made his reputation describing and classifying little-known invertebrate types. By the late 1850s he was a professor at the Royal School of Mines and had begun to work on vertebrate anatomy and paleontology. Until he read the *Origin of Species* Huxley had no interest in explaining how living things had evolved over time. As an exponent of the position sometimes known as "scientific naturalism" he certainly didn't believe in divine creation, but he could see no way of approaching the topic along lines that would satisfy the demand for a scientifically credible mechanism of transformation. Unlike Darwin he dismissed the Lamarckian mechanism of the inheritance of acquired characteristics. He had also been discouraged by his reading of Robert Chambers' *Vestiges of the Natural History of Creation*, realizing that its vague idea of a divinely preordained law of development left the topic in the hands of natural theology (see Ruse, Chapter 4, this volume).

Huxley was one of several naturalists to be informed about Darwin's theory before it was published (see van Wyhe, Chapter 10, this volume), but only when he read the *Origin of Species* did he appreciate the full scope of Darwin's new vision. As he later wrote, his first reaction was to say "How extremely stupid not to have thought of that."[2] He saw that natural selection depended for its explanatory power solely on observable causes – variation, heredity and the balance between reproduction and the limitations of the environment – and was thus open to experimental testing. He also appreciated that it did not imply a built-in progressive trend. For adaptive features such as the beaks of the Galapagos finches, it was a plausible and almost certainly valid explanatory hypothesis. This was the kind of theory required by the philosophy of scientific naturalism, thus opening the field to investigation and checkmating the claim that design by the Creator was the only reason why species are as they are.

It was because the exponents of natural theology sought to discredit Darwin's initiative that Huxley took on the task of defending the theory in public, necessitated by Darwin's reluctance to participate in open debate. His activities would eventually earn him the title "Darwin's bulldog," although John van Wyhe has shown that this phrase was not used until after his death in 1895.[3] It does, however, convey an appropriate sense of how Huxley was regarded in the early phases of the debate.

---

[2] Huxley, "On the Reception of the 'Origin of Species'" in Francis Darwin, ed. *The Life and Letters of Charles Darwin* (London: John Murray, 1887, 3 vols.), II, pp. 179–204, see p. 197.
[3] John van Wyhe, "Why There Was No 'Darwin's Bulldog'," *The Linnean*, 35 (2019): 26–30.

Fortunately for the Darwinians, Huxley was asked to write a review for the London *Times* which appeared on Boxing Day, 1859. In February of the following year he lectured on the theory at the Royal Institution, using varieties of pigeons to emphasize Darwin's point about the power of artificial selection. His main defense came in the *Westminster Review* in April 1860, a substantial article which introduced the term "Darwinism" and included several passages that have been quoted endlessly by historians. It explained the theory of natural selection and the various lines of evidence in its favor and declared that "every philosophical thinker hails it as a veritable Whitworth gun in the armoury of liberalism." He did, however, express doubts about the power of natural selection (Section 12.5 Last Words). Turning to the opposition from religious thinkers he noted that "Extinguished theologians lie about the cradle of every science as the strangled snakes beside that of Hercules."[4]

The episode that has come to symbolize Huxley's determination to resist the clerical attacks came on Saturday June 30 at the annual meeting of the British Association for the Advancement of Science in Oxford. Here the Bishop of Oxford, Samuel Wilberforce, primed by Huxley's rival Richard Owen, attacked the theory as bad science and worse philosophy. Huxley rose in defense and savaged the Bishop's effort to ridicule the idea of a link between humans and the great apes. We don't know his exact words, but they implied that he would rather be descended from an ape than from a man of great influence who attacked a theory he didn't understand. His remarks provoked an outcry from both sides in the audience, but in this instance we have to issue our first warning about the tendency to exaggerate Huxley's role. Modern studies show that the widely accepted image of Huxley demolishing the Bishop was constructed long after the event, when the triumph of evolutionism was assured (see Brooke, Chapter 13, this volume).[5] At the time, as shown by contemporary reports, many were not impressed by Huxley's performance and it was the botanist Joseph Hooker who did most to swing the meeting in Darwin's favor.

Late in 1862 Huxley got the opportunity to promote the theory to a very different audience. The last of a series of lectures on biology to working men was devoted to a defense of the *Origin of Species*. Huxley didn't publish the lectures directly, but they were taken down in shorthand and issued as a cheap pamphlet, with his approval, making the information widely available to

---

[4] Several of Huxley's commentaries are reprinted in his *Collected Essays*, especially volume 2, *Darwiniana* (London: Macmillan, 1894); for his reviews of the *Origin* see pp. 1–21 and 22–79; quotations from p. 23 and p. 78.

[5] E.g. Frank A. L. James, "An 'Open Clash between Science and the Church'? Wilberforce, Huxley and Hooker at the British Association, Oxford, 1860" in David M. Knight and Matthew D. Eddy, eds., *Science and Beliefs: From Natural Theology to Natural Science* (Aldershot: Ashgate, 2005): 1751–93.

ordinary people.[6] He defended Darwin vigorously in a controversy that erupted when he was awarded the Royal Society's Copley Medal in 1864.[7] He continued to publish supportive articles in formal periodicals, including a savage attack on St. George Jackson Mivart in the *Contemporary Review* for 1871.[8] Originally a supporter of Darwin, Mivart had now developed a very different view of how evolution works in which many parallel lines of development advanced along predetermined lines. But it was Mivart's efforts to make his theory compatible with Roman Catholic theology that really sparked Huxley's anger. In 1880 he gave an appreciative talk about Darwin's achievements in a Royal Institution lecture to celebrate the *Origin*'s coming of age. When Darwin died he wrote a substantial obituary and later spoke at the unveiling of a statue at the new Natural History Museum in South Kensington.[9]

12.3   Extending the Theory

Huxley also bolstered the theory by applying it in areas that Darwin had avoided, either for fear of provoking greater public outcry or because he genuinely did not appreciate how the science could be pushed forward. Most controversial was the topic of human origins, which Darwin had only alluded to briefly in the *Origin of Species*. But Huxley was also a pioneer in the attempt to reconstruct the history of life on earth by explaining how the major groups of animals could have arisen from previously existing forms. Although widely seen as promoting the Darwinian vision, this project also exposed the ways in which that vision could be modified by different perceptions of how evolution works.

Darwin knew that giving the human race an animal ancestry was anathema to traditional Christians, which is why Wilberforce chose to taunt Huxley on the topic. Huxley was already embroiled in a controversy with Richard Owen over the closeness of the relationship between humans and apes, a debate with obvious implications for the question of an evolutionary link. The popular press highlighted the issue with caricatures of gorillas displaying human behavior (Figure 12.2), although the Darwinian theory only implied an ape-like common ancestor which would not have been identical to any of its descendants. The controversy with Owen centered on the degree of anatomical similarity between the great apes and humans, with Owen stressing the differences and Huxley the similarities. Much attention focused on the structure of the brain, with the question of whether the apes had an organ called the hippocampus minor symbolizing the disagreement. Most historical studies

---

[6] Reprinted in Huxley, *Darwiniana*, pp. 303–475.
[7] This episode is described in Janet Browne, *Charles Darwin: The Power of Place* (London: Jonathan Cape, 2002), pp. 244–7.
[8] Ibid., pp. 120–86.   [9] Ibid., pp. 248–52 and 253–302.

Figure 12.2 *Punch*, May 18, 1861, "Monkeyana." (Wellcome Collection. Attribution 4.0 International (CC BY 4.0))

assume that Huxley's demonstrations carried the day and left Owen discredited.

Huxley's published his arguments in his *Man's Place in Nature* of 1863 and they were widely seen as a vindication of the view that humans have an ape ancestry.[10] They certainly undermined the assumption that the gulf between humans and apes is so great that our species must have been divinely created. In fact, though, Huxley's position was quite restricted in its evolutionary implications. He made no effort to imagine what the common ancestor would have looked like and said nothing about how it could have been transformed into the earliest humans. The best-known hominid fossil of the time, the Neanderthal

---

[10] Reprinted in vol. 7 of the *Collected Essays*, Huxley, *Man's Place in Nature and Other Anthropological Essays* (London: Macmillan, 1894).

cranium, was shown to have so large a capacity that it could not be the "missing link." There was no mention of the adaptive pressures that might have forced our ancestors to walk upright and gain a larger brain. Darwin himself would address these questions in his book *The Descent of Man* of 1871, but in the early 1860s Huxley was not prepared to tackle them.

Huxley soon became active in the attempt to reconstruct other major steps in the evolution of life on earth, but his inspiration for entering the field was provided not by Darwin but by the German evolutionist Ernst Haeckel (see Richards, Chapter 20, this volume). Darwin was reluctant to speculate about the origins of the main animal classes in part because he was well-aware of the imperfection of the fossil record. Huxley too had dismissed the idea that the earliest known fossils of any group would give any clue about the group's origin, which almost certainly lay much further back in time. He changed his viewpoint when Haeckel's *Generelle Morphologie* of 1866 showed that it was possible to use evidence from comparative anatomy and embryology to produce hypotheses about common ancestors and their derivation from previous forms.[11]

Haeckel encouraged Huxley to take seriously the implication embedded in Darwin's theory that the shared characters that identify a particular grouping of species have been inherited from a common ancestor. Each natural group is monophyletic, because the same basic character is unlikely to have evolved twice, allowing evolution to be depicted as a "tree of life" with divergent branches. Once that point is accepted, we can ask how the common ancestor evolved from a single branch within a previously existing group. Huxley's paleontological interests allowed him to address these issues and his first project centered on the origin of birds from reptiles, as outlined in an address to the Royal Institution in 1868. The fossil *Archaeopteryx*, discovered in 1861, showed that creatures with both reptilian and avian features had existed, checkmating critics who claimed that the classes were completely distinct from one another. But Huxley doubted its relevance to the question of the birds' origin because he thought it came too late in the geological sequence to be an evolutionary link. He focused instead on a small dinosaur, *Compsognathus*, which had hind-limbs identical to those of birds and could thus be regarded as the "missing link" between the classes.

To fill in the sequence Huxley suggested that the first birds had been terrestrial like the modern ostrich and had only later developing wings for flying. His theory was taken seriously for some time, although it faced the serious problem of explaining how a bird adapted to running could ever grow

[11] For details see Peter J. Bowler, "Thomas Henry Huxley and the Reconstruction of Life's Ancestry," in Alan P. Barr, ed., *Thomas Henry Huxley's Place in Science and Letters* (Athens: University of Georgia Press, 1997), pp. 119–39.

wings large enough to get it into the air. Huxley's focus on the anatomical resemblance to dinosaurs led him to put aside the question of how the adaptive transition to flight could have taken place. The alternative theory that the earliest birds had climbed trees and developed wings to glide from branch to branch eventually prevailed, making *Archaeopteryx* a good example of the intermediate forms.

Huxley also worked on the evolution of crocodiles, but it was his involvement in the study of horse fossils that attracted most attention. He had tried to identify plausible ancestors to the modern horse in the European fossils, but when he visited America in 1876 he was shown a far better series discovered there by Othniel Charles Marsh. This displayed the specialization of the limbs with a steady reduction in the digits ending with the hoof of the modern horse. Huxley's popular lectures, published as his *American Addresses*, included a diagram of the sequence that was widely reproduced in later surveys of evolutionism.[12] At one level, the demonstration of specialization for the adaptive strategy of running on open plains was an important vindication of Darwinism. But the simple linear pattern made up of a few fossil specimens made the process seem so regular that some naturalists thought it could not be the result of a process feeding on "random" variations. Marsh's rival Edward Drinker Cope used the linearity of such sequences to claim that the Lamarckian process of use-inheritance offered a more plausible explanation of such trends. Later fossil discoveries would show that the evolution of the horse family is better seen as an irregularly branching tree rather than a straight line, a pattern far more compatible with Darwin's vision.

12.4   Huxley's Reservations

We have seen that when Huxley first learned about the theory of natural selection he said: "How extremely stupid not to have thought of that!"[13] This comment has encouraged the view that he saw natural selection as the obvious explanation of how evolution works. In fact, however, his endorsement of the selection mechanism was highly constrained. He welcomed Darwin's theory as an important step toward a naturalistic theory of evolution, but a survey of his efforts to explore its applications suggests that he found it difficult to follow Darwin's assumption that adaptation was the key to all evolutionary changes. As a morphologist, not a field naturalist, he was far more interested in the deep internal structures of the different forms of life, which were difficult to see as the

---

[12] Reprinted in vol. 4 of the *Collected Essays*, Huxley, *Science and Hebrew Tradition* (London: Macmillan, 1894), pp. 46–138; for the diagram see p. 130.

[13] Huxley, "On the Reception of the 'Origin of Species'" in Francis Darwin, ed. *The Life and Letters of Charles Darwin* (London: John Murray, 1887, 3 vols.), II, pp. 179–204, quote from p. 197.

accumulated product of small adaptive modifications. In his early morphological work he had rejected the view that all of a form's characters must have some adaptive value. Darwin convinced him that many superficial features are shaped by adaptation, but even then Huxley could not accept natural selection as a complete explanation of evolution. His tendency to focus on internal structural constraints led him to think along lines that bear at least some resemblance to the ideas of naturalists who looked for alternative mechanisms of change, including saltationism and orthogenesis.

For all the support he offered after the *Origin of Species* was published, Huxley made it clear that there were issues that needed to be resolved before Darwin's theory could be confirmed. These were laid out in detail in his contribution to the *Westminster Review*. One issue centered on how the theory could be reconciled with the traditional view that distinct species could not interbreed or hybridize. Darwin showed that artificial selection could produce significant structural differences, as in the varieties of pigeons, but these were still interfertile and thus could not be seen as analogous to natural species. To resolve this, Darwin would need to show that continued selection could eventually produce new forms that did not interbreed. Darwin did not think the barriers to interbreeding were so clear-cut, and modern biologists would agree.

Huxley was also reluctant to accept Darwin's belief that all evolutionary changes must be slow and gradual. Early in his review he discussed cases where new varieties had appeared abruptly from individuals born with a significant difference to the norm for their species. These included the so-called Ancon sheep, a short-legged variety founded by a single ram born with reduced limbs. Toward the end of the review Huxley returned to the topic and insisted: "Mr. Darwin's position might, we think, have been even stronger than it is if he had not burdened himself with the aphorism, '*Natura non facit saltum*,' . . . . Nature does make leaps now and then."[14] These leaps – sports of nature or saltations – could obviously found new varieties within a domesticated species so they might also be a source for transformations in nature. Any new forms created in this way would not have been shaped by gradual natural selection. We also know that he had reservations about Darwin's theory of heredity, pangenesis.[15]

Huxley raised this possibility again in his response to Albert Kölliker's objections to Darwinism, making it clear that the pressure of the environment might not be as relentless as Darwin supposed: "If . . . the new variety is by no means perfectly adapted to its conditions, but only fairly well adapted to

[14] Huxley, *Darwiniana*, p. 77.

[15] Huxley's letter to Darwin expressing his reservations has not been found, but its tone is obvious from Darwin's response on July 12, 1865 and Huxley's reply four days later; see *The Correspondence of Charles Darwin*, vol. 13 (Cambridge: Cambridge University Press, 2015), pp. 196–7 and 202–3.

them, it will persist, so long as none of the varieties it throws off are better adapted than itself."[16] Evidently he did not believe the struggle for existence to be so intense that species must be maintained in a state of perfect adaptation to the environment. There was often some leeway in which new characters produced by forces within the organism could appear and reproduce.

Perhaps because of these reservations, Huxley often promoted the wider principles of evolutionism without focusing on natural selection. An early example of this tactic can be found in his Royal Institution lecture of 1860, of which Darwin himself complained "He gave no just idea of *natural* selection."[17] Huxley's doubts were still in play in 1878 when he provided an article on "Evolution in Biology" for the *Encyclopaedia Britannica*.[18] Here he cautioned: "How far Natural Selection suffices for the production of new species remains to be seen" and went on to argue for the possibility that "variability is definite, and is determined in certain directions rather than in others, by conditions inherent in that which varies." Far from being plastic in the hands of the environment, species could only evolve in directions predetermined by the variations that the internal constitution of the organisms permitted. Selection could favor some of these directions and oppose others, but Huxley seemed to imply that there might be adaptively neutral saltations or variation-trends driving the species in certain directions. Darwin's belief that species change only in response to pressure from the environment was too restrictive – some new characters might appear and survive even though they conferred no adaptive benefit.

Huxley was certainly not moving toward the sort of position adopted by the growing number of naturalists who were actively suspicious of the selection theory. Unlike Mivart he did not imagine variation trends that could drive multiple lines of evolution in the same direction, thereby undermining the principle of monophyletic descent. Nor would he follow the advocates of "orthogenesis" or evolution by definitely directed variation, who argued for trends that could drive a species in a harmful direction and precipitate its extinction. But he was willing to imagine internal forces that restricted, if they did not actually control, variation and might thus play a role in shaping evolution in ways that did not represent a straightforward response to environmental pressures. He was not prepared to endorse all the principles of Darwinism as Darwin himself understood them, because he saw natural selection as a process that limited but did not necessarily control the direction of evolution.

---

[16] For the response to Kölliker see Huxley, *Darwiniana*, pp. 80–106.

[17] On the lecture and Darwin's response see Desmond, *Huxley: The Devil's Disciple*, pp. 267–9.

[18] Reprinted in Huxley, *Darwiniana*, pp. 187–226; quotations from p. 223.

12.5   Last Words

Huxley died in 1895. In his later years, as social tensions multiplied, he found himself increasingly at odds with his one-time friend Herbert Spencer. The result was his 1893 Romanes Lecture "Evolution and Ethics" which seems to project a very harsh, Darwinian vision of nature driven by the struggle for existence.[19] The emphasis on struggle was, however, a product of his desire to oppose what was starting to be known as Spencer's "social Darwinism" – his insistence that since nature was driven by individual competition, the same process should be taken as the basis for social progress. Huxley only stressed the harshness of nature in order to argue that civilization required the development of a morality that turned its back on nature.

Spencer knew that any species adapted to live in social groups must develop cooperative instincts, but by the 1890s his emphasis on free enterprise was increasingly taken to imply that social progress requires ruthless competition analogous to the struggle for existence in nature. To oppose this position, it suited Huxley to portray nature in the harshest possible light to make the point that a civilized society must repudiate its ways. "Evolution and Ethics" is full of references to the struggle for existence as the basis for a "gladiatorial theory of existence." It even harps on the "survival of the fittest" – actually Spencer's term (see Depew, Chapter 15, this volume), although he also endorsed Lamarckism on the grounds that that struggle could promote self-improvement as well as the elimination of the unfit.

At first sight Huxley's emphasis on struggle seems to imply that he had at last become more conscious of the power of natural selection. But there is clear evidence that he had not abandoned his earlier reservations. In 1894 he read William Bateson's revival of the saltationist view of evolution in his *Materials for the Study of Variation*. He wrote Bateson a letter saying: "I see you are inclined to advocate the possibility of considerable 'saltus' on the part of Dame Nature in her variations. I always took the same view, much to Mr. Darwin's disgust."[20] To support his claim that most new varieties have arisen by abrupt saltations, Bateson openly rejected the Darwinian view that the pressure of the environment allows only adaptive variations to flourish (Radick, Chapter 18, this volume).

That Huxley could endorse Bateson's position suggests that he was still unwilling to abandon his long-standing reservations about the power of natural selection. To the end, he thought internal factors could shape evolution independently of environmental pressure.

---

[19] Reprinted in vol. 9 of the *Collected Essays*, Huxley, *Evolution and Ethics and Other Essays* (London: Macmillan, 1894).

[20] Huxley to Bateson, February 20, 1894, in Leonard Huxley, ed., *Life and Letters of Thomas Henry Huxley* (London: Macmillan, 1900, 2 vols.), II, p. 372.

Myth 13   That Huxley Defeated Wilberforce, and Ridiculed His Obscurantism, in the 1860 Oxford Debate

John Hedley Brooke

13.1   The Legend

The story has been told several times:

Huxley is best known for his public exchange in 1860 with Bishop Samuel Wilberforce. The bishop, a clever, witty debater, opened himself to attack by making a gentle joke about Huxley's ancestry. Huxley, furious, replied famously to the effect that he would rather be descended from an ape than a bishop.[1]

Or

Wilberforce and Huxley clashed in a debate over evolution that was to go down in history, not least for the jibes they traded ... Although it would take some time for evolution to become fully accepted by science, the debate is widely regarded as the turning point for the acceptance of Darwin's theory, not least because the theory was considered significant enough to be grounds for serious debate in the first place.[2]

Oxford, June 30, 1860. The annual meeting of the British Association for the Advancement of Science is taking place. The discussion of Darwin's theory takes an infamous turn when Darwin's protagonist, Thomas Henry Huxley, is cheekily asked by Samuel Wilberforce, Bishop of Oxford, whether he would prefer to think of himself descended from an ape on his grandfather's or grandmother's side (Figure 13.1). Huxley whispers to a neighbor: "The Lord hath delivered him into mine hands," and then, replying to the provocation, declares that he would rather have an ape for an ancestor than a bishop. Many years later, Isabella Sidgwick recalled that "no one doubted [Huxley's]

My thanks to Kostas Kampourakis for his assistance in the preparation of this chapter, to Nick Spencer for alerting me to Richard England's essay, "Censoring Huxley and Wilberforce" (2017) on which I have drawn, and to the editors of *Science and Christian Belief* in which my earlier essay "The Wilberforce-Huxley Debate: Why did it Happen?" was published in vol.13 (2001), 127–41.

[1] pbs.org, Huxley, Darwin's Bulldog www.pbs.org/wgbh/evolution/library/02/2/l_022_09.html.

[2] Michelle Starr, "The Day the Theory of Evolution Levelled Up," 2016 www.cnet.com/science/huxley-wilberforce-the-day-the-theory-of-evolution-levelled-up/.

(a)   (b)

Figure 13.1 *Vanity Fair* caricature of Samuel Wilberforce, Bishop of Oxford (1869, left (a)) and Thomas Henry Huxley (1871, right (b)). Reproduced with permission from John van Wyhe ed. 2002–. *The Complete Work of Charles Darwin Online*. (http://darwin-online.org.uk/)

meaning, and the effect was tremendous. One lady fainted and had to be carried out; I, for one, jumped out of my seat." According to another report,

the room was crowded to suffocation long before the protagonists appeared on the scene, 700 persons or more managing to find places . . . the very windows by which the room was lighted down the length of its west side were packed with ladies, whose white handkerchiefs, waving and fluttering in the air at the end of the Bishop's speech, were an unforgettable factor in the acclamation of the crowd.

In yet another report Soapy Sam got what he deserved; for he had spoken for no less than half an hour with "inimitable spirit, emptiness and unfairness."[3] Huxley's riposte was a victory for science over religion, or as a mid-twentieth

---

[3] Isabella Sidgwick, "A Grandmother's Tales," *Macmillan's Magazine*, 78, no. 468, Oct. 1898, 433–4; Leonard Huxley, *Life and Letters of Thomas Henry Huxley*, 2 vols., London, 1900, 1, 179–89; 2nd edition, 3 vols., London 1908, 1, pp. 263–5; J. R. Lucas, "Wilberforce and Huxley: A Legendary Encounter," *The Historical Journal*, 22 (1979), 313–30.

century commentator put it, "Huxley had committed forensic murder with a wonderful artistic simplicity, grinding orthodoxy between the facts and the supreme Victorian value of truth-telling."[4]

13.2   Reconsidering the Legend

It is a great story, the kind that would have to be invented even if the event it describes never happened. In this case, a startling confrontation did occur as the Bishop and Huxley voiced their divergent assessments of Darwin's science (but see Bowler, Chapter 12, this volume). Determining exactly what was said and how their altercation should be interpreted has not been so easy. Contemporary reports in newspapers, and by contemporaries who were present, lack unanimity. Some accounts of the meeting, as in the *Athenaeum,* scarcely mention the celebrated exchange at all, arguably sanitizing a report that would have otherwise exposed ungentlemanly behavior by both parties.[5] Generalizations about a head-on conflict between science and the Anglican Church also turn out to be simplistic. From Leonard Huxley's *Life* of his father, we learn that a man in Holy Orders "joined in – and indeed led – the cheers for the Darwinians."[6] Wilberforce must not be assumed to typify a monolithic Christian reaction to Darwin.

One of the most distinguished of the Darwinians was Joseph Hooker, Assistant Director of Kew Gardens. But to read his account of the proceedings is to meet the view that Huxley had caused hardly a stir. He had not even had the strength of voice for his stinging reply to carry. According to Hooker, the person who really won the day for the Darwinians was . . . Hooker! In fact, the more closely we look at the legend the more slippery it becomes. The idea that Huxley won a famous victory for evolutionary science was not even countenanced in Leonard Huxley's *Life*. The result of the encounter, though a check to the anti-Darwinian sceptics, could not be represented as an "immediate and complete triumph for evolutionary doctrine." This was precluded by the "character and temper of the audience, most of whom were less capable of being convinced by the arguments than shocked by the boldness of the retort."[7] One of Huxley's most respected biographers, Adrian Desmond, declares that talk of a victor is ridiculous.[8] From the noisy reaction of the audience, as reported in the *Oxford Chronicle*, there was "prolonged cheering" for Wilberforce, laughter and "loud applause" for Huxley.[9]

---

[4] William Irvine, *Apes, Angels, and Victorians: A Joint Biography of Darwin and Huxley*. London: Weidenfeld and Nicolson, 1956, 5–6.

[5] Richard England, "Censoring Huxley and Wilberforce: A New Source for the Meeting that the *Athenaeum* 'Wisely Softened Down'," *Notes and Records of the Royal Society* 71 (2017), 371–84.

[6] Leonard Huxley (ed.), *Life and Letters of Thomas Henry Huxley*, 3 vols., London 1908, 1, 263.

[7] Ibid., 273–4.   [8] Adrian Desmond, *Huxley: The Devil's Disciple*. London: Penguin, 1994.

[9] *Oxford Chronicle and Berks and Bucks Gazette*, July 21, 1860, p. 3, in England, "Censoring Huxley and Wilberforce," 375–8.

There is an additional reason not to speak of victors. Instead of anti-Darwinians being converted either by Huxley or Hooker, we know that at least one Darwinian was de-converted in the debate. This was Henry Baker Tristram, one of the first to apply Darwin's principle of natural selection. Tristram had been fascinated by the phenomenon of camouflage – how the desert larks of North Africa, for example, were of a darker hue than those of more favored districts. Competition between lighter and darker birds gave him the answer, as the darker would be less visible to desert predators. Tristram had been converted by another naturalist, Alfred Newton, whose own conversion to Darwinism reminds us that conversion is not an experience confined to the religious. Newton recalled that "it came to me like the direct revelation of a higher power; and I awoke . . . with the consciousness that there was an end of all the mystery in the simple phrase 'Natural Selection'." But Newton also tells us that his one convert, Tristram, soon sank into apostasy. The occasion was the Wilberforce–Huxley debate. Apparently Tristram "waxed exceedingly wroth as the discussion went on and declared himself more and more anti-Darwinian."[10] So much for Huxley's victory.

In the three decades that followed and in the relative absence of coverage in other than ephemeral print sources, the event largely disappeared from public awareness until it was resurrected in the 1890s as an appropriate tribute to a recently deceased hero of scientific education.[11] That (too?) delicious remark, "the Lord hath delivered him into mine hands," which Huxley claimed for himself in a letter to Francis Darwin of June 27, 1891, only came to prominence with Leonard Huxley's *Life and Letters* in 1900. Once references to the Oxford altercation began to gather momentum, as a result of the *Life and Letters* (of Darwin and Hooker as well as Huxley) it took on the character of a foundation myth – one of the defining moments of an emerging scientific professionalism. For this reason, the original event became important because in retrospect one could see that serious issues had been involved. There were questions about the freedom to speak one's mind and about the autonomy of the sciences. There were also broader issues, such as divisions within the Anglican Church.

13.3   The Freedom to Speak One's Mind

Leonard Huxley denied that his father had scored a victory, but he concluded his account with an upbeat message: "The importance of the Oxford meeting lay in the open resistance that was made to authority, at a moment when even

---

[10] I. Bernard Cohen, "Three notes on the reception of Darwin's ideas on natural selection (Henry Baker Tristram, Alfred Newton, Samuel Wilberforce)," in D. Kohn (ed.), *The Darwinian Heritage*. Princeton: Princeton University Press, 1985, pp. 589–607

[11] F. A. J. L. James, "An 'Open Clash between Science and the Church'? Wilberforce, Huxley and Hooker on Darwin at the British Association, Oxford, 1860," in D. M. Knight and M. D. Eddy (eds.) *Science and Beliefs: From Natural Philosophy to Natural Science, 1700–1900*. Aldershot: Ashgate, 2005, pp. 171–93.

a drawn battle was hardly less effectual than acknowledged victory. Instead of being crushed under ridicule, the new theories secured a hearing, all the wider, indeed, for the startling nature of their defence."[12] Darwin would have concurred. He wrote to Huxley some three weeks after the event. "I am indeed most thoroughly contented with progress of opinion. From all that I hear from several quarters, it seems that Oxford did the subject great good. – It is of enormous importance the showing the world that a few first-rate men are not afraid of expressing their opinion."[13] There is a certain poignancy here given Darwin's debilitating anxieties about public engagement. In another letter to Huxley, Darwin wrote: "I honour your pluck; I would as soon have died as tried to answer the bishop in such an assembly."[14] There were social pressures that could lead to repression. It was not merely that to speak out on matters of religion was to risk ostracism. It was part of the culture of a scientific gentleman – certainly earlier in the century – that one would not press one's heterodoxy if by so doing one injured the faith of more sensitive brethren. The risks were still real in 1860.

Huxley himself was not insensitive on the subject of what it was appropriate to say in public. Part of him cautioned restraint. On June 28, two days before his encounter with Wilberforce, Huxley had been present at another session of the "British Asses" as they were affectionately called. He had heard Oxford's Professor of Chemistry, Charles Daubeny, deliver a paper on "the final causes of the sexuality of plants," with particular reference to Darwin's work on the origin of species. Huxley had been invited to enter the discussion but had shown no enthusiasm to do so on the ground that "a general audience, in which sentiment would unduly interfere with intellect, was not the public before which such a discussion should be carried on."[15]

But there was also a part of Huxley that could not be suppressed – especially when provoked by England's leading anatomist, Richard Owen (see Rupke, Chapter 14, this volume), from whom he had experienced condescension. At that Thursday meeting Owen had argued that the brain of a gorilla was so different from the brain of a man that a continuity premised on the action of natural selection was unwarranted. Not so for Huxley whose brain had been making a special study of brains. He had found himself, after all, on his feet, flatly contradicting his adversary. This battle over brains was to become fiercely acrimonious over the next couple of years. Perceptions of what

---

[12] Leonard Huxley (ed.), *Life and Letters of Thomas Henry Huxley*, 3 vols., London, 1908, 1, 274.

[13] Darwin Correspondence Project, "Letter no. 2873," accessed on September 4, 2022, www.darwinproject.ac.uk/letter/?docId=letters/DCP-LETT-2873.xml.

[14] Darwin Correspondence Project, "Letter no. 2854," accessed on September 4, 2022, www.darwinproject.ac.uk/letter/?docId=letters/DCP-LETT-2854.xml.

[15] Leonard Huxley (ed.), *Life and Letters of Thomas Henry Huxley*, 3 vols., London, 1908, 1, 261.

happened on the Saturday meeting of the British Association cannot be detached from what had occurred on the Thursday.

Among the inner circle of Darwinians, it was supposed that Owen and Wilberforce were in league and that the bishop had been coached by England's Cuvier. "Hooker tells me," Darwin wrote to Huxley, "you fought nobly with Owen . . . and that you answered the B. of O. capitally."[16] Note that Huxley had *answered* the bishop but that his *fight* had been with Owen. The confrontation between Huxley and Wilberforce cannot be reduced to a simple clash between science and religion. The bishop enrolled eminent scientists of the day in his critique of Darwin's theory.

13.4   The Autonomy of Science and Its Formation as a Profession

In a classic study, historian Frank Turner pointed out that the Victorian conflict between science and religion was in an important respect an epiphenomenon.[17] It reflected a social transformation in the organization and practice of the sciences. Whereas natural history and the life sciences had often been a favorite study of the English clergy, their essentially amateur approach was being overtaken by new standards of professionalism. In the eyes of the young professionalizers, the clerical scientists, such as the geologists Adam Sedgwick (Cambridge) and William Buckland (Oxford) came to epitomize an old guard, whose science was informed by a natural theology in which nature was redolent of divine design. As a student in Cambridge, the young Darwin had assimilated and appreciated the writings of William Paley on that subject of design and was never to completely renounce the idea that the *laws* of nature had been designed (see Ruse, Chapter 4, this volume). But there were increasing problems for clerical scientists. One was identified by Darwin's mentor, Charles Lyell, who questioned the wisdom of trying to combine two demanding loyalties. As the sciences moved rapidly toward specialization it was too much to expect that an enthusiast, whose primary responsibilities lay with his Church, could find time to keep up to speed in natural history. There is a sense in which Wilberforce himself fell into this long established but now threatened category of the clerical naturalist. He was not a scientific ignoramus. Ten years before his faux pas with Huxley he had been assiduously attending Richard Owen's celebrated Hunterian Lectures.

The person who perhaps best epitomizes the arrival of a younger generation of professional scientists was none other than Thomas Henry Huxley – and professional in the sense of aspiring toward earning a living from science as well as seeking to ring-fence new standards of rigor that the clergy would soon be unable

---

[16] Darwin Correspondence Project, "Letter no. 2854," accessed on September 4, 2022, www.darwinproject.ac.uk/letter/?docId=letters/DCP-LETT-2854.xml.

[17] Frank M. Turner, "The Victorian Conflict between Science and Religion: A Professional Dimension," *Isis*, 69 (1978), 356–76.

to meet. There was no privilege in Huxley's background, and he was so impecunious that he had to defer his marriage for five years. His lament was that he could get honor in science but no income. He had his FRS before the age of twenty-six, recognition for his papers, and a Royal Medal in 1852. But it had been a dreadful struggle. Exasperated, he had told his future wife that "to attempt to live by any scientific pursuit is a farce."[18] Owen's scientific and political ascendancy on the one hand and Oxbridge privilege on the other were twin irritants.

In retrospect we can see that there was a trend toward the exclusion of clerics from the sciences. In the period 1831–65 no fewer than forty-one Anglican clergy had presided over the various sections of the British Association. Wilberforce himself had been a Vice-President of the organization. Between 1866 and 1900 the number dropped to three. In the collision between Wilberforce and Huxley scientific and religious interests were certainly involved, but also in contention were contrasting styles and methodologies of science. Wilberforce could protest that, because of its intrinsically hypothetical structure, Darwin's theory of natural selection marked a dangerous departure from sacrosanct Baconian induction. In his study of *The Post-Darwinian Controversies*, James Moore observed that, before 1872, Darwin's hypothetico-deductive model of scientific explanation was "one of the most offensive aspects of his work."[19] For his part, one of the reasons why Huxley was so attracted to Darwin's vision of evolution was that it could be easily incorporated into a naturalistic worldview that rendered a clerical natural theology superfluous for science. Of Huxley's performance at the British Association meeting, the *Oxford Chronicle* reported that he had protested against the subject of evolution being dealt with by amateurs in science and made the occasion of appeals to passion and feeling.[20]

13.5   Divisions Within the Church

There were internal as well as external threats to the unity of the established Church that also placed it on the defense. A keen advocate of the sciences in Oxford was Baden Powell, whose uphill struggle and the personal animosity he experienced led him to a high degree of disillusionment with Oxford theology. It drove him to the radical position that natural scientists should have complete freedom and autonomy in exploring the causes of natural phenomena; but with the proviso that the moral sphere should remain the preserve of the theologian. It was an elegant way of avoiding further retrenchment in territorial squabbles.

[18] Leonard Huxley (ed.), *Life and Letters of Thomas Henry Huxley*, 3 vols., London, 1908, 1, 196.
[19] James R. Moore, *The Post-Darwinian Controversies*. Cambridge: Cambridge University Press, 1979, 196.
[20] *Oxford Chronicle and Berks and Bucks Gazette*, July 21, 1860, p. 3, in England, "Censoring Huxley and Wilberforce," 376.

But there was irony in his religious odyssey because his ultra-liberal theology came to resemble the Unitarianism he had vigorously contested in his youth. Powell, like Huxley, was censured by Wilberforce for views the bishop described as "scarcely-veiled atheism." This was Wilberforce's response to an essay by Powell in *Essays and Reviews* – that book which rocked the Church in 1860 more than Darwin's *Origin*.[21]

Essays and Reviews contained divisive essays from Oxford divines among others on how the Scriptures should be read. They were divisive because they reflected advances made in the contextualization and historical criticism of the sacred text. A presupposition common to most if not all of the contributors was that the Bible had been written by ordinary men whose beliefs reflected the age in which they had lived and who were fallible in their understanding of nature. Wilberforce was to be angered and saddened in almost equal measure because one of his own ordinands, Frederick Temple, now headmaster of Rugby school (and a future Archbishop of Canterbury), was a contributor. When Wilberforce wrote his review of the book, he could not contain himself:

that men holding such posts should advocate such doctrines; that the clerical head of one of our great schools, ... two professors in our famous University of Oxford, one of whom is also tutor of one of our most distinguished colleges, ... that such as these should be the putters forth of doctrines which seem at least to be altogether incompatible with the Bible and the Christian Faith as the Church of England has hitherto received it.

This was all too much to swallow. It was a paradox, "rare and startling"; it was not Anglicanism but capitulation to German metaphysics. "The English church," he continued, "needs in her posts of trust such men as his past career has made us believe Dr. Temple to be. We lament with the deepest sorrow the presence of his name among these essayists."[22] Wilberforce even pleaded with Temple to renounce the association. His review was published in January 1861, after his skirmish with Huxley; but it exposes a deep division between conservative forms of Anglicanism and the liberalizing trends that Temple now personified. This division is also part of the context, part of the background tension against which the Darwinian debates were played out. It is often said that Darwin called into question the historicity of the Adam and Eve narrative. The truth, as Wilberforce knew, is that the biblical critics had got there first. He wanted *Essays and Reviews* condemned in the Convocation of Canterbury.

We can perhaps see how in the battle of wits between Wilberforce and Huxley there might be churchmen happy to see the bishop put down. We have already heard reference to one person in Holy orders rooting for the

---

[21] Samuel Wilberforce, "Essays and Reviews" (1861), in Samuel Wilberforce, *Essays Contributed to the Quarterly Review*, 2 vols., London: Murray, 1, 1874, 104–83, p. 108.
[22] Ibid., 106–9.

Darwinians. The Vice-Chancellor of the University certainly took the view that the bishop got no more than he deserved. When Joseph Hooker claimed that he had been more effective than Huxley he said that he had been "congratulated and thanked by the blackest coats and whitest stocks in Oxford."[23]

13.6   Wilberforce No Ignoramus

Whether the bishop's baiting of Huxley should be seen as anything more than a bit of ad-libbing to brighten a couple of hours in a stuffy room, it was certainly a joke in poor taste that badly backfired. Huxley was angered by such flippancy when the appraisal of Darwin's science was a serious matter. Wilberforce expressed surprise that Huxley had taken umbrage, partly because he believed that, in the main body of his speech, he had offered a serious assessment. It had clearly been premeditated. Wilberforce was confident that the best science and the best philosophy were on his side. This is clear from one of the most revealing texts of the day: his long, formal review of Darwin's *Origin* for the *Quarterly Review*.[24] This was published only days after the fireworks in Oxford. When he spoke, he had all the resources of this review on which to draw. It makes interesting reading. It pulls no punches when proclaiming, understandably but unwisely, that the notion of a "brute origin" of humankind was "utterly irreconcilable" with human uniqueness as understood in Christian theology.[25] Unwisely because forward-looking Anglican clergy, such as Frederick Temple and Charles Kingsley, were among the first of Darwin's sympathizers, refusing to slam the door on rapprochement.[26]

Those expecting in Wilberforce's review a sustained theological diatribe will be disappointed. Darwin is not set up for ridicule. His writings are said to be "unusually attractive"; the book is "most readable," its language so "perspicuous" that it sparkles. He is evidently impressed by the interdependence of all of nature as Darwin has described it. Indeed, it is a wonder Wilberforce has not been hailed as a new age prophet! He speaks of the "golden chain of unsuspected relations which bind together all the mighty web which stretches from end to end of this full and most diversified earth." Darwin's argument is then contested; but the bishop identified moves made by Darwin that could easily

---

[23] Sheridan Gilley, "The Huxley-Wilberforce Debate: A Reconsideration," in Keith Robbins (ed.), *Religion and Humanism (Studies in Church History)*, 17, Cambridge: Cambridge University Press, 1981, 333.

[24] Samuel Wilberforce, "Darwin's Origin of Species" (July 1860), in Samuel Wilberforce, *Essays Contributed to the Quarterly Review*, 2 vols., London: Murray, 1, 1874, 52–103.

[25] Ibid., 94.

[26] John Hedley Brooke, "Wilberforce, Huxley, and Genesis," in Michael Lieb, Emma Mason and Jonathan Roberts (eds.), *The Oxford Handbook of the Reception History of the Bible*. Oxford: Oxford University Press, 2011, 397–412, 409.

produce incredulity. It was one thing to argue that all living things might have descended from a few original forms; but Darwin had been lured further by the quest for unity: "Analogy would lead me one step further, namely, to the belief that all animals and plants have descended from some one prototype." For Wilberforce that extra step would strain credulity even if no other did. We might expect him to cavil at Darwin's references to self-acting powers in nature. They could surely be taken to imply the autonomy of a natural order? But no: Wilberforce is content to say that there is a self-acting power in nature, continuously working in all creation. What is this power? Surprisingly perhaps, it turns out to be natural selection. Darwin is even said to have established the law of natural selection. To be sure the bishop assigned limits to its action; but he did not deny the real effects of a struggle for life. Such a struggle, he wrote, "actually exists, and that it tends continually to lead the strong to exterminate the weak we readily admit." But then we detect the limits of his tolerance. It is in this law of natural selection that we see a "merciful provision against the deterioration, in a world apt to deteriorate, of the works of the Creator's hands." Natural selection prevents the deterioration of existing species rather than effecting new ones.[27]

In his critique, Wilberforce pounced on the analogy Darwin had drawn between the selective breeding of domesticated species and what he believed natural selection could achieve, given an enormity of time. The problem was that, although the domestic breeder could accentuate and accumulate variation to produce fancy pigeons and the like, the evidence suggested that, in the wild, the progeny of variant forms would soon revert to type. This was not a ridiculous objection. It had been used by Charles Lyell against the evolutionary hypothesis of Lamarck. A second difficulty was the seeming absence of transitional forms in the fossil record. Here Darwin had appealed to Lyell's reasoning that, since fossilization required specific physico-chemical preconditions, the record would inevitably be incomplete. Moreover, transitional forms, precisely because they would have fewer representatives, were less likely than stabilized species to leave a record. But was that not like using a theory to explain why there was so little direct evidence for it? Wilberforce censured the logic, judging it unsatisfactory.[28] Darwin had been worried about the degree to which he was exploiting the imperfection of the fossil record, seeking reassurance from Lyell on that very point. His own reaction to Wilberforce's review is arresting: "it is uncommonly clever; it picks out with skill all the most conjectural parts, and brings forward well all the difficulties. It quizzes me quite splendidly."[29]

[27] Wilberforce, "Darwin's Origin of Species," 52–3, 58–61.   [28] Ibid., 64, 70–4.

[29] Darwin Correspondence Project, "Letter no. 2875," accessed on September 22, 2022, www.darwinproject.ac.uk/letter/?docId=letters/DCP-LETT-2875.xml.

The myth of Huxley's victory enjoys longevity, partly because of its enter-tainment value, partly because it fits so snugly within encapsulating narratives that present scientific progress as the primary vehicle of secularization. But Wilberforce was no obscurantist. He attacked both the launch of biblical proof-texts against innovative science and the zealous appropriation of new science for the defense of Scripture.[30] Coached by Richard Owen, he knew there were serious scientists among Darwin's critics. Nor was Huxley always quite the person the mythology suggests. He was certainly anticlerical, a vehement critic of Roman Catholicism and orthodox Anglican theology; but one can be dissenter without being antireligious. Science, including evolutionary biology, was for Huxley neither theistic nor antitheistic. He even confessed that "the antagonism between science and religion about which we hear so much, appears to me to be purely factitious." It was a fabrication, "by short-sighted religious ... and equally short-sighted scientific people."[31]

[30] Wilberforce, "Darwin's Origin of Species," 92–3.
[31] T. H. Huxley, "The Interpreters of Genesis and the Interpreters of Nature" (1885), in *Science and the Hebrew Tradition: Essays*, London, 1904, 139–63, 160–1; Bernard Lightman, "Victorian Sciences and Religions: Discordant Harmonies," *Osiris* 16 (2001), 343–66, 348.

Myth 14   That Darwin's Critics Such as Owen Were Prejudiced and Had No Scientific Arguments

Nicolaas Rupke

14.1   Summary of the Argument

Some of the criticism that followed the publication of the *Origin of Species*[1] was generated by prejudice. Few experts in the field would deny or doubt that. After all, many creationists, then and now, were and are motivated by religious belief in objecting to Darwin and his *magnum opus*. They tend to be quite open about it.[2] Yet Darwin and many in the "Darwin industry" have turned this openly admitted fact into a double-pronged self-serving myth, one prong explicit, the other implicit. What was the main mythical component in the *Origin of Species* and in the later adulatory commemorations, especially of 1959 (centenary) and, more recently, 2009 (sesquicentenary)?

The main myth was, and has continued to be, that opposition to Darwin's theory equates to creationism, and that, by contrast, Darwinism is equal to evolution theory, "the only game in town."[3] Explicitly, Darwin and his followers lumped their critics together, collectively painting them with the tar brush of creationism. A case in point has been the enduring misrepresentation – from Darwin till Dawkins – of Richard Owen (Figure 14.1) and several of his allies, among whom George Campbell, Eighth Duke of Argyll, and St. George Jackson Mivart.

Yet Owen had been an adherent of evolution well before his critique of Darwin's 1859 book. He himself was severely censured by creationists when he put forward his famous vertebrate archetype as palpable evidence for the

Warm thanks are due to Wolfgang Boeker, Kostas Kampourakis and Bob Richards for valuable *addenda et corrigenda*.

[1] My primary Darwin sources are Charles Darwin, *On the Origin of Species by Means of Natural Selection; or, The Preservation of Favoured Races in the Struggle for Life* (London: John Murray, 1859); Charles Darwin and Alfred Russel Wallace, *Evolution by Natural Selection,* with a foreword by Gavin de Beer (Cambridge: Cambridge University Press, 1958). The latter book contains Darwin's early expositions of his theory and the Darwin-Wallace papers of 1858 presented at the Linnean Society.

[2] As was I, during my creationist high school and undergraduate years.

[3] Richard Dawkins, *The Greatest Show on Earth. The Evidence for Evolution* (New York, Free Press, 2009), p. 426.

Figure 14.1 Richard Owen. Photograph by Maull & Polyblank. Credit: Wellcome Collection. Public Domain Mark

evolutionary origins of backboned animals. Owen's brilliant work on the vertebrate archetype was, however, in direct touch with, and an improvement upon, European continental work in idealist morphology. Owen joined a continental elite of evolutionists whose writings preceded Darwin's *Origin of Species* by more than a century.

This brings us to the second prong, the implicit part of Darwin's myth building. Gillian Beer, in her classic *Darwin's Plots*, has recounted the influence of the *Origin of Species* on the narrative imaginations of George Eliot, G. H. Lewes, Thomas Hardy and others. In the book's third edition, she significantly expands her scope.[4] The first of Darwin's plots was his own *magnum opus*. I agree and argue that the *Origin of Species* was a "plot" to keep the predominantly continental evolution theory, which was chiefly structuralist, out of Great Britain in order to preserve a Darwin-Malthusian functionalist ethos.

---

[4] Gillian Beer, *Darwin's Plots: Evolutionary Narrative in Darwin, George Eliot and Nineteenth-Century Fiction* (3rd ed., Cambridge, Cambridge University Press, 2009), see especially "Darwinian Myths," pp. 79–136.

The main representatives of the continental tradition were neither creationist nor forerunners of Darwin, but spokesmen of a more all-inclusive theory of evolution than Darwin's. Its representatives, among whom Immanuel Kant, Johann Friedrich Blumenbach, and Alexander von Humboldt, tried to grasp the totality of evolution, cosmic evolution, combining the inorganic with the organic.

In this chapter, I briefly rework parts of the familiar Owen versus Darwin controversy. In addition, I broaden and deepen the story of their animosity by examining Darwin's political agenda in misrepresenting Owen and the others.

14.2   Owen Was Not a Creationist

Although today the explicit myth of Owen's presumed creationism is still alive and kicking, a demythologizing has taken place ever since Roy MacLeod's classic paper of 1965.[5]

Owen's conversion from Paleyan natural theology to scientific naturalism and evolution theory took place in the course of the 1840s and 1850s, roughly at the time that Darwin formulated his theory of evolution by means of natural selection. The two naturalists conducted a courteous, even friendly correspondence, and Darwin came to be aware of Owen's evolutionary views, while he himself kept his views under wraps until 1858–9 (but see van Wyhe, Chapter 10, this volume).

The change of view that Owen underwent took place in the context of his detailed and extensive homological research program. By his own account, he began working systematically on problems of transcendental morphology in 1841, as part of his curatorial task to arrange the osteological collection of the Hunterian Museum.[6] The catalogue was not published until 1853, but in the intervening years various spin-offs from this basic museum work appeared in print. First, Owen's Hunterian lectures of 1841 and also the series covering the

---

[5] Roy M. MacLeod, "Evolutionism and Richard Owen, 1830–1868: An Episode in Darwin's Century," *Isis*, vol. 56, 1965, pp. 259–80. For further details and revisionist literature on Owen see Rupke, *Richard Owen: Victorian Naturalist* (London and New Haven: Yale University Press, 1994); *Richard Owen: Biology without Darwin* (Chicago and London: University of Chicago Press, 2009); "Richard Owen's Vertebrate Archetype," *Isis*, vol. 84, 1993, pp. 231–51. See also Phillip R. Sloan, "Whewell's Philosophy of Discovery and the Archetype of the Vertebrate Skeleton: The Role of German Philosophy of Science in Richard Owen's Biology," *Annals of Science* 60 (2003): 39–61.

[6] "My duties in the Museum of the Rl. College of Surgeons, especially in preparing the Catalogue of Comparative Osteology, led me to test the validity of homologizing in a higher or more abstract degree than the 'special' route; and, with concomitant study of development of skeletons, enforced an advance of thought beyond the law of adaptation or conditions of existence." Owen's autobiographical sketch of the road that led him to the theory of organic evolution, BM(NH), L, OC 90, R. Owen, *On the Archetype and Homologies of the Vertebrate Skeleton* (John van Voorst, 1848), 7.

years 1844–8 significantly benefitted from his cataloguing labor. The part that dealt with fishes was published in 1846 as the *Lectures on the Comparative Anatomy and Physiology of the Vertebrate Animals*.

Second, and more importantly, the catalogue work served Owen in composing his extensive account of transcendental osteology, presented to the British Association for the Advancement of Science (BAAS) in the form of a major Report in 1846. It was enormously detailed, densely packed with specifics, loaded with technical terms, and tedious to read. The report, with some additions, was published in book form in 1848 under the title *On the Archetype and Homologies of the Vertebrate Skeleton*. The following year, 1849, Owen expanded upon some parts of his BAAS Report in a lecture at the Royal Institution, published as *On the Nature of Limbs*. It was less overloaded with anatomical detail and nomenclature than his Report and more accessible to a wider audience. The most popular rendition of this research appeared in 1854 as part of W. S. Orr's *Circle of the Sciences*; it went through at least a dozen different British and American printings, most of which carried the title: *The Principle Forms of the Skeleton and the Teeth*.

In the process, Owen gave a precise definition to the often synonymously used terms "homology" and "analogy." A homologue was defined as: "The same organ in different animals under every variety of form and function." An analogue, on the other hand, is: "A part or organ in one animal which has the same function as another part or organ in a different animal."[7]

With respect to the relative validity of "form" (homology) versus "function" (analogy), Owen habitually illustrated the inadequacy of the functionalist method by citing the development of the skull, because it was the interpretation of the skull with which transcendental anatomy had been most publicly associated. The skull of the human fetus, at the time of birth, consists of some twenty-eight separate pieces which ultimately unite into an unyielding whole along a series of rigid sutures (Figure 14.2). The functional purpose of the loose, dissembled nature of the fetal skull was believed to be that it facilitates childbirth by making possible a change of shape of the head when it has to pass through the birth canal. However, this function is present only in placental mammals; in various other vertebrates no supple and adjustable cranium is required by the process of birth, and yet their crania are ossified from the same number of points as occur in the human embryo. For example, the tiny, "prematurely" born kangaroo, even though at the time of birthing it is as

[7] Richard Owen, *On the Archetype and Homologies of the Vertebrate Skeleton* (London, Longman, 1847), 7. See also the glossary appended to Owen's *Lectures on the Comparative Anatomy and Physiology of the Invertebrate Animals, Delivered at the Royal College of Surgeons, in 1843*, (London: Longman, 1843), pp. 374 and 379.

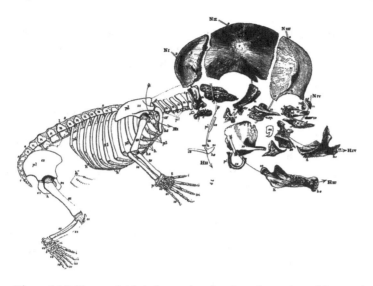

Figure 14.2 Human fetal skeleton, showing Owen's version of the vertebral
theory of the skull in support of his structuralist approach to morphology
(from Owen, *Archetype and Homologies*, fig. 25)

minuscule as a thimble, nevertheless exhibits uncoalesced skull bones
which do not serve the function of making parturition safer as in higher
mammals. The chick when it breaks through its eggshell also has
a composite cranium, not yet coalesced into a single, solid whole but
made up of a variety of movable pieces which serve no such function as
in the human foetus. Owen concluded: "These and a hundred such facts
force upon the contemplative anatomist the inadequacy of the teleological
hypothesis."[8]

A more inclusive view had to be taken of what governs the fact that in all
vertebrates, as a rule, the skull is composed of the same number of pieces
arranged in the same general way. The answer – Owen believed – is provided
by the vertebrate theory of the skull. It had become a commonplace that the
sacrum can be seen as a series of "metamorphosed" vertebrae. Controversial
was the notion that, at the other end of the vertebral column, the skull, too, was
little more than a number of metamorphosed vertebrae. Lorenz Oken's theory
that the head must be interpreted as a recapitulation of the rest of the body was
to a large extent intuitively produced and based on limited and selected

---

[8] Owen, *Archetype and Homologies*, p. 73.

examples.[9] By contrast, Owen's was the most extensive comparative study of available vertebrate material, and he presented his theory as the end-product of a process of inductive labor. What's more, that theory involved the natural origin of species, publicly enunciated by Owen well before Darwin did a similar thing.

Also, the people who attended Owen's Hunterian lecture courses, the last of which was delivered in 1855, had not come away with the impression that Owen was a creationist; quite the opposite. The Duke of Argyll recalled: "I had heard many of his lectures, and all these had left on my mind an impression that he himself believed that somehow or another different races of animals had descended from each other by ordinary generation."[10] Mivart, too, confirmed that his revered teacher had in no way been averse to the idea of evolution; "on the contrary, he was its decided advocate."[11] Imagine Owen's feelings when in November 1859 the *Origin of Species* appeared, in which Darwin cast Owen in the role of a leading advocate of the immutability of species, lumping him together with outspoken anti-evolutionary creationists such as Agassiz and Sedgwick. Owen was out-raged. Whereas he had passed up the opportunity to review the anonymously published *Vestiges of the Natural History of Creation* (an amateur rendition of largely Continental-Scottish evolution theory), this time he forsook caution and wrote an essay on the *Origin of Species* for the *Edinburgh Review*. It was a condemning review.

Some have cited Owen's critique as evidence that he opposed evolution,[12] but it shows nothing of the sort. In fact, the review was most of all a forceful reiteration of Owen's own evolutionary stance, to which was added a critique of natural selection. In addition to Darwin's book, nine other publications were included in this review, among them the French translation of Owen's own work on the archetype, his presidential address to the BAAS, and his *Palaeontology*, which had just come off the press. In Owen's view, Darwin made the fundamental mistake of confusing the issue of the origin of species by natural law with that of the nature of such a law.

One only [Owen wrote in reference to himself], in connexion with his palaeontological discoveries, with his development of the law of irrelative repetition and of homologies, including the relation of the latter to an archetype, has pronounced in favour of the view of the origin of species by a continuously operative creational law; but he, at the same

[9] Ibid., 73–7.
[10] Dowager Duches of Argyll (ed.), *George Douglas, Eighth Duke of Argyll, K.G., K.T. (1823–1900), Autobiography and Memoirs* (London: John Murray, 1906), vol. 1, 581.
[11] [St. George Jackson Mivart], "A century of science," *Living Age* 205: 396 (1895).
[12] For example, Gavin de Beer, *Charles Darwin: A Scientific Biography* (New York: Doubleday, 1963), 164.

time, has set forth some of the strongest objections or exceptions to the hypothesis of the nature of that law as a progressively and gradually transmutational one.[13]

Owen quoted from his *Palaeontology* "that perhaps the most important and significant result of palaeontological research has been the establishment of the axiom of *the continuous operation of the ordained becoming of living things.*" With respect to the mode of operation, various ideas had been put forward by the French (Buffon, Demaillet, Lamarck, Geoffroy; see Corsi, Chapter 2, this volume) and by the author of *Vestiges*; but – asked Owen rhetorically – is the entire series of phenomena of parthenogenesis or "alternation of generations," on which he and others had worked, of no relevance with respect to the introduction of new species on this planet? This was an unambiguous indication that Owen had visualized the evolutionary process as saltational heterogeny, analogous to the cycle of metagenesis. As this and other possibilities had by no means been exhaustively explored by the scientific community, why consent to Darwin's additional speculation?[14]

The assumption that Owen wrote this critique of the *Origin of Species* in subservience to the powers-that-be, or to his own or his patrons' religious belief in Genesis,[15] is incorrect. "We have no sympathy whatever" – Owen openly declared – "with Biblical objectors to creation by law, or with the sacerdotal revilers of those who would explain such law."[16] Owen's defamed review of 1860 was not what in Darwinian mythology it has become, namely an obscurantist attack on an honest and persecuted prophet of truth. The self-portrayal by Owen as a misrepresented evolutionist was a true likeness, and Darwin's depiction of his adversary a myth.[17] Let us go further yet, as currently done by Richard Delisle, in "Deconstructing the Origin of Species."[18]

14.3   The *Origin of Species* Was a Gillian Beer-Type Plot

Owen's approach to origins was, as stated above, more inclusive than Darwin's. The structuralist view had antecedents from as far back as the year 1755, when Kant published his *Allgemeine Naturgeschichte und Theorie des Himmels*

---

[13] [Richard Owen], "Darwin on the origin of species," *Edinburgh Review*, 111, pp. 503–4.
[14] Ibid., p. 503.
[15] G. Himmelfarb, *Darwin and the Darwinian Revolution* (New York, NY: Doubleday, 1959), p. 277 ("perhaps in subservience to the powers-that-be").
[16] Owen, "Darwin on the Origin of Species," p. 511.
[17] In later years, Owen further developed his evolutionary "derivative hypothesis." See Rupke, *Owen: Biology without Darwin*, 2009, pp. 170–81.
[18] See for example Richard G. Delisle, Maurizio Esposito, and David Ceccarelli (eds.), *Unity and Disunity in Evolutionary Biology* (Cham: Springer Nature, 2024).

(*General History of Nature and Celestial Theory*). Kant's unified theory of evolution had many and eminent followers, not least Humboldt with his first two volumes of *Kosmos*.[19] The history of the cosmos was portrayed as an integrated process of material complexification that followed natural laws, while the origin of life and species – organic evolution – was understood as determined by molecular processes. The mechanism of evolution centered on affinity and bonding (from chemical to cultural, under changing environmental conditions, with Malthusian competition playing a subordinate role).[20]

Structuralism was primarily Continental, more specifically German. Most leading structuralists were Germans; Johann Friedrich Blumenbach in Göttingen, and his students Lorenz Oken and Gottfried Reinhold Treviranus are just a few of the early influential names. Perhaps most prominent was Johann Wolfgang von Goethe, active in Jena and Weimar – the cultural heartland of the emerging nation-state of Germany – as well as the Dresden polymath and Goethe biographer Carus, author of the structuralist classic *Von den Ur-Theilen des Knochen- und Schalengeruste* (*On the Fundamental Components of Endo- and Exo-Skeletons*), published in 1828, which advanced the architectural approach to organic form in the context of German Romanticism and Idealist philosophy. Throughout the nineteenth century, Jena remained a hotbed of the morphological method, some major representatives of which were Karl Gegenbaur and Ernst Haeckel.[21]

Let me add some of the characteristic points and great names of the pre- and post-*Origin of Species* structuralist tradition. From the start – the late eighteenth century – the structuralist approach to the origin of species was a more all-embracing one than Darwinian theory ever was or is, if only because the question of the origin of organic diversity was treated as intrinsically related to the question of the origin of life – abiogenesis. This was believed to be a natural process – spontaneous generation – engendered by a constellation of material conditions and molecular forces. It connected the evolutionary history of life with the evolution of the earth, the solar system, our galaxy, and the elements. The literary epitome of this grand synthesis was Alexander

---

[19] Bogdana Stamencović, "Humboldt, Darwin and Theory of Evolution," *History and Philosophy of the Life Sciences*, in press.

[20] See also Gerrit Jan Mulder, *Het streven der stof naar harmonie* (Rotterdam: H.A. Kramers, 1843–44).

[21] Also Scotland, in the person of Wentworth D'Arcy Thompson, kept the non-Darwinian, structuralist/formalist tradition alive. Gillian Beer cites Thompson as a Darwin critic. See his *On Growth and Form*, originally published in 1917, followed by both abridged and expanded editions. The early twenty-first century is exhibiting a growth of interest in the history of "deviations from Darwin." See, for example, Abigail Lustig, Robert J. Richards, and Michael Ruse, eds., *Darwinian Heresies* (Cambridge: Cambridge University Press, 2004); George S. Levie, Kay Meister, and Uwe Hossfeld, "Alternative Evolutionary Theories: A Historical Survey," *Journal of Bioeconomics* 10 (2008): 71–96; Jerry Fodor and Massimo Piatelli-Palmarini, *What Darwin Got Wrong* (New York: Farrar, Straus and Giroux, 2010); Peter J. Bowler, *Darwin Deleted: Imagining a World without Darwin* (Chicago: University of Chicago Press, 2013).

von Humboldt's *Cosmos: A Sketch of a Physical Description of the Universe* (five volumes, published between 1845 and 1862), whereas a more amateur and outspoken rendition was Robert Chambers' *Vestiges of the Natural History of Creation*, published in 1844.[22] They saw the history of the cosmos as an integrated process of material complexification that followed natural laws, and they understood the origin of life and species – organic evolution – as a process driven by molecular forces, "molecular evolution" in today's terminology. Owen referred to the formative forces as "inner tendencies." A century after Owen, Erwin Schrödinger, who made groundbreaking contributions to quantum mechanics, inspired the DNA revolution by arguing along similar structuralist lines in his *What Is Life? The Physical Aspect of the Living Cell*, published in 1943, while today Simon Conway Morris speaks of "deep structure" and Keith Bennett writes "that macroevolution may, over the long-term, be driven largely by internally generated genetic change, not adaptation to a changing environment."[23]

Darwin ignored the Kant-Humboldtian theory of progressive development in the physical, nonliving sphere. He belittled or altogether dismissed the fundamental question of the origin(s) of life, in part because his evolutionary mechanism of natural selection can only operate once living systems exist. Toward the very end of the *Origin of Species* he inserted a couple offhand remarks that seemed more to dismiss than seriously address the question. Darwin wrote: "I should infer from analogy that probably all the organic beings which have ever lived on this earth have descended from some one primordial form, into which life was first breathed," and further he referred to life "having been originally breathed into a few forms or into one."[24]

A long-held view is that Darwin avoided the issue because "[t]his enabled him to work out a theory of descent without bogging down in unanswerable questions."[25] To put it another way: it allowed him to make natural selection seem more efficacious than it was, because natural selection has no obvious grip on the transition from lifeless matter to primitive life (even though later attempts have been made to apply natural selection to stages of the primordial

---

[22] See a combined review of Humboldt's *Cosmos* and *Vestiges*: John Crosse, "The Vestiges etc.," *Westminster Review*, vol. 44, 1845, pp. 484, 490.

[23] Rupke, "Myth 13: that Darwinian natural selection has been 'the only game in town'" in Ronald L. Numbers and Kostas Kampourakis, eds., *Newton's Apple and Other Myths about Science* (Cambridge, MA: Harvard University Press, 2015), pp. 103–11, 244–6. Simon Conway Morris, ed., *The Deep Structure of Biology: Is Convergence Sufficiently Ubiquitous to Give a Directional Signal?* (West Conshohocken, PA: Templeton Foundation Press, 2008); Keith Bennett, "The Chaos Theory of Evolution," *New Scientist* 208, October 2010, pp. 28–31, at 31.

[24] Darwin, *Origin of Species*, 484, 490.

[25] Neal C. Gillispie, *Charles Darwin and the Problem of Creation* (Chicago: University of Chicago Press, 1979), 130.

abiogenesis process, famously by the Glasgow organic chemist Alexander Graham Cairns-Smith).[26]

Here I argue that there was more to Darwin's avoidance of the origin of life than the fact that his proposed mode of evolution would fare better without it. His ignoring the issue, unprofessional as this may have been given its fundamental significance, proved tactically clever in that by so doing Darwin averted a confrontation with representatives of the then leading theory of the origin of species – for the most part men whom he was anxious to get on his side and was intellectually ill-equipped to counter. In order to succeed, Darwin could not afford alienating the many powerful and prestigious representatives of the theory of autogenesis.[27]

The origin of life was removed from the Darwinian research agenda, even more so after the *Bathybius* debacle, when Huxley, Haeckel,[28] and many other eager converts to Darwinism misidentified albuminous coagulates in ocean floor samples from the *Challenger* expedition as incipient life.

14.4   Concluding Observations

The mythical part of the Darwinian account of evolution is based on several of its narrative ploys, such as forgetting, ignoring, and misrepresenting.

There is and was nothing new about putting Darwin's views forward as the one and only scientific theory of the origin of species by presenting it as the alternative to the biblical belief in special creation. Contrasting Darwin and Moses – so to speak – was in fact initiated by Darwin himself when he added to the third edition of the *Origin of Species* "an historical sketch" of evolutionary thought. "Until recently," he wrote, "the great majority of naturalists believed that species were immutable productions, and had been separately created Some few naturalists, on the other hand, have believed that species undergo modification, and that the existing forms of life are the

---

[26] Alexander Graham Cairns-Smith, *Seven Clues to the Origin of Life* (Cambridge: Cambridge University Press, 1985). By contrast, Darwin narrowed the scope of evolutionary biology by belittling the problem of how life had originated. Both Ernst Mayr and Richard Dawkins have reasserted the separateness and uniqueness of biology, keeping the conundrum of life's origins at arm's length; both men have also doubted the validity of astrobiology with its implications of multiple spontaneous origins of life. On Darwin's stance, see Rupke, "Darwin's Choice," in *Biology and Ideology from Descartes to Darwin*, ed. Denis R. Alexander and Ronald L. Numbers (Chicago: University of Chicago Press, 2014), 139–64.

[27] On autogenesis, see Rupke, "Neither Creation Nor Evolution: The Third Way in Mid-Nineteenth Century Thinking About the Origin of Species," *Annals of the History and Theory of Biology*, vol. 10, 2005, pp. 143–72.

[28] On the break-up between Darwin and Haeckel, see Rupke, "The Break-up between Darwin and Haeckel," in Uwe Hossfeld, Georgy Levit and Ulrich Kutchera (eds.), Special Issue Ernst Haeckel (1834–1919): The German Darwin and his impact on modern biology. *Theory in Biosciences*, vol. 138 (2019) (1), pp. 113–17.

descendants by true generation of pre-existing forms"[29] (see also Corsi, Chapter 2, this volume). These few naturalists, however, were said to be mere forerunners of Darwin, imperfectly grappling with evidence; and according to one admirer, Darwin provided "the magnificent synthesis of evidence, all known before, and of theory, adumbrated in every postulate by his forerunners – a synthesis so compelling in honesty and comprehensiveness that it forced men such as Thomas Henry Huxley to say: How stupid not to have realized that before!"[30]

To repeat: the equating of evolution with Darwinian theory, as initiated by Darwin himself and perpetuated today by his followers, is a myth, presented in the form of a historical narrative of evolutionary biology that in part defined itself in opposition to creationism while ignoring notions of structuralist (or formalist) evolutionary theory – that is, the attribution of the origin of life, and of the many forms it has taken, primarily to the operation of physico-chemical and mechanical forces, and less to natural selection, the impact of which, however, is not denied.

The sacral veil covering Darwin's *Origin of Species* should be lifted. Darwin was not a modern-day Moses who replaced metaphysical creationism with naturalistic evolution theory.

Two major issues separated Owen and Darwin. The first concerned Owen's most considerable accomplishment, the establishment of the British Museum of Natural History in South Kensington, perhaps the greatest architectural embodiment of Victorian science anywhere. Darwin was an active part of the attempt at a political assassination of this accomplishment. He and his cronies lost, Owen won, with the help of Gladstone and other Liberal science supporters. "[T]he Darwinian camp acted with the same jealous duplicity of which Owen habitually is accused." "Owen's attack on Darwin's crowning glory appeared in April 1860, nearly a year and a half after Huxley, Darwin and others had attempted to sabotage Owen's vision of what was to become his most monumental accomplishment."[31]

---

[29] Charles Darwin, "An Historical Sketch on the Progress of Opinion on the Origin of Species Previously to the Publication of the First Edition of This Work," in *The Origin of Species*, 4th ed. (London: John Murray, 1866), xiii. See also Curtis N. Johnson, *Darwin's "Historical Sketch". An Examination of the "Preface" to the Origin of Species* (New York: Oxford University Press, 2020).

[30] Bentley Glass, preface to *Forerunners of Darwin: 1745–1859*, ed. Bentley Glass, Owsei Temkin, and William L., Straus Jr. (Baltimore: Johns Hopkins University Press, 1959), vi.

[31] See my essay "The Road to Albertopolis: Richard Owen (1804–1892) and the Founding of the British Museum of Natural History," in Nicolaas A. Rupke (ed.), *Science, Politics and the Public Good* (London: Macmillan Press, 1988), 63–89; quotation on 85. Further details on Owen and the Victorian Museum Movement are to be found in Nicolaas A. Rupke, *Richard Owen: Victorian Naturalist* (New Haven: Yale University Press, 1994), chapters 1 and 2, and N. A. Rupke, *Richard Owen: Biology without Darwin* (Chicago: University of Chicago Press, 2009), chapter 2.

All Darwin had left in his effort to defeat Owen was misrepresentation, and dishonestly characterizing him as a creationist. Yet Owen preceded his scheming adversary as a scientifically better informed paleobiologist and supporter of the Kantian unified theory of evolution. Regrettably, the Darwin industry has persisted in putting forward the mythology that Owen & Company were prejudiced men lacking scientific facts.[32]

[32] Disappointingly, even Curtis N. Johnson's comprehensive *Darwin's "Historical Sketch." An Examination of the "Preface" to the Origin of Species*, misses the political context of the Owen–Darwin dispute and perpetuates a conventional Darwin idolatry.

Myth 15 That Natural Selection Can Also Be Accurately Described As the Survival of the Fittest

David Depew

15.1   How "Survival of the Fittest" Got into *Origin of Species*

Darwin had reason to be anxious when the *Origin of Species* was published on November 24, 1859. The book did not introduce "transformism" to the British reading public. It intervened in an already fraught debate about it that was being conducted in public because it carried implications for theology and politics as well as natural history (see Corsi, Chapter 2; Fara, Chapter 3; Ruse, Chapter 4, this volume). As Darwin feared, this thick context shaped how his version of transformism would be interpreted – and misinterpreted.

A few years after he returned from his youthful voyage on the *Beagle*, Darwin settled down to formulate and empirically support his double-barreled hypothesis that all organisms on earth, extinct as well as extant, evolved from a common ancestor and that the exquisite functional fit between organisms and environments – adaptation – does not result from "special creation" by a divine maker. Nor, he believed, does the adaptation that leads to new species arise from the inheritability of acquired characteristics, as the journalist Robert Chambers, following the lead of the French evolutionist Jean-Baptiste Lamarck, maintained in his widely read but also widely panned *Vestiges of the Natural History of Creation*, published in 1844.[1] Instead, it comes from a process Darwin called natural selection. He was still working at proving these things twenty years later when in June 1858, he read a manuscript Alfred Russel Wallace, a field biologist then collecting specimens in Malaysia, had sent to him. In it, Wallace proposed a scenario about how new species displace less well adapted ones.[2] Wallace's argument was

I owe many improvements to this chapter to the learning and generosity of James Moore. I have also taken to heart advice from James Lennox, Kostas Kampourakis, Phillip Sloan, Frank Sulloway, Robert Richards, Michael Ruse, and John van Wyhe. Any remaining errors are mine.
[1] James Secord (2000). *Victorian Sensation: The Extraordinary Publication, Reception, and Secret Authorship of Vestiges of the Natural History of Creation.* Chicago: University of Chicago Press.
[2] Alfred Russel Wallace (1858). "On the Tendency of Varieties to Depart Indefinitely from the Original Type." *Journal of the Proceedings of the Linnean Society (Zoology)* 3: 53–62. Darwin

close enough to his own to panic Darwin into thinking that, after decades of research – and with more left to do – he had been "forestalled."[3]

Darwin's allies resolved the issue of priority by arranging for Wallace's paper to be read at a meeting of the Linnean Society together with an unpublished sketch of his own theory of natural selection that Darwin had written in 1844 and an excerpt from a letter explaining it to the American botanist Asa Gray, which he had sent before he read Wallace's paper. These documents appeared together in the Society's transactions. So in the end Darwin and Wallace shared the priority of simultaneously publishing on evolution by natural selection (see Ruse, Chapter 11, this volume). Darwin then set about composing an "abstract" of the evidence he had been collecting in a "large species book" on which he had long been working. After the publication of the first edition of *Origin of Species* he continued to address issues raised by his theory in five revised editions.

In Darwin's hands, natural selection says, first, that all living beings labor under the struggle for existence implied by Malthus's postulate that resources lag behind the natural rate of reproduction. Darwin goes on to claim that slight differences between organisms are always arising by chance in the sense that they are not hatched in order to help an organism in the struggle for existence, as Lamarck had supposed. If they happen to be both helpful and inheritable, however, chance variations will tend to be passed to offspring and gradually shaped into adaptations so fitted to their purpose that one might well believe that God created them.[4]

Darwin's argument is guided by a comparison between natural selection and the "artificial selection" of plant and animal breeders who "pick from the litter" and cross what for their purposes are the best. The idea is that if what breeders achieve is marvelous what natural selection can do is even more wonderful, not least because nature works for the good of the organism rather than the breeder (*Origin of Species*, Ch. IV). The analogy seems to imply a selective agent. To some it might suggest that natural selection is a secondary cause that the Creator deploys. To others, it might mean that the evolutionary process is not natural enough to count as the subject of a natural science. Initially, Darwin may have tried to have it both ways, partly to attract clerical support that never came. In

---

had been favorably impressed by an earlier piece by Wallace showing that new species arise at times and in places where those most closely resembling them have already existed ("On the Law Which Has Regulated the Introduction of New Species." *The Annals and Magazine of Natural History*, Series II: 16 [1855]: 184–96). The new piece explained how that can happen in ways close to, but upon closer inspection, not identical with Darwin's account (see Ruse, Chapter 11, this volume).

[3] Charles Darwin to Charles Lyell, June 18, 1858, LETTER 2285. Darwin Correspondence Project (www.darwinproject.ac.uk/). Henceforth DCP

[4] Darwin thought Lamarck's implication that one species turns into another is unable to account for the extent of extinctions in the fossil record.

Figure 15.1 Herbert Spencer. Photograph, 1889 (Wellcome Collection. Public Domain Mark)

Origin's third and later editions, however, he made it clear that "When I speak of natural selection I mean only the aggregate action of many natural laws." Sure, it's a metaphor. But "who objects to chemists speaking of elective affinities" or to physicists when they talk about the "attraction" of gravity? These are metaphors too, but people have got used to them as expressing literal truths (*Origin of Species*, Ch. IV).

That didn't reassure some members of a circle of professional biologists to whom the retiring and often ailing Darwin turned to help him fight his public battles. One of these was Wallace, who had returned to England in 1862. In 1866, he wrote Darwin to say, "'Natural selection' is a stumbling block" to nonprofessional readers and "is not the best adapted [phrase] to impress . . . the general naturalist public."[5] He suggested that in the forthcoming fourth edition of *Origin* Darwin should consider replacing it with "survival of the fittest," a phrase used by the philosopher Herbert Spencer (Figure 15.1) in his 1864 *Principles of Biology* in connection with linking his own theory to Darwin's. "Survival of the fittest" seemed purely a matter of natural causes and effects, not of anyone's purposes.[6] "This survival of the fittest," wrote Spencer, "which

---

[5] Alfred Wallace to Charles Darwin, July 2, 1866, DCP-LETT-5140.

[6] It is intriguing to consider why, beyond its rhetorical use in batting down the objection about intentional agency, Wallace commended Spencer's phrase to Darwin. Darwin's anxiety when he read Wallace's 1858 paper may have led him to overlook differences in how the two men treat natural selection (see Ruse, Chapter 11, this volume). Neither in that paper nor subsequently did Wallace use Darwin's comparison of natural to artificial selection. His view was that species modified by artificial selection, including our own self-domesticated kind, are disanalogous to wild types because they are so coddled that if released into the wild they will die or revert to type. Darwin's analogy implies that just as breeders select individual organisms individuals are

I have here sought to express *in mechanical terms*, is that which Mr. Darwin has called 'natural selection', or the preservation of favored races in the struggle for life."[7]

A few days later Darwin agreed to insert "survival of the fittest" here and there into the almost completed proofs of the new edition.[8] "I have called this principle, by which each slight variation, if useful, is preserved, by the term Natural Selection," we read in the fifth edition, "in order to mark its relation to man's power of selection. But the expression often used by Mr. Herbert Spencer of the Survival of the Fittest is more accurate, and is sometimes equally convenient."[9] Darwin said this not because he had thought the matter through but because, like Wallace, other naturalists he respected were also worrying about the intentionalistic implications of "natural selection."[10] He was more chagrined than he should have been by Wallace's charge that "natural selection" harbors an ambiguity. The process of natural selection, Wallace argued, "does not so much select special variations as exterminate the most unfavorable ones," which fail in the "struggle for existence."[11] Darwin's metaphorical comparison to the work of breeders calls attention instead to the transgenerational outcome of this process. It's a big difference.

It was probably with relief that Darwin reported to Wallace that it was too late to make a wholesale change. In any case, he never did it when he published *Variation of Animals and Plants Under Domestication* in 1868 or when he amended *Origin*'s fifth and sixth editions. The reason is that "survival of the fittest" obscures what he wished to highlight. A few years earlier he had told the botanist Joseph Hooker, another member of his circle and a close confidant, "My enormous difficulty for years was to understand adaptation and this made me I cannot but think rightly insist so much on natural selection."[12] He tells

the units between which natural selection picks. Varieties ("races") are its transgenerational fruits. By contrast, even after he realized he had to explain where varieties come from and professed to accept Darwin's stress on individuals, varieties remained for Wallace the basic units between which selection discriminates. Those that displace their neighbors and predecessors are the fittest and best adapted. So Spencer's phrase made sense to Wallace in his own framework. In addition, his views about society were closer to Spencer's than to Darwin's; he even named his first child after him. Over time, Wallace drew out the implications of his approach, leading Darwin to complain, "I fear we will never quite understand each other" (Darwin to Wallace, September 23, 1868, LETTER 6386, DCP).

[7] Herbert Spencer (1864) *Principles of Biology. Vol. 1, para.* 165. London: Williams and Norgate. My italics.

[8] Charles Darwin to Alfred Wallace, July 5, 1866, DCP-LETT-5145.

[9] Charles Darwin (1869) *On the Origin of Species by Means of Natural Selection* (5th ed.), p. 72. London: John Murray.

[10] See Diane Paul, "The Selection of 'Survival of the Fittest'." *Journal of the History of Biology* (1988): 411–24.

[11] Alfred Wallace to Charles Darwin, July 2, 1866, DCP-LETT-5140; Darwin to Wallace, July 5, 1866, DCP-LETT-5145.

[12] Charles Darwin to Joseph Hooker, November 26, 1862, DCP-LETT-3834.

Hooker that he adopted the analogy to artificial selection to make this point. If he had it to do over again, he says, he might consider replacing "natural selection" with "natural preservation," a phrase that appears in *Origin*'s subtitle, but he doesn't mention anything like "survival of the fittest." At the beginning of Chapter IV Darwin defines "natural selection" to include both "the preservation of favorable and the rejection of injurious variations." Generally, he stresses rejection when he is referring to what happens in one generation and to preservation when he is talking about what takes place over a sequence of generations. Clearly, Darwin's metaphorical comparison of natural to artificial selection was intended to stress the slow transgenerational process of accumulating small heritable variations as the mother of adaptations and to downplay the eliminative means by which it is in part accomplished.

Over time Wallace, Darwin, and Spencer recognized the consequences of their different ways of understanding natural selection, but the sources of these differences often remained elusive. At root, they are subtle differences about how correctly to assign the roles in adaptive natural selection of chance, environmental pressure, and functionality, goal-directedness, or generally biological teleology (Lennox, Chapter 16, this volume). A closer look at Spencer's interpretation will show that Darwin's invitation to substitute "survival of the fittest" as a synonym for natural selection is at cross-purposes with his theory of adaptation. The issue is worth going into because to this day when the "general naturalist public" thinks of Darwinism the phrase "survival of the fittest" often springs to the lips.

15.2   Spencer and "Survival of the Fittest"

It tells you something about the Victorian public sphere that Herbert Spencer could make a good living and acquire fame by writing philosophy books. His model of how both nature and society work was inspired by the discovery of developmental biologists that embryos invariably move from an initially undifferentiated state to an integrated system of differentiated and specialized organs that equip them to deal with the challenges posed by their environments. In 1852, Spencer applied this model to social systems.[13] He was out to refute Malthus. He didn't deny that population pressure is real but, in contrast to the anti-utopian pessimism with which Malthus confronted apologists for the French Revolution, he proposed that competitive markets follow nature itself by spontaneously dividing labor into finer and finer specializations. As a result, wealth, leisure, and civilization increase and, because acquired habits and skills are inheritable in the way Lamarck postulated, the burden of making a living

---

[13] Herbert Spencer, "A Theory of Population, Deduced from the General Law of Animal Fertility," *Westminster Review* 57 (1852): 468–501.

grows lighter. As Spencer had already been claiming, it's an all-around bad idea, then, and an unnecessary one, to use the state to constrain the market in the name of charity. Spencer was a liberal in the nineteenth-century sense of the term.

From the start, Spencer grounded these claims in evolutionary biology. As early as 1851, he responded to the debate about Chambers' *Vestiges* by arguing that there is some evidence for what he called "the development hypothesis" but none for special creation.[14] A year later he was arguing that Malthusian population pressure does not affect all organisms, including humans, equally. There are winners and losers:

those prematurely carried off [in the struggle for existence] must, in the average of cases, be those in whom the power of self-preservation is the least. It unavoidably follows that those left behind to continue the race are those in whom the power of self-preservation is the greatest. [They] are the select of their generation. So that by ... the ceaseless exercise of the faculties needed to contend with the dangers of nature ... and the death of all men who fail to contend with them successfully, there is ensured a constant progress in the species.[15]

In 1864, Spencer deduced these and other claims from his reading of *Origin of Species* in the light of the principles of what he called his "synthetic philosophy": "From the continual destruction of the individuals least capable of maintaining their equilibria in the presence of a new incident force there must eventually be reached an altered type completely in equilibrium with altered conditions ... [It is] what Mr. Darwin calls 'natural selection'."[16]

I come now to the key point. Darwin and Spencer both thought that gradual evolutionary progress is at work in the natural world and in human history, not only within species but also in the evolution of new, more sophisticated ones. They also concurred that the division of labor in ecology and economics is the principle means by which progress of both sorts is achieved. Darwin, too, was a political liberal, even if he doubted Spencer's upbeat expectation that over time the discipline of the market could be relaxed without weakening incentives to work. But these agreements conceal disagreement about adaptive natural selection. For Spencer, organisms that adjust to environmental pressures, whether *in utero* or after, are adapting, if not yet fully adapted. In fact, they may well acquire their helpful characteristics by their own effort, pulling themselves up by the bootstraps of their "life-preserving power," as it were.

Not so for Darwin. Ever the pluralist about evolutionary agencies, he did not deny that passing acquired characteristics to offspring can result in adaptation.

[14] Herbert Spencer, "The Development Hypothesis." *The Leader*, March 20, 1851.

[15] Spencer, "A Theory of Population," para. 15. By "exercise of faculties needed to contend with the dangers of nature" Spencer meant in part that evolutionarily advanced nations spontaneously develop ways to regulate reproduction. (They do.)

[16] Herbert Spencer (1864). *Principles of Biology*, para. 160, 165. London: John Chapman.

What he denied is that Spencer's Lamarckian scenario counts as adaptation by natural selection. For Darwin, that happens only when a variant trait arises by chance rather than orientation to an end and when that trait is inheritable enough to enhance the reproductive success of those possessing it. Under the Malthusian struggle for existence, competitors lacking the helpful variant will tend to be eliminated. By contrast, as a reproduction-enhancing trait spreads over many generations it will no longer exist by chance but because it performs a life-preserving function. Spencer takes himself to have satisfied "Mr. Darwin's" transgenerational requirement by saying that "eventually an altered type arises that is completely in equilibrium with altered conditions." But lacking Darwin's analogy to artificial selection, he stresses environmental pressure over chance both as the source of variation and as the shaper of adaptations, thereby portraying the various stages of Darwin's triad of chance, environmental pressure, and function as more causally homogeneous and less exposed to contingencies than they are.

15.3 "Survival of the Fittest" and "Social Darwinism"

In the years between his promise at the end of *Origin of Species* that natural selection would throw "light on the origin of man and his history" and the publication of *The Descent of Man* in 1871, Darwin became aware that, unless more fecund but less civilized nations and races are to be judged fitter than advanced peoples, he would have to amend his way of measuring success merely by counting offspring. His account of natural selection's role in human evolution in the *Descent of Man* is far less lyrical than the *Origin of Species*. Although he recognizes that the species is on an upward trajectory, he stresses extermination of less developed nations and races by those that are more advanced as the means of progress. But in explaining how progress occurs he also frequently calls on evolutionary agencies other than natural selection: Lamarck's and Spencer's idea that the adaptive efforts of one generation can be passed to another (Burkhardt, Chapter 8, this volume); "group selection" that favors cooperative over uncooperative communities; and sexual selection of the sort that highlights the favors bestowed by and norms of behavior inculcated by the presumably gentler sex (Hamlin, Chapter 21, this volume). Positive, adaptive natural selection's diminished presence in *Descent of Man* shows that Darwin was sticking to his guns on what it actually is: an agency that works on initially chance differences between individual organisms and shapes adaptations by comparative reproductive success protracted over many generations.

In the late nineteenth century, many influential Europeans went all in for nationalism, imperialism, racism, and eugenic proposals for retaining their supposed superiority over the colonized and keeping their own populations from degenerating. Read in the context of the turn by professionals toward eliminative natural selection, Darwin's invitation to substitute "survival of the fittest" and his appeal to the negative side of natural selection in *The Descent of Man* may have encouraged superficial objections and misunderstandings. In public discourse to this day, we often hear that "Darwinism" is self-refuting since "survival of the fittest" is an empty and thus nonexplanatory tautology. Isn't the survival of the fittest the survival of those that survive? "Survival of the fittest" also narrows the difference between the struggle for existence and adaptive natural selection by putting a premium on physical robustness, sexual prowess, and aggression. It's tempting to blame Spencer or Darwin or Wallace or all three. On the whole, however, it seems likely that by mindlessly recycling phrases like "struggle for existence" and "survival of the fittest" ideologists with little or no knowledge of evolutionary biology, not professional evolutionary theorists, bear most of the responsibility for the erasure of Darwin's Darwinism in late nineteenth-century public discourse.[17]

This is certainly true of the American sociologist William Graham Sumner, who taught at Yale in the Gilded Age. Convinced that an unconstrained competitive market was both necessary and sufficient to weed out the unfit, Sumner declared that "A drunkard in the gutter is just where he ought to be . . . The law of survival of the fittest was not made by man, and it cannot be abrogated by man."[18] The arch-capitalist Andrew Carnegie softened Sumner's claim that the classes owe nothing to each other by piously calling on philanthropy, but not, God forbid, legislation to handle any collateral damage.[19] Neither Carnegie nor Sumner knew a thing about biology and whatever ideas about it they helped disseminate bear little resemblance to those of evolutionary experts. Still, the picture of evolution they circulated made it easier to believe that sentimentality and sympathy were interfering with the laws of nature by allowing allegedly unfit races to outbreed what in 1916 the eugenicist Madison Grant called "The Great Race" and were threatening "Nordic" supremacy by tolerating interracial marriage and letting the mentally handicapped, sickly, and morally degraded procreate.[20]

---

[17] Gregory Claeys, "The 'Survival of the Fittest' and the Origin of Social Darwinism." *Journal of the History of Ideas* 61 (2000): 223–40; and Paul, "Survival." Darwinism was still alive during this period, if not well. Faithful Darwinian "biometricians" in England were quietly at work finding cases of trans-generational adaptive natural selection nature by counting offspring. This was easier to do than it might otherwise have been because industrial pollution was causing rapid environmental change, creating strong selection pressures.

[18] William Graham Summer (1883). "The Forgotten Man." In William Graham Sumner (1918). *The Forgotten Man and Other Essays.* Keller, Albert G. (ed), pp. 465–98. New Haven: Yale University Press.

[19] Andrew Carnegie, "Wealth," *North American Review* 191 (1889). Later retitled *The Gospel of Wealth.*

[20] Madison Grant (1916). *The Passing of the Great Race.* New York: Charles Scribner's Sons.

"Pop Darwinism" of this sort permeated the background of the notorious Scopes trial. In 1925, John Scopes defied a Tennessee law prohibiting teaching evolution ("Darwinism") in that state in order to challenge its constitutionality. The populist hero and three-time Democratic nominee for the American presidency William Jennings Bryan took up the case against the accused less because he was a Biblical literalist than because he opposed the influence of elite educational reformers on local school boards, rejected imperialism abroad so heartily that he resigned as Woodrow Wilson's Secretary of State to protest the United States' entry into World War I, and was deeply disturbed by eugenics, which he associated with German militarism, Friedrich Nietzsche, and capitalists like Carnegie. Scopes was duly convicted, but his opponents lost the battle for public opinion. The journalist H. L. Mencken ridiculed the townspeople of Dayton, Tennessee as "boobs." In the popular mid-twentieth-century play and film *Inherit the Wind*, a thinly disguised account of the trial, there is no hint that the textbook Scopes used offered health advice to high school students laced with 1920s-era eugenics.[21]

It didn't help that in 1944 the Columbia University historian Richard Hofstadter retraced the history of what he was the first to call Social Darwinism.[22] The phrase caught on. Many historians of evolutionary theory have deplored its inaccuracy. Sometimes they suggest that Hofstadter should have talked about "Social Spencerism" instead. But that is just as misleading.

Eventually Spencer became aware that Darwin's view of natural selection differed from his. He didn't deny that it is a real phenomenon but only that it deserved the primacy Darwin claimed for it. He allowed that it could explain defensive and feeding mechanisms that lie at the competitive interface between organisms and environments, such as the bioluminescent fishing rod that rises from forehead of the angler fish, but could not account for the integration of the many traits of organisms into a coherent whole.[23] For his part, Darwin remained unfailingly pluralistic about evolution's factors and, arguably, overly eager to acknowledge mechanisms touted by others. He was so obliging about use inheritance, a form of Lamarckism, and the direct effect of environments in molding the traits of developing individuals, which was close to Spencer's model, that it became increasingly difficult to say why he still assigned primacy to trans-generational adaptive natural selection.[24]

---

[21] George William Hunter (1914). *A Civic Biology*. New York: American Book Co. See also Edward Larson (1997). *Summer for the Gods*. New York: Basic Books.

[22] Richard Hofstadter (1944). *Social Darwinism in American Thought*. New York: Columbia University Press.

[23] Herbert Spencer (1891). *The Factors of Organic Evolution*. New York: Appleton and Co.

[24] If he hadn't mentioned the direct effect of environments in early editions of *Origin*, the aged Darwin says, it was only because the best evidence for it was not yet secure (Darwin to Editors of *Nature*, November 5, 1880, DCP-LETT-12800). As early as the manuscript he mined in writing *Origin*, Darwin wrote: "Seeing how absolutely necessary whiteness is in the snow-covered arctic

Still, the history of science is as full of ironic reversals as political history. In the early 1880s, Spencer was at the height of his influence, but within a decade he had been dethroned by the embryologist August Weismann's demonstration, as it was widely taken to be, that acquired traits are not heritable. The rediscovery of Mendel's laws of inheritance in 1900 drove another nail in the Lamarckian coffin, but early genetics was almost as fatal to Darwinism. Mendelism's advocates favored single-generation macro-mutations as the origin of new species rather than multi-generational selection (Radick, Chapter 18, this volume). By the 1920s, however, Mendelian genetics had been integrated with Darwinian natural selection by setting development aside and instead using newly devised statistical and probabilistic methods to track changes in gene frequencies in populations. If not already dead, Spencerism was finally killed off because it was based on models of development and didn't have a statistical bone in its body. One of the many revealing implications of the shift to "population thinking" is that Malthusian population pressure is not a universal force constantly bearing down on all living beings but a limiting case. This meant that the sudden elimination implied by "survival of the fittest" was replaced by tracing the long-term effects of slightly lower rates of reproduction and that Darwin's cherished idea of gradual adaptation by amplifying slight, initially chance variations finally came into its own.

Under the aegis of the Modern Evolutionary Synthesis, which unified genetics and natural selection, evolutionary biologists have for over three-quarters of a century been accumulating a vast body of empirically secure knowledge. Since the 1940s they have gone to great lengths to introduce this sort of Darwinism into school curricula and to inform the public that traces of "survival of the fittest" that linger in the public memory are scientifically worthless. Still, it is surprising how tenacious this myth is.

regions ... we might attribute the absence of color [in a white bird] to a long course of selection. But it may be that whiteness is the direct effect of cold and that the struggle for life has come into play only insofar as colored animals in arctic regions live under a great disadvantage" R. C. Stauffer, ed. [1975]. *Charles Darwin's Natural Selection: Second Part of his "Big Species Book."* Cambridge: Cambridge University Press, p. 377.

Myth 16 That Darwin Banished Teleology from Biology

James G. Lennox

16.1   Introduction

It has long been believed that one of Charles Darwin's most important accomplishments was to have banished teleology from biology. Here is what Friedrich Engels wrote to Karl Marx on December 11 or 12, 1859: "Darwin, by the way, whom I am just now reading, is quite splendid. There was one aspect of teleology that had not yet been destroyed, but now that has been done. Never before has such a wonderful attempt been made to prove historical development in nature, and certainly never with such success."[1] And here is what we read in a widely read textbook on evolution.

His [Darwin's] alternative to intelligent design was design by the completely mindless process of natural selection, according to which organisms possessing variations that enhance survival or reproduction replace those less suitably endowed, which therefore survive or reproduce in lesser degree. This process cannot have a goal, any more than erosion has the goal of forming canyons, for *the future cannot cause material events in the present.* Thus the concepts of goals or purposes have no place in biology (nor in any other of the natural sciences).[2]

But as this chapter will demonstrate, Darwin thought deeply about the relationship between natural selection and teleology, and concluded that his theory, far from banishing teleology, provided a profoundly original foundation for teleological explanation.

A teleological explanation is one in which some property, process or entity is said to exist or be taking place *for the sake of* a certain result or consequence. For example, if someone were to ask me why I swim a few times a week, I might respond, "In order to keep fit," thus explaining my swim by pointing to a (beneficial) *consequence* of swimming, keeping fit.

[1] Quoted in Chapter II of Richard Weikart, *Socialist Darwinism: Evolution in German Socialist Thought from Marx to Bernstein*. San Francisco: International Scholars, 1998.
[2] D. J. Futuyma, *Evolution.* 4th ed., Sunderland, MA: Sinauer Associates, Inc, 2005, p. 250.

Such explanations have been an integral part of the scientific study of life at least since the biological works of Aristotle in Ancient Greece.[3] The large, hooked beaks and powerful talons of raptors, for example, might be explained by showing how they are especially suitable for capturing and eating animals that they prey on for food. Similarly, the coordinated actions of a wolf pack might be explained by showing their value for hunting prey. Explanations of this sort were first denominated "teleological" in the eighteenth century, formed from two Greek words – *telos* (meaning "end" or "goal") and *logos* (meaning "reason" or "account"). That is, these explanations cite *the end or purpose* of a biological structure, behavior or characteristic as the *reason why* the organism has it.

It is, however, very common to read that, thanks to Charles Darwin, biology no longer has any need or use for such explanations. For example:

> ... a myth has grown up, partly the work of [Asa] Gray, partly the work of Darwin's son and biographer, Frances Darwin, that Darwin somehow "brought teleology back into biology". In any nontrivial sense of that word, he did the exact opposite, getting rid of teleology and replacing it with a new way of thinking about adaptation or at least show how it could be accomplished.[4]

Even Charles Darwin's friend and staunch defender, Thomas Huxley, occasionally seems to make much the same point: At one point he says that the impression he came away with on first reading the *Origin of Species* was that "Teleology, as commonly understood, had received its deathblow at Mr. Darwin's hands."[5] But the crucial words here are "as commonly understood," for after saying this he goes on to say that "there is a wider Teleology, that is not touched by the doctrine of Evolution, but is actually based upon the fundamental proposition of Evolution."

What did Charles Darwin think about teleology? Did he think he had delivered it a death blow? As we shall see, he did not. He certainly did give biology a new way of thinking about adaptation, which entailed a new way of thinking about how it is that identifying the ends served by organic parts, processes and activities explains them. And that may well be what Huxley was referring to when he said there was a wider teleology, which not only doesn't conflict with Darwin's ideas but is actually based on them. As we shall

---

[3] Since Darwin often refers to such explanations as providing the "final cause" of the characteristic he is explaining, it is important to know the background of this expression. It originates in Medieval discussions of Aristotle's four types of cause. One of those causes identifies what a part or process is for: Aristotle's expression is *to hou heneka*, which literally translated is "that for the sake of which." Medieval philosophers translated it into Latin as *causa finalis*, final cause. When the term "teleology" came into scientific use in the nineteenth century, it was understood as an explanation that picks out the final cause of a biological structure or behavior.

[4] M. T. Ghiselin, Foreword to *The Various Contrivances by which Orchids are Fertilised by Insects, by Charles Darwin (1863)*. Chicago: University of Chicago Press, 1984, p. xi–xix.

[5] T. H. Huxley, *Darwiniana Collected Essays*. London: Macmillan, 1893, p. 82.

see, Charles Darwin would strongly object to the claim that he "got rid of teleology."

It will be noted that evolutionary biologist Michael Ghiselin, in the above quotation, aims to dispel a contrary myth – that Darwin brought teleology back into biology! He blames the creation of *this* myth on Charles Darwin's friend and foremost American defender, Harvard botanist Asa Gray, and on his son Francis Darwin, who was the first person to publish some of his father's letters and unpublished manuscripts. In this chapter, then, my aim is to show that what Michael Ghiselin claims to be a myth is in fact reality, and that his assertion that Darwin rid biology of teleology is itself a myth.

16.2   Darwin and Gray: Teleology Wedded to Morphology

An ideal place to begin is with the historical source of the statement that Charles Darwin brought teleology back into biology. The source is indeed Harvard botanist and Fisher Professor of Natural History, Asa Gray (Figure 16.1). In a brief appreciation of Darwin, published in the science journal *Nature* in June of 1874, Gray praised Darwin for his "great service to Natural Science in bringing back to it Teleology: so that instead of Morphology versus Teleology, we shall have Morphology wedded to Teleology."[6] Now if Darwin were in the business of ridding biology of teleology, you would expect him either to remain respectfully silent about Gray's claim, or to insist that his friend has misunderstood him.

Darwin quickly responded: "What you say about Teleology pleases me especially and I do not think anyone else has ever noticed the point. I have always said you were the man to hit the nail on the head."[7] Darwin had apparently forgotten that in an 1862 review of his book on the fertilization of orchids by insects, Gray had said much the same thing, although restricted to having "brought back teleological considerations to botany."[8] Be that as it may, Darwin here goes out of his way to say that Gray's claim about bringing Teleology back to Natural Science *especially* pleased him. (By the way, Darwin's son Francis is implicated in the creation of Ghiselin's "myth" because he included this letter in an edited collection of his father's correspondence – he "outed" Darwin the teleologist!)

[6] Asa Gray, "Natural Selection Is Not Inconsistent with Natural Theology," in A. Hunter Dupree (ed.), *Asa Gray: Darwiniana*. Cambridge, MA: Harvard University Press, 1963.

[7] Correspondence Vol. 22 [1874], 278. Darwin Correspondence Project, "Letter no. 9483," accessed on 3 January 2024, https://www.darwinproject.ac.uk/letter/?docId=letters/DCP-LETT-9483.xml.

[8] In A. Gray, "Fertilization of Orchids Through the Agency of Insects," *American Journal of Science and Arts* 34 (1862), 420–29, 428.

Figure 16.1 Asa Gray, the Harvard botanist with whom Darwin had the longest running and most significant exchange of correspondence dealing with the subjects of design in nature and religious belief. Credit: Public Domain

In order to understand Gray's comment, we need to understand what he means by contrasting "Morphology versus Teleology" with "Morphology wedded to Teleology," and this will require a brief excursion into a debate (often referred to as a debate about the priority of form vs. function) that raged between two French naturalists at the Natural History Museum in Paris during the years just before and during Darwin's years in college and on HMS Beagle: Georges Cuvier and Ettiene Geoffroy St. Hilaire. Darwin refers to this debate explicitly in the closing paragraphs of chapter VI of the *Origin of Species* – what Gray terms "morphology" Darwin refers to as Saint-Hilaire's doctrine of Unity of Type, and what Gray terms "teleology" Darwin refers to as Cuvier's doctrine of Conditions of Existence:

It is generally acknowledged that all organic beings have been formed on two great laws – Unity of Type, and the Conditions of Existence. By unity of type is meant the fundamental agreement in structure, which we see in organic beings of the same class, and which is quite independent of their habits of life. On my theory, unity of type is explained by unity of descent. The expression of conditions of existence, so often insisted on by the illustrious Cuvier, is fully embraced by the principle of natural selection.[9]

Darwin's thought here is that the shared characteristics that allow us to identify (say) the many species of Kangaroos and Wallabies as the same "type" of animal are accounted for by the fact that all of them can be traced back to a common ancestor; while their distinctive adaptations, which Cuvier designated their "conditions of existence," are accounted for by natural selection "now adapting the varying parts of each being to its organic and inorganic conditions of life."[10] Once one accepts Darwin's view of descent with modification due to natural selection, the role of morphology is to understand the similarities and differences between related species in evolutionary terms as reflecting genealogical relationships, and to explain the specific adaptive modifications of related species teleologically – to identify what these adaptations are *for*. A Darwinian need not choose between morphology and teleology. To use Gray's metaphor, in Darwinian biology they are "wedded." We see much the same point in this comment by Thomas Huxley: "The apparently diverging teaching of the Teleologist and of the Morphologist are reconciled by the Darwinian Hypothesis."[11]

As it turns out, Darwin had been worrying about how to understand the central role of teleological concepts in the study of living things from the very beginning of his quest to understand the origin of species. His "Species Notebooks," in which he struggles to solve the mystery of the origin of new species, span a period from July of 1837 to July of 1839, shortly after the Beagle returned to England. Early in the first notebook he uses the expression "final cause of life" when discussing the fact that many organisms seem to be alive for the sole purpose of reproducing. This is a theme that returns late in Notebook D: "The final cause of sexes is to obliterate differences. Final cause of this because the great changes of nature are slow. If animals became adapted to every minute change, they would not be fitted to the slow great changes really in progress."[12] Late in the third of these notebooks, Darwin reports that he hit on the idea that he would later call "natural selection" while reading economist Thomas Malthus's *Essay on the Principle of Population*. Comparing the struggle for existence resulting from species reproducing more offspring each generation than can possibly survive to forcing wedges into gaps in nature,

---

[9] C. Darwin, *On the Origin of Species*. London: John Murray, 1859, 206. In this quote one can see that Darwin gives explanatory priority to function over form.
[10] Darwin, *On the Origin of Species*, 206. G. Cuvier, *La Règne Animal*. Paris: 1816. About this principle peculiar to natural history, Cuvier wrote: "c'est celui des conditions d'existence, vulgairement nommé des causes finales."
[11] Huxley, *Darwiniana*, 86.
[12] P. Barrett et al., *Charles Darwin's Notebooks, 1836–1844: Geology, Transmutation of Species, Metaphysical Enquiries*. Cambridge: Cambridge University Press, 1987, Notebook D, 167; Cf.

he says: "The final cause of all this wedging, must be to sort out proper structure and adapt it to change"[13] (see also Love, Chapter 9, this volume).

16.3   Darwin's Context: Natural Theology

There is an important reason why someone educated in England in the early nineteenth century would spend considerable time thinking about this topic. Permeating the thought of most of Darwin's teachers and professional colleagues were the philosophical premises underlying what had come to be known as Natural Theology. This way of thinking has ancient roots in Plato's idea that the natural world is the creation of a divine Craftsman, whose goal was to craft a material world as good as it could possibly be. This idea, through Plato's influence on Christianity, heavily influenced leading figures of the scientific revolution such as Francis Bacon, Isaac Newton, Robert Boyle and John Ray.[14] The core idea of Natural Theology is that the scientist, by studying nature and discovering its "laws," is in fact learning about God's plan for his creation, for the natural world is full of evidence of divine design. Darwin read Reverend William Paley's *Natural Theology* while at Cambridge, and in his autobiography he admitted that when he studied it he didn't question Paley's premises and admired the logic of his argument (see Ruse, Chapter 4, this volume).

Paley famously argued that if you were walking in a field and you came across a rock, you might well think that its shape and size were simply due to chance interactions of physical forces such as heat and erosion. But if you were to come across a pocket watch, and you examined it, you would see a complicated mechanism and would naturally conclude that there was an intelligent craftsman who designed it for the purpose of keeping time. Paley went into great biological detail in order to show that the complicated "contrivances" in organisms, adapting them for seeing, or swimming, or digesting, etc., were even more obviously the product of an intelligent craftsman.

every indication of contrivance, every manifestation of design, which existed in the watch, exists in the works of nature; with the difference, on the side of nature, of being greater and more, and that in a degree which exceeds all computation. I mean that the

---

Notebook E, 48: "My theory gives great final cause ≪I do not wish to say only cause, but one great final cause – nothing probably exists for one cause≫ of sexes ≪in separate≫ ≪animals≫ . . . ". The words in double angle brackets are later additions.

[13] In quoting from these notebooks I am correcting various grammatical, spelling and punctuation errors in the original.

[14] For more details on the history of teleology before Darwin, see J. G. Lennox, "Teleology," in Evelyn Fox Keller and Elisabeth Lloyd (eds.), *Keywords in Evolutionary Biology*. Cambridge, MA: Harvard University Press, 1992, 324–33 and J. G. Lennox and Kostas Kampourakis, "Biological Teleology: The Need for History," in Kostas Kampourakis (ed.), *Philosophy of Biology: A Companion for Educators*. Dordrecht: Springer, 2013, 421–54.

contrivances of nature surpass the contrivances of art, in the complexity, subtility, and curiosity of the mechanism.[15]

And since this craftsman is clearly a being of much greater power and intelligence than a watchmaker, since he is responsible for all of the amazing adaptations of all the world's creatures, this must, declared Paley, be God.

Shortly after Darwin found in Malthus the key to how there could be in nature something that brought about results analogous to those produced by breeders of domestic animals and plants, he decided to test out his new idea by making an abstract, with his own critical commentary, of a recent Natural Theology tract by John Macculloch, entitled *Proofs and Illustrations of the Attributes of God from the Facts and Laws of the Physical Universe, being the Foundation of Natural and Revealed Religion*.[16] This process leads him to question whether he should continue to use the language of final causes, since it has come to be so closely associated with the kinds of explanations he finds so vacuous in works like Macculloch's. One of his notes makes the point explicitly: "The Final cause of innumerable eggs is explained by Malthus. – [is it anomaly in me to talk of Final causes: consider this!] consider these barren Virgins."[17] The reference to final causes being "barren Virgins" is likely borrowed indirectly from Francis Bacon's *De Augmentis Scientiarum* III.5, thanks to a reference in another work of Natural Theology, the third in the series of Bridgewater Treatises, William Whewell's *Astronomy and General Physics Considered with Reference to Natural Theology*,[18] which Darwin had recently read.

As we have seen, Darwin clearly decided that there was nothing anomalous about him discussing final causes. But already in the late 1830s, Darwin is worrying about this issue, and he is doing so while in the process of comparing natural selection's ability to explain adaptations with that of a work of Natural Theology.

16.4   Teleology after the *Origin of Species*: Darwin on Contrivances in Orchids

What prompted Gray's appreciative remarks about Darwin's use of teleology? It might have been the *Origin of Species*: after all, Gray would no doubt have been delighted to read that natural selection is "daily and hourly scrutinizing throughout the world, every variation, however slight; rejecting that which is

[15] William Paley, *Natural Theology: Or Evidences of the Existence and Attributes of the Deity*, 12th ed. London: J. Faulder, 1809, 18.
[16] Barrett et al., *Notebooks*, 631–41.   [17] Ibid., 637.
[18] W. Whewell, *Astronomy and General Physics Considered with Reference to Natural Theology*. London: William Pickering, 1833.

bad, preserving and adding up all that is good";[19] and that "natural selection can act only through and for the good of each being."[20] However, Gray's specialty was botany, and as we've seen, his praise for Darwin's integration of morphology and teleology first appears in an 1862 review of Darwin's *The Various Contrivances by which Orchids are Fertilized by Insects*. Darwin's choice of the word "contrivance" in this work might well have encouraged readers such as Gray to read his teleology through a Natural Theological lens, given William Paley's characterization of biological adaptations as *contrivances of nature*. But according to Darwin's theory, adaptation results from the differential survival of variations that arise by chance, and in his investigation of adaptations for fertilization in orchids, Darwin gives the following example of a case where a very simple solution to an adaptive problem appears to be passed over in favor of a more complicated solution, in a species of the genus *Malaxis*. He supposes that at one point in the past its ovarium was oriented so that the labellum hung downward, but at a certain point in its history it became advantageous to have the labellum in the more typical, upward position.

[T]his change, it is obvious, might be simply effected by the continued selection of varieties which had their ovarium a little less twisted; but if the plant only afforded varieties with the ovarium more twisted, the same end could be attained by their selection until the flower had turned completely round on its axis: this seems to have occurred with the *Malaxis*, for the labellum has acquired its present upward position, and the ovarium is twisted to excess.[21]

Notice again that Darwin sees himself engaged in a teleological inquiry, a search for the end to be served by this adaptation. But selection must attain that end by making use of whatever chance variations are actually present, which leads to this oddly complicated *contrivance*. Asa Gray good-naturedly responded that "we believers in real design make the most of your 'frank' and natural terms, 'contrivance, purpose,' etc., and pooh-pooh your endeavors to resolve such contrivances into necessary results of certain physical processes."[22] Note Gray's self-description: he is among the believers in *real* design, Darwin a mere pretender.

A valuable source for furthering our understanding of Charles Darwin's thinking about teleology is, then, his correspondence with his friend and American advocate, Asa Gray. For Gray was *both* an advocate of Natural Theology and of Natural Selection. He was open to the idea that Darwin had discovered the "intermediate causes," or natural laws instituted by God, to create new species adapted to their ever-changing conditions of existence.

---

[19] Darwin, *On the Origin of Species*, 84.   [20] Ibid., 86.

[21] Charles Darwin. *The Various Contrivances by which Orchids Are Fertilized by Insects*. London: 1862, 349–50.

[22] Correspondence vol. 11 (1999), 253. Darwin Correspondence Project, "Letter no. 4056," accessed on 3 January 2024, https://www.darwinproject.ac.uk/letter/?docId=letters/DCP-LETT-4056.xml.

Darwin suggests, in his autobiographical remarks about his changing religious views, that when he wrote the *Origin of Species* he shared this view, and its frontispiece facing its title page has two quotes that confirm this, one from Francis Bacon's *Advancement of Learning* and one from William Whewell's *Bridgewater Treatise on Astronomy and General Physics*: both quotations stress that the scientific search for natural laws is in harmony with a belief in God as the author of those laws.

16.5   Teleology without Natural Theology

Over the decade following the publication of the *Origin of Species*, Darwin and Gray came to see just how different their views about teleology were. This years-long dialogue on chance and design helped Darwin to see that his understanding of teleology was fundamentally different from, and in the end incompatible with, that derived from Natural Theology.[23] And, in corresponding with others, Gray shows himself to be aware that he and Darwin grounded their teleology in very different ways. In a letter to Alphonse de Candolle in 1863, for example, Gray admits he recognizes that Darwin does not accept the inference from the presence of ends in nature to an unnatural designer: "Under my hearty congratulations of Darwin for his striking contributions to teleology, there is vein of petite malice, from my knowing well that he rejects the idea of design, while all the while he is bringing out the neatest illustrations of it!"[24] Darwin was equally aware that there is a deeper disagreement behind their common endorsement of teleological reasoning. In an 1861 letter to Gray, Darwin reports on Sir John Herschel's first public response to the *Origin of Species*, in a new edition of Herschel's *Physical Geography*.

[He] agrees to certain limited extent; but puts in a caution on design, so much like yours that I suspect it is borrowed. – I have been led to think more on this subject of late, & grieve to say that I come to differ more from you. It is not that designed variation makes, as it seems to me, my Deity "Natural Selection" superfluous; but rather from studying lately domestic variations & seeing what an enormous field of undesigned variability there is ready for natural selection to appropriate for any purpose useful to each creature.[25]

This exchange reveals deep disagreement over whether a significant element of chance, in the form of that "enormous field of undesigned variability," is compatible with a teleological account of adaptive modification. Darwin

---

[23] For more details about this long-distance dialogue, see J. G. Lennox, "The Darwin/Gray Correspondence 1857–1869: An Intelligent Discussion about Chance and Design," *Perspectives on Science* 18.4 (Winter 2010), 456–79.

[24] Jane Loring Gray, ed., *The Letters of Asa Gray*. Boston: 1893, Cambridge: The Riverside Press, 498.

[25] *Correspondence* vol. 9 (1994), 162: 3176.

thought it was – notice his closing expression: "ready for natural selection to appropriate for any purpose useful to each creature." Darwin is here stressing that the constant appearance of undersigned – that is, random – variation is a primary tenant of his theory. Thinkers such as Gray, Herschel, and Whewell could not accept this. For them it had disturbing implications that were deep and far-reaching – among them that the appearance of humans on this earth was a matter of chance, not something preordained by God.

In 1862, the same year he published the first edition of his *Orchid* book, Charles Darwin presented the results of his research on sexual dimorphism in the genus *Primula* (the genus that includes Primroses and Cowslips) to the botanical section of the Linnean Society.[26] In the species of Primula, individual plants have two distinct forms of their reproductive parts, and these parts are structured in such a way to increase the chances of cross-fertilization. In the concluding section of the paper, he wrote:

> The meaning or use of the existence in Primula of the two forms [of flowers] in about equal numbers, with their pollen adapted for reciprocal union, is tolerably plain; namely, to favour the intercrossing of distinct individuals. With plants there are innumerable *contrivances for this end*; and no one will understand *the final cause of the structure of many flowers* without attending to this point.[27]

Once again, Darwin describes himself as engaged in a *teleological* enquiry, a search for the *final cause* of a particular feature of these two varieties of plants. And he refers to the different mechanisms to encourage intercrossing in different plants as contrivances that are present *for the sake of that end*. It is unlikely that Darwin would have used such loaded expressions unreflectively; as we have seen, this has been a topic of reflection for him since the 1830s. And, indeed, there is a clear sense in which Darwin *has* identified the "final cause" of the trait in question. For he has identified the advantageous consequences of intercrossing that explain why it is selectively favored.

Allow me to illustrate Darwin's teleology with one last example, which Darwin found so compelling that he published a review of the journal article in which the example appears. It was written by Henry Walter Bates, and the example is one of a particular kind of "mimicry," now known as "Batesian Mimicry" (Figure 16.2), which Bates studied in the butterfly populations of the Amazon Basin. Darwin first presents an overview of the meticulous observations made by Bates, after which he says:

> Mr. Bates has given to these facts the requisite touch of genius, and has, we cannot doubt, hit on *the final cause of all this mimicry*. The mocked and common forms [of

---

[26] This was the first of five such publications between 1862 and 1868, which were republished as a single book, *The Different Forms of Flowers on Plants of the Same Species*, in 1880. In all, Darwin studied 144 species in 40 genera.

[27] Barrett, *Notebooks*, 2:59; emphasis added.

Figure 16.2 Batesian Mimicry: The butterflies labeled 1 (upper left) and 2 (upper right) are mimics adapted by natural selection to look like those labeled 1a (bottom left) and 2a (bottom right), which predators find distasteful and avoid. Credit: Bates, Henry Walter (1862). "Contributions to an insect fauna of the Amazon Valley – Lepidoptera: Heliconidae." Transactions of the Linnean Society (23): plate LV, after p. 865. Linnean Society of London

butterfly] must habitually escape, to a large extent, destruction, otherwise they could not exist in such swarms; and Mr. Bates never saw them preyed upon by birds and certain large insects which attack other butterflies; he suspects that this immunity is owing to a peculiar and offensive odour that they emit. The mocking forms, on the other hand, which inhabit the same district, are comparatively rare, and belong to rare groups. . . . Now if a member of one of these persecuted and rare groups were to assume a dress so like that of a well-protected species that it continually deceived the practiced eyes of an ardent entomologist, it would often deceive predacious birds and insects, and thus escape entire annihilation. This we fully believe is *the true explanation of all this mockery*.[28]

Notice the parallelism in the two italicized phrases, with the substitution of "true explanation" for "final cause." How exactly does this true explanation identify the final cause of mimicry? There is a *valuable consequence* for any individuals in the "persecuted and rare" species appearing like individuals in one of the "well-protected" species – they will then largely avoid predation and will pass on their "disguise" to their offspring. *It is this valuable consequence of mimicry that explains its selective advantage* – that is the end achieved by mimicry. One can see why Charles Darwin went out of his way to draw the

[28] "A review of H. W. Bates, 'Contributions to an Insect Fauna of the Amazon Valley'," *Transactions of the Linnean Society* 23 (1862), 495–566. Originally published in *Natural History Review: Quarterly Journal of Biological Science*, 1863, pp. 219–24. Reprinted in Barrett et al., *Notebooks*, 2:87–92 (emphasis added).

attention of naturalists to this article: it is a perfect example of selection-grounded teleology, including a treasure trove of evidence of the requisite "enormous field of undersigned variability" for natural selection to sort through.

16.6   Conclusion

As I have argued elsewhere,[29] the view that Darwin must be opposed to teleology rests on two beliefs, often left implicit: first, that teleological explanations either appeal to God's plan (as in Natural Theology) or to some long discredited vital force; and second, that Darwin's theory of natural selection appeals to neither, nor did Darwin think it did. And as Darwin's disagreement with Gray over what his use of teleology commits him to shows, the second premise is clearly true. The myth that Darwin was an opponent of teleology depends on the first premise. But that premise is false, both philosophically and historically. Darwin explicitly endorsed and used teleological explanations, most obviously in his many botanical publications written in the last two decades of his life. That he opposed the use of such explanations in biology is a myth; that he gave us a new way of understanding them is not.

---

[29] J. G. Lennox, "Darwin Was a Teleologist." Biology and Philosophy, 8 (1993), 409–21; Lennox, "The Darwin/Gray Correspondence," 456–79.

Myth 17 That Darwin's Success Depended on Undermining "Aristotelian Essentialism"

James G. Lennox

17.1 Introduction

It has long been assumed by historians and philosophers that Darwin and Aristotle were in opposite camps. For instance, Ernst Mayr in 1970 wrote that:

Owing to its belief in essences this philosophy [Typology] is also referred to as essentialism and its representatives as essentialists (typologists). [They] were influenced by the idealist philosophy of Plato and the modifications of it by Aristotle. ... The concept of unchanging essences and of complete discontinuities between every eidos (type) and all others make genuine evolutionary thinking well-nigh impossible.[1]

Elliott Sober seemed to agree:

Aristotle is typical of exponents of the Natural State Model in holding that variation is introduced into a population by virtue of interference with normal sexual reproduction. ... The essentialist attempts to understand variation within a species as arising through a process of deviation from type. ... evolutionary theory undermined the essentialist's model of variability [and] removed the need for discovering species essences.[2]

I begin this chapter with a caveat: Charles Darwin's theory of evolution by natural selection certainly did conflict with a certain kind of essentialism about species that was dominant among naturalists, even among Darwin's supporters, in the nineteenth century; and he did argue strenuously in the *Origin of Species* and elsewhere to undermine it. The myth I shall debunk in this chapter is that this essentialism is Aristotelian in nature.

17.2 The Creation of a Myth

I will begin with a quotation from a philosopher well known for his defense of the centrality of "falsification" to the scientific method, Sir Karl Popper. In *The Open Society and Its Enemies*, published near the end of World War II, he had

---

[1] Erns Mayr, 1970. *Populations, Species and Evolution*. Cambridge MA: Harvard University Press.

[2] Elliott Sober, 1994. "Evolution, Population Thinking, and Essentialism." In *From a Biological Point of View*. Cambridge: Cambridge University Press, 226.

Figure 17.1 Aristotle's statue, in Aristotle's park, Stagira, Halkidiki, Greece.
Source: sneska/E+/Getty Images

particularly harsh things to say about Aristotle's (Figure 17.1) influence on the history of science.

> Every discipline, so long as it used the Aristotelian method of definition has remained arrested in a state of empty verbiage and barren scholasticism, and the degree to which the various sciences have been able to make any progress depended on the degree to which they have been able to get rid of the essentialist method.[3]

These critical remarks about "the Aristotelian method of definition" and "the essentialist method" are not specific to biology – but biology is certainly to be included.

Under the influence of Popper's historical claim about Aristotle's influence, David Hull published an influential paper in 1965 entitled "The effects of essentialism on taxonomy: 2000 years of stasis." In it he writes: "Aristotelian definition had to be abandoned both for species names and for 'species' [i.e. the species concept]. Typologists could ignore the actual untidy distribution of properties among living organisms and the variety of methods of reproduction used to perpetuate species. Evolutionists could not."[4]

Note the use of the concept "typologist" in this passage, apparently substituted for purveyors of "Aristotelian definition" in the previous sentence. The distinction Hull makes between "typologists" and "evolutionists" had, by 1960,

---

[3] Sir Karl Popper, 1945. *The Open Society and Its Enemies*, Vol. 1. London: Routledge and Kegan Paul, 206.

[4] David Hull, 1965. "The Effects of Essentialism on Taxonomy: 2000 Years of Stasis," *BJPS* vol. 15, no. 60, 314–26.

become common in the founding documents of the so called "Neo-Darwinian Synthesis." In *Systematics and the Origin of Species*, for example, Ernst Mayr stresses that a "typological" approach to the species concept is incompatible with evolutionary thinking, and in *Animal Species and Evolution*, published in 1963 and its abridged 1970 edition, he cites Plato and Aristotle as its source. Part of this worldview, according to the myth, is that variations within a species are deviations from type, a view Elliott Sober refers to as the Natural State Model, and names Aristotle as a "typical exponent."

It is time to summarize the key elements of this myth:

* Aristotle should be considered a reformed Platonist
* According to Plato's Theory of Forms, knowledge is of eternal and changeless types of which the things of this world are imperfect images
* Aristotelian forms are Platonic forms embedded in matter
* All perfect members of a species have an identical essence
* Variations among species members are deviations from type
* The essence is unchanging and eternal
* Knowledge, in the form of definitions, is of these essences
* Reproduction passes on the species essence to offspring
* This was the "paradigm" that Darwin needed to overturn

17.3   Aristotle on Kinds and Their Forms

Let's begin with getting clear on Aristotle's approach to the study of living things. We fortunately have ample resources for doing so, since more that 25 percent of the works of Aristotle are zoological in focus. Given the purpose of this chapter, I will focus exclusively on Aristotle's views about animal kinds, how they are defined and how their attributes are explained, and the important role that continuous variations play in distinguishing one form of a kind from another.

One preliminary point is necessary: Aristotle's "taxonomic" vocabulary is limited to two terms, *genos* ("kind") and *eidos* ("form"). The same group of animals can be referred to by either term, but when Aristotle is discussing subgroups within a wider class, the wider class will typically be referred to as the *genos* (kind), and the narrower groupings as *eide* (forms). Charles Darwin uses the word "form" in much the same way, as in this justly famous closing sentence of the *Origin of Species*:
                                                                          ˋ
There is grandeur in this view of life, with its several powers, having been originally breathed into *a few forms* or into one; and that, whilst this planet has gone cycling on according to the fixed law of gravity, from so simple a beginning *endless forms* most beautiful and most wonderful have been, and are being, evolved.[5]

---

[5] Charles Darwin, 1859. *On the Origin of Species by Natural Selection*. London: Charles Murray, 490.

In what follows, I will present a highly compressed and simplified account of Aristotle's biology[6] that will focus on just three questions:

1. How does Aristotle approach the problem of identifying biologically significant kinds?
2. What role does his understanding of variations within kinds play in his biology?
3. Is Aristotle's biology "essentialist," and if so, is it anything like what they mythologists refer to as "typological essentialism"?

17.4   Identifying Kinds and Their Many Forms

Let us begin with a passage from Aristotle's *On the Parts of Animals* (*PA*):

> In those cases where people refer to kinds in a clearly defined manner and where the kinds have [i] a single common nature and [ii] forms in them not too distinct from each other we should refer to these animals in common as a single kind, as we do with the birds and the fish. And we should do the same with any other currently unnamed group that embraces – as kinds do – many forms within it. (644b1–6)[7]

Aristotle here lays down two conditions that should be met before we demarcate a group of animals as a "kind": they need to have a shared "common nature," and distinct forms within the kind that are not to be "too distinct" from each other. He apparently thinks these conditions have already been met by the kinds "bird" and "fish" – but he advises us to identify other such groups that have not yet been named. (Later, I will provide an example of one such group that Aristotle was the first to properly identify, the cephalopods.)

To grasp what Aristotle is up to here, we need a little context. In the pages leading up to this quotation, he has introduced a method for organizing species into more general groupings. This method involves taking very general kinds – for example, Birds and Fish – identifying all their shared features ("general differences"), and then subdividing those differences into more and more specific, determinate forms. To cite a single example: all birds have beaks, which differ markedly from the mouths of other kinds of animals; but they can be divided into hooked or straight, wide or narrow, short or long, and so on. Aristotle recommends tracing down such differences for *all* the shared traits simultaneously, identifying more and more determinate forms of the

---

[6] For a somewhat richer, but still introductory, presentation, see my entry on "Aristotle's Biology," in *Stanford Encyclopedia of Philosophy* (https://plato.stanford.edu/search/searcher.py?query=Aristotle%27s+Biology). Much of the research on which that entry is based can be found in James G. Lennox, 2001. *Aristotle's Philosophy of Biology: Essays on the Origins of Life Science*. Cambridge: Cambridge University Press.

[7] All translations are from James G. Lennox, 2001. *Aristotle: On the Parts of Animals I–IV*. Oxford: Oxford University Press. The numbers refer to the pagination in the standard Bekker edition of Aristotle.

general kind in this way. This method is based, however, on an undefended assumption: that researchers can identify these general kinds. Aristotle is well aware that this is not an obvious step: he notes, for example, that if you start with the general category "flyer," you will be starting with a kind consisting of some (but not all) insects, some (but not all) mammals, and most (but not all) birds (see *PA* IV.14, 697b14–26 on ostriches)! That is, you will be breaking up kinds with multiple shared characteristics and uniting forms that share only one. The above passage is part of Aristotle's solution to the problem of identifying kinds that are *appropriate* starting points for biological inquiry – what I elsewhere have dubbed "entry level kinds."[8]

But how do we identify, at this preliminary stage of inquiry, kinds with a "common nature"? And how do we determine whether forms are "not too distinct from one another"? These concerns take us to my second question.

17.5   Differing by More and Less

In his *Posterior Analytics*, Aristotle articulates and defends an epistemology that allows for stages of "knowing": Aristotelian inquirers can have compelling reasons for thinking that a number of forms share a common nature prior to knowing what that common nature is – in fact getting to that stage provides them with clues in the search for that nature. Aristotle recommends that in order to identify such kinds, one needs to study *continuous variations* among similar forms of animal. He makes the point immediately after the passage we discussed above, where he outlines norms for identifying biologically useful kinds.

it is by finding groups where the figures of the parts and of the whole body *bear a likeness* that the kinds have been demarcated; for example, members of the kind "bird" are related in this way to one another, as are members of the kind "fish," the soft-bodied animals [*cephalopods*] and the hard-shelled animals [*mollusks*]. For their parts are not merely analogous to one another, as bone in mankind is analogous to fish-spine in fish, but rather *their parts are of the same kind and differ only by being larger or smaller, softer or harder, smoother or rougher and so on – speaking generally, they vary by more and less.* (644b8–16; emphasis added)

With sufficient experience of birds, researchers are able to develop a profile of their common features – feathers, beaks, two fleshless legs and toes, wings, and so on: they will notice that, from one form of bird to the next, these shared features vary continuously along many axes: feathers, for example, vary in length, color patterns, density and length of the barbs; beaks vary in hardness, length, depth, curvature. One takes note of these differences in distinguishing

---

[8] James G. Lennox, 2005. "Aristotle on Entry Level Kinds," in G. Wolters and M. Carrier eds., *Homo Sapiens und Homo Faber: Festschrift für Jürgen Mittelstrass*. Berlin: de Gruyter, 87–100.

forms of bird. There is a shared overall body plan ("Unity of Type" as Darwin might say), but each part, and each activity associated with it, differs in innumerably many measurable ways as you go from one form of bird to the next.

On the very first page of the *History of Animals*, after distinguishing the general categories of likeness he will use to organize this work, he turns to the ways in which animals alike in kind (he again uses Bird as his example) differ from each other.

Practically speaking, most of their parts differ by way of the oppositions among their affections, for example by color or configuration, by them being affected some more and some less; and again, by being greater or lesser, larger or smaller, and generally by excess and defect. ... Generally speaking, most of the parts from which the entire bulk of the animal is constituted are either the same or differ by opposition and according to excess and defect; for one can treat 'the more and less' as a sort excess and defect. (*HA* I.1, 486a25–b17)

Aristotle eventually identifies nine Great Kinds (*megista genê*), within which the parts and overall bodily plan of the different forms vary by "more and less" along many dimensions: five "blooded" kinds (vertebrates) – birds, fish, live-bearing quadrupeds (terrestrial mammals, minus humans), egg-laying quadrupeds (reptiles and amphibians), and cetaceans (aquatic mammals); and four "bloodless" kinds (invertebrates) – insects, soft-shelled animals (crustaceans) hard-shelled animals (testaceous mollusks), and cephalopods (non-testaceous mollusks). He observes that many animals do not fall into any of these categories (human beings, for example!) and others that seem to have some traits that are typical of one kind and some that are typical of another. In fact, he seems positively fascinated by these animals, which he calls "dualizers": apes, monkeys, and baboons, for example, dualize between the human, biped form and quadrupeds. Apes in particular, he notes, are "human-form" (*anthrôpoeideis*) in many respects: he mentions likenesses in the nose, ears, teeth, eyelashes, arms and legs, fingers, fingernails, male nipples, and female vagina, but notes that their feet are more like hands than those of humans (HA II.502a16–b27).

It is Aristotle's views about how one kind of bird (or fish, or insect, or cephalopod) differs from another that will be especially important for the question at the heart of this chapter: was it in fact Aristotle's views about animal kinds and their "essences" that Darwin had to overcome, or something quite different?

17.6   Kinds, Forms, and Continuous Variation in Practice

To this point, our focus has been on Aristotle's *theory* of how different forms of animals vary and the role that ought to play in identifying kinds with common natures. To see these ideas in practice, I will focus on one of those "great kinds" that he was the first to identify as a unified kind, the cephalopods – *ta malakia*, "the softies." HA IV begins by summarizing the discussion of the blooded animals in the first three books, and then lists the four "bloodless" kinds he is about to discuss, giving one or two examples of each. He then turns to the first kind on his list:

> Among the animals called "softies" [cephalopods] these are the external parts: first, the so-called feet; second, the head, continuous with the feet; third, the sac, containing the internal organs, which some mistakenly call the head; and fourth the fin, which encircles the sac. In all of the softies the head turns out to be between the feet and the belly. Moreover, all have eight "feet," and all have two rows of suckers, except for one kind of octopus. The cuttlefish, and the large and small squids, have a distinctive feature, two long tentacles, the ends of which are rough with two rows of suckers, by which they capture food and convey it to their mouth and fasten themselves to a rock when it storms, like an anchor. (HA IV.1, 523b22–32)

Aristotle starts with a number of features that all cephalopods share, and with the distinctive way the parts are organized – for example, that the head is contiguous with the "arms" (which, he notes, are referred to as "feet" – *hoi podes*).[9] He notes two exceptions, an unusual octopus in which the arms only have a single row of suckers (he later names it the *helodonê*), and the two additional tentacles which distinguish squids and cuttlefish from the other cephalopods: I'll return to them in discussing Aristotle's form of essentialism. The remainder of the chapter on cephalopods discusses first the external and then internal parts of each form, closing with a discussion of variations in several forms of octopus.

  Throughout the entire discussion of these four great kinds, the pattern is the same: a brief discussion of the features that are common to all the forms within that kind, followed by a much longer description of the ways in which those features vary continuously from one form to the next, along with features distinctive to the different forms.

  Following Aristotle's general methodological recommendations in his philo-sophical works, his study of animals distinguishes the fact-collecting and organ-izing investigation from the search for causal explanations. *HA* is a contribution to the former task. I will now turn to those treatises which report on his causal inquiries, for that will provide insight into Aristotle's essentialism.

17.7   Aristotelian Essences

For Aristotle, there is an intimate connection between discovering what is *causally basic* to being a certain kind of thing and what it is *to be* that thing – that is, what is its essence.[10] With respect to animals, it is their peculiar way of life,

---

[9] Cephalopod is a conjunction of the Greek words for head and foot.

[10] There is no Aristotelian term that corresponds to our word "essence." The words so translated are all variants on the Greek verb meaning "to be."

and the activities that way of life requires, that explain why each animal has the parts it has with those more and less variations characterized in *HA*. Every animal has those fundamental capacities that Aristotle collectively refers to as their *psychê* (misleadingly translated "soul") – their abilities to reproduce, nourish themselves, perceive, and locomote. But what fascinates Aristotle is the wildly varied ways in which animals do all these things, and the ways in which all these activities are integrated.

To illustrate this "causal essentialism" we will look at Aristotle's causal explanations of the single row of suckers on the arms of one form of octopus, and the two additional tentacles found only in squids and cuttlefish. Both are explained in *Parts of Animals* IV.9.

Now while the other octopuses have two rows of suckers [on each arm], one kind has a single row. This is because of the length and thinness of their nature; for it is necessary that the narrow arm have a single row of suckers. It is *not, then, because it is best* that they have this feature, but because it is *necessary owing to the distinctive account of their being* (οὐσία). (685b13–17, emphasis added)

There are two aspects of this explanation that are initially surprising: (1) the narrowness of the arms or the whole body is identified as essential and (2) this explanation is *not* teleological – they don't have a single row of suckers because it is best to be that way. But this explanation accords well with Aristotle general account of biological explanation in book I:

Hence it would be best to say that, since this is what it is to be a human being, on account of this it has these things; for it cannot be without these parts. If one cannot say this, one should say the next best thing, i.e. either that in general it cannot be otherwise, or that at least it is good thus. (640a32–b1)

Part of what it is to be a *heledonê*, its essence, is to have narrow arms, which necessitates a single row of suckers.

The possession, by cuttlefish and squids, of two long tentacles, the ends of which are rough with two rows of suckers, which distinguishes them from other cephalopods, is explained in the same chapter.

Since the cuttlefish and squids have small 'feet' that are useless both for taking hold . . . of the rocks when there are waves and storms, and for feeding . . . they have two long tentacles by which they moor themselves . . . and by which they hunt down prey from afar and bring it to their mouths. The octopuses, on the other hand, don't have these, because their "feet" are useful for these activities. (*PA* IV.9 685a33–b2).

Unlike octopuses, cuttlefish and squid "feet" are useless for feeding or mooring – for those functions they have tentacles. Here we have a teleological explanation, but one that is (1) tied to a deficiency in another part and (2) connected directly to distinctive aspects of a cephalopod's way of life.

The facts and the explanations we've looked at so far concern features of kinds and subkinds of cephalopods, rather than variations *within* these kinds. But they highlight how attuned Aristotle is to *variations within wider kinds* – differences displayed by a form of a kind are to be explained by a more fundamental distinctive feature of that form of animal. If one species of octopus has arms with only one row of suckers, it is either essential or explained by some other essential feature, such as the narrowness of its arms. That is the essence of Aristotelian essentialism.

But what about *individual variations* – has Aristotle really no interest in them, or does he see them as mere deviations or accidents, as the myth claims? That idea is hard to reconcile with the following passage, from Aristotle's account of inheritance in *Generation of Animals* IV, which is entirely concerned with variations among members of what we would refer to as "species," starting with the most obvious and important one, sex differences between males and females. In chapter 3 he turns to puzzles about offspring resembling one parent with respect to one trait and the other with respect to another trait; or looking more like a grandparent than a parent; or female offspring resembling the male parent in nonsexual characteristics. The general framework for the discussion is summarized in the following passage:

> In relation to generation the distinctive and particular features are always stronger. For Koriskos is both a man and an animal; but the man is nearer to what is distinctive [to Koriskos] than the animal. And both the particular and the kind generate, but more so the particular; for this is the being (οὐσία) [of the animal], and what is coming to be is, while a certain *sort* of animal, also *this* animal here, and this is the being [of the animal]. (767b31–768a1)

Aristotle insists that it is in the distinctive characteristics of the individual parents that being resides, not in the general characteristics. However, and this is what even most scholarly commentary on this passage misses, the causes aimed at reproducing Socrates' distinctively *snub* nose will of necessity also produce a *human* nose and an *animal* nose. You can describe a particular person's nose in a way that focuses on its uniquely *individuating* features, its distinctively *human* features, or those features which make any nose a nose. Individual variations are not, for Aristotle, "deviations due to interference with the animal's nature" – quite the opposite! It is in those features that the being of the animal exists.

Aristotle does, however, stress that typically *the individual variations make no functional difference* to the life of the animal, no *essential* difference – to cite one of his examples, it is essential that eagles have distinctively eagle eyes, but not that they have blue or brown eyes. You can ignore the variations at that level when considering the being or essence of (say) bald eagle, even though every distinctive feature of bald eagles will vary "by more and less." And *that* is one important difference between Aristotle and Darwin – because of the way natural selection works, those tiny variations between one bald eagle and the next *may* turn out to be crucial to evolution, because they may provide a slight competitive advantage to the life and reproductive potential of one individual relative to another.

To summarize then:

1. Aristotle is not a typological essentialist
2. Aristotle is an explanatory essentialist
3. Aristotle considers variation within and between kinds natural and important and in need of explanation.
4. Individual variations are most real, general similarities are consequences – the characteristics of a species only exist as distinctive features of individual organisms.

Notice that there is nothing in this sort of essentialism that would "make genuine evolutionary thinking well-nigh impossible," as the myth claims.

17.8   The Essentialism Darwin Was Up Against

If "typological" essentialism is not, either historically or philosophically, Aristotelian, there was, nevertheless, an essentialism that Darwin had to challenge if his ideas were to gain a fair hearing. It was an essentialism much more akin to the "typological" variety.[11] It receives a very clear statement and defense in the book that was Darwin's "bible" during the Beagle Voyage and had a profound impact on him: Charles Lyell's *Principles of Geology*. Volume 2 characterizes the view in these terms: "the majority of naturalists agree with Linnaeus that all individuals propagated from one stock have certain distinguishing characters in common, which will never vary, and which have remained the same since the creation of each species."[12]

Essentially the same view appears in a number of negative reviews of the *Origin of Species*, such as this, from a review by William Hopkins: "Every natural species must by definition have had a separate and independent origin, so that all theories ... which assert the derivation of all classes of animals from one origin, do, in fact, deny the existence of natural species at all."[13] Note the concern about creation and origins: each species creation must be separate and independent, and once created the species cannot vary.

---

[11] How akin is up for debate; cf. Ron Amundson, 1998. "Typology Reconsidered: Two Doctrines on the History of Evolutionary Biology," *Biology and Philosophy*, 13, 153–77.

[12] Charles Lyell, (1832) 1990. *Principles of Geology*, Vol. II. Chicago: University of Chicago Press, 3.

[13] David Hull, 1973. *Darwin and His Critics*. Chicago: University of Chicago Press, 241.

This is the essentialism that Charles Darwin deals with in passages like the following:

Systematists will have only to decide ... whether any form be sufficiently constant and distinct from other forms, to be capable of definition; and if definable, whether the differences be sufficiently important to deserve a specific name. .... Hence ... we shall be led to weigh more carefully and to value higher the actual amount of difference between two forms.[14]

In this passage, Darwin is stressing *degree of constancy and difference* in deciding what counts as a species – that greater value should be given to what Aristotle would call "more-and-less differences" between forms. Doing so allows conceptual space for the gradual transformation that he has argued is the source of new species.

17.9   Darwin (Finally) Reads Aristotle

Evolutionary thinking was a response to the need for a comprehensive explanation of a vast range of information about fossils and the biogeographic distribution of hundreds of thousands of species that only became available in the eighteenth century. It is hard to imagine how Aristotle would have responded to that information – and that is not our concern.

We do, however, know how Darwin responded to what Aristotle had accomplished. In January of 1882, three months before his death, he received by mail a copy of a new translation of *On the Parts of Animals*, a gift from its translator, William Ogle. Darwin writes a brief thank you note at that time. But over a month later, on February 22, he writes a second note to Ogle, in which he says he has read all of Ogle's long introduction and "not more than a quarter of the book proper." From other things he says we know he read all of book one, and some of book two.[15] His note concludes as follows: "From quotations which I had seen I had a high notion of Aristotle's merits, but I had not the most remote notion what a wonderful man he was. Linnaeus and Cuvier have been my two gods, though in very different ways, but they were mere school-boys to old Aristotle." In very different ways, his two gods were also purveyors of the kind of essentialism that made it so hard for some readers to understand what Darwin was up to in the *Origin of Species*. Aristotle's essentialism it was not.

---

[14] Darwin, *On the Origin of Species by Natural Selection*, 484–5.
[15] For a detailed and insightful discussion of this correspondence, see Allan Gotthelf, 1999. "Darwin on Aristotle," *Journal of the History of Biology*, 32, 3–30. Reprinted in Allan Gotthelf, 2012. *Teleology, First Principles, and Scientific Method in Aristotle's Biology*. Oxford: Oxford University Press, 345–69.

Myth 18 That Darwin's Theory Would Have Become More Widely Accepted Immediately Had He Read Mendel's 1866 Paper

Gregory Radick

18.1 Introduction

Many have wondered: What would have happened if Darwin had read Mendel's work? Nowadays an easily encountered answer runs as follows:

He would have been overjoyed, because it solved the greatest weakness of natural selection: it did not work under the theories of inheritance at the time. The most common theory was "blended inheritance" and natural selection cannot work under it ... Mendelian genetics is one of [the] pillars of [the] Modern Synthesis (neo-Darwinism) formulated in the 1940s.[1]

Here is another instance, referring to the Victorian engineer Fleeming Jenkin, who first identified this supposed weakness in Darwin's theory:

There was no denying Jenkin's inescapable logic: to salvage Darwin's theory of evolution, he needed a congruent theory of heredity ... [For Darwin, reading Mendel's] study might have provided the final critical insight to understand his own theory of evolution. He would have been fascinated by its implications, moved by the tenderness of its labor, and struck by its strange explanatory power. Darwin's incisive intellect would quickly have grasped its implications for the understanding of evolution.[2]

The claim that Darwin's theory would have become more widely accepted immediately had he read Gregor Mendel's (Figure 18.1) paper is actually two myths wrapped together:

- The myth that Darwin's theory of natural selection was unworkable and so unconvincing until integrated with Mendel's theory of inheritance.
- The myth that, had Darwin read Mendel's 1866 paper on his experiments with crossbred peas, the needed integration would have taken place around 1870 rather than around 1940.

Acknowledgements: Many thanks to Kostas Kampourakis, Shruti Santosh, and Anya Plutynski for their helpful comments on a draft version of this chapter.

[1] *Quorum.com*, top answer in 2022, www.quora.com/What-would-have-happened-if-Darwin-had-read-Mendels-work (accessed September 8, 2022).
[2] Siddhartha Mukherjee, *The Gene: An Intimate History* (London: Bodley Head, 2016), p. 46.

GREGOR MENDEL

About the year 1862

Figure 18.1 Gregor Mendel, 1862. Wellcome Collection. Attribution 4.0 International (CC BY 4.0)

Let us call these the "Darwin needed Mendel" myth and the "decades wasted" myth, respectively. In what follows I shall consider each in turn, taking them in that order because, for the most part, the idea that Darwin's theory had a Mendel-sized gap in it (the "Darwin needed Mendel" myth) is what has prompted some Darwinians to lament the fact that Darwin never read Mendel, on the view that, had Darwin done so, he would have plugged the gap himself and so sped up acceptance of his theory by decades (the "decades wasted" myth). But the historical record, always full of surprises, shows that in fact a version of the "decades wasted" myth predates the "Darwin needed Mendel" myth. Even more surprisingly, this if-only-Darwin-had-read-Mendel lament came not from a champion of Darwinian natural selection but from an *opponent*, who made his lament not in the wake of the successful integration of natural selection theory and Mendelian genetics – the "Modern Synthesis" of the 1930s and 40s, after which Darwinian theory went fully mainstream – but in the early twentieth century, when there was so much controversy over natural selection that the period was later dubbed the "eclipse of Darwinism."[3]

---

[3] The phrases "Modern Synthesis" and "eclipse of Darwinism" both come from Julian Huxley, *Evolution: The Modern Synthesis* (London: Allen & Unwin, 1942).

The Partners 1907

Figure 18.2 William Bateson (right) with his collaborator Reginald
Punnett (left). Credit: John Innes Archives, courtesy of the John Innes
Foundation

Our counterfactually minded anti-Darwinian was the most fervent
Mendelian who ever lived, the Cambridge biologist William Bateson (Figure
18.2). "Had Mendel's work come into the hands of Darwin," wrote Bateson in
Mendel's Principles of Heredity: A Defence, published in 1902, just two years
after Mendel's paper had become an unexpected talking point in European
botany, "it is not too much to say that the history of the development of
evolutionary philosophy would have been very different from what we have
witnessed." Bateson here emphatically did not mean: if only Darwin had read
Mendel, then biologists would have embraced the theory of natural selection
more strongly and swiftly. Before 1900, Bateson was best known for arguing
that evolutionary theorizing had taken a wrong turn with Darwin's *Origin of
Species*, since, in Bateson's view, the evidence showed that new species evolve
from existing ones not, as Darwin had thought, by gradual adaptive change
driven by natural selection, but by non-adaptive jumps from one stable form to
another. For Bateson, reading Mendel's paper would have alerted Darwin to his
error. His corrected theory would then have redirected his followers toward
Mendel-style experimental hybridizing with all-or-nothing "unit" characters as
the best way to understand the discontinuous nature of the origin of species.
When Bateson imagined Mendel's paper coming into Darwin's hands, it was to

help twentieth-century biologists rid their science of natural selection theory for good.[4]

18.2   The "Darwin Needed Mendel" Myth

It is by no means clear that Darwin needed *any* theory of inheritance, whether Mendel's or anyone else's. Consider that, on Darwin's presentation of his theory, natural selection will take place whenever three conditions are satisfied: first, there is variation among individual organisms; second, there is a Malthusian "struggle for existence" sufficiently intense that only some of those organisms will survive to reproduce; and third, the offspring of the survivors on the whole inherit the adaptive variations that caused their parents to become survivors. In Darwin's view, the evidence in support of nature's fulfilling these conditions was overwhelming. Indeed, in the first chapter of the *Origin of Species*, in the sole extended discussion of inheritance per se in the book, he wrote that it was only "theoretical writers" who had ever doubted that "like produces like." By contrast, he went on, the two kinds of practical men whose occupations gave them the largest scope to observe parents and off-spring up close – namely, breeders and doctors – took the principle utterly for granted. For Darwin's own theoretical purposes, then, he needed only for offspring to inherit their parents' distinctive characteristics; and he took it to be an incontrovertibly well-evidenced fact that, in general, that was what happened.[5]

The theory of natural selection nevertheless came to acquire a reputation for needing a theory of inheritance – indeed, a better theory than Darwin came up with in his ignorance of Mendel's paper – because of a review of the *Origin of Species* that appeared in 1867 in the *North British Review* by Jenkin. In this review, Jenkin drew attention to what he declared to be a fatal problem for Darwin's theory, in the form of an imperial fantasy. Imagine, wrote Jenkin, that a male colonist ends up on an island and begins reproducing prodigiously with the native women. Imagine further that, in terms of the struggle for existence, the colonist is superior to the natives. According to Jenkin, the children will

---

[4] William Bateson, *Mendel's Principles of Heredity: A Defence* (Cambridge: Cambridge University Press, 1902), pp. 37–9, quotation on 39. For Bateson on species, see his *Materials for the Study of Evolution Treated with Especial Regard to Discontinuity in the Origin of Species* (London: Macmillan, 1894). For further discussion of Bateson's counterfactual see Gregory Radick, *Disputed Inheritance: The Battle over Mendel and the Future of Biology* (Chicago: University of Chicago Press, 2023), pp. 293, 295–6.

[5] Charles Darwin, *On the Origin of Species by Means of Natural Selection* (London: John Murray, 1859), pp. 12–14, quotations on 12. For further discussion see Gregory Radick, "*Origin*'s Chapter I: How Breeders Work Their Magic." In *Understanding Evolution in Darwin's "Origin": The Emerging Context of Evolutionary Thinking*, ed. Maria Elice Brzezinski Prestes (Cham: Springer, 2023), pp. 205–19.

inherit not the colonist's advantageous variations but, because of the native mothers, a watered-down version of them, so that whatever advantage the variations brought the colonist will, on average, be halved. Jenkin went on to show that, even taking into account the tendency of the colonist-native offspring to survive disproportionately, and so to be represented disproportionately among the parents of the next generation, within a short while, the colonist's advantageous variations will be utterly swamped. On Jenkin's calculations, then, it appeared that, far from inheritance preserving the advantageous variations which natural selection then amplified and accumulated, as Darwin had supposed, inheritance blended advantageous variations away into nothing.[6]

For anyone impressed with Jenkin's swamping argument, the theory of natural selection appeared to be a bust, due to Darwin's failure to bolster it with a theory of inheritance that explained (as the Mendelian theory would eventually do) why advantageous variations will not be blended into oblivion. But for anyone who had read – let alone written – the *Origin of Species*, there was an obvious objection. As an empirically well-attested matter of fact, under artificial selection, advantageous variations *do* accumulate, with the result that wild progenitor species become modified. On the farm, in the garden, and in the aviary, human breeders consistently avoid the potential for swamping identified by Jenkin by iteratively mating the best males not with the average females, but with the best ones. For Darwin, in nature, a comparable situation obtains, since the struggle for existence – the natural counterpart to the human breeder – ruthlessly culls all potential mates except for those that vary most fully in the directions favored under the prevailing conditions of life. In nature as under domestication, then, the scenario under which Jenkin-style swamping is a threat rarely if ever actually arises. Accordingly, Darwin regarded Jenkin's critique not as landing a devastating blow to the theory of natural selection but, when it came to variation and inheritance, as providing an occasion to state more clearly what Darwin already believed. Previously Darwin had doubted that what he called "sports" – individuals that, by the accidents of birth, differ dramatically from the rest of their generation – could be anything like as evolutionarily consequential as individuals that were undramatically superior thanks to inborn variations that made them incrementally faster, taller, smarter, and so on. Now, thanks to Jenkin's mathematical underscoring of the point, Darwin saw that even the minimal role he had grudgingly assigned to sports in adaptive evolution was unnecessary. In the next, fifth edition of the *Origin of*

---

[6] H. C. F. Jenkin, "The *Origin of Species*," *North British Review* 46 (1867): 277–318, reprinted in David L. Hull, *Darwin and His Critics* (Cambridge, MA: Harvard University Press, 1973), pp. 302–44.

Species, he wrote sports out of the picture entirely, acknowledging his debt to Jenkin for the intellectual favor.[7]

At the time, Jenkin's critique was regarded as but one of many challenges raised against the theory of natural selection.[8] Notwithstanding such challenges, neither the theory nor the idea of the branching tree of life – which Darwin represented in the *Origin of Species* as a consequence of natural selection – went into "eclipse." When Bateson expressed his wish that Darwin counterfactually had read Mendel, it was precisely because he was exasperated by the prevalence of Darwinian theorizing around him in the early twentieth century. A notable irritation for Bateson was the Oxford biologist Walter Frank Raphael Weldon (Figure 18.3), whose 1902 critique of Mendel had provoked Bateson into writing his book-length defense of Mendel's work. By then, Weldon's enthusiasm for Darwinian natural selection had extended to pioneering empirical studies that sought, via statistical analysis and

Figure 18.3 Walter Frank Raphael Weldon. Courtesy of Special Collections & Galleries University of Leeds Libraries

[7] For further discussion of Darwin's response to Jenkin, see, e.g., Susan W. Morris, "Fleeming Jenkin and *The Origin of Species*: A Reassessment," *British Journal for the History of Science* 27 (1994): 313–43; Tim Lewens, "Natural Selection Then and Now," *Biological Reviews* 85 (2010): 829–35.

[8] The best survey of the challenges to the theory of natural selection remains Peter J. Bowler, *The Eclipse of Darwinism: Anti-Darwinian Evolution Theories in the Decades around 1900* (Baltimore: Johns Hopkins, 1983/1992).

other means, to catch selection in action in the changing dimensions of crab shells and snail shells. At his death in 1906, Weldon left unfinished a manuscript setting out an alternative to Mendelian theory stressing the extent to which bits of chromosome have variable effects on bodies depending on internal and external contexts. This went against the emphasis in Mendelian theory on dominance as a property that certain character versions have or do not have, categorically (in pea seeds, for example, yellowness and roundness have it and greenness and wrinkled-ness do not). Even after Weldon's death, Oxford remained a place where natural selection theory, often wedded to Weldonian emphases, thrived.[9]

So how did Jenkin's review nevertheless come to be remembered as stopping Darwinism in its tracks for decades? The answer lies with a book by the greatest theorist of natural selection in the generation after Weldon's, the English mathematician Ronald Fisher. Fisher's *The Genetical Theory of Natural Selection*, published in 1930, was, along with contemporary work by the English geneticist J. B. S. Haldane and the American geneticist Sewall Wright, what laid the foundations for the Mendelized natural selection of the Modern Synthesis. The book begins with a chapter entitled "The Nature of Inheritance" in which Fisher identified Darwin's commitment to blending inheritance as a weakness built into the theory of natural selection when, with seemingly no other option available, Darwin signed up to the general consensus in his era about how inheritance worked. Fully aware of the problem he thus created for himself in explaining how advantageous variations might be preserved and so accumulated by natural selection, Darwin was forced, on Fisher's reconstruction, into the very unsatisfactory position of supposing that advantageous variations are not so much preserved as constantly generated anew thanks to environmental changes inducing high levels of variability. As Fisher wrote to Darwin's son Leonard in 1932, "I do not believe that your father would ever have ascribed the great variability of domesticated races to the effect of their environment on their mutation rates, had he not thought that variations were continually dissipated by blending." In the *Genetical Theory*, Fisher even speculated that had Darwin or anyone else only thought harder about the possibility that inheritance might be nonblending or "particulate," they could have arrived at Mendel's correct theory of inheritance with no need of the experiments that Mendel did – and so, at a stroke, could have resolved the difficulty that blending inheritance posed for the theory of natural selection.[10]

---

[9] On Weldon and his work and legacies, see Radick, *Disputed Inheritance*. On the Oxford tradition in theoretical and empirical studies of adaptive evolution by natural selection, from the era of Weldon and E. B. Poulton to that of E. B. Ford and beyond, see Michael Ruse, *Monad To Man: The Concept of Progress in Evolutionary Biology* (Cambridge, MA: Harvard University Press, 1996).

[10] Ronald Fisher, *The Genetical Theory of Natural Selection* (Oxford: Clarendon Press, 1930), ch. 1; letter from Fisher to L. Darwin, October 14, 1932, in *Natural Selection, Heredity, and Eugenics*, ed. J. H. Bennett (Oxford: Clarendon Press, 1983), pp. 154–5, quotation on 155.

By the early 1950s, the Fisherian notion that Jenkin had exposed a disastrous flaw in the pre-Mendelian theory of natural selection was becoming a commonplace.[11] And so was born the familiar version of our second myth.

18.3   The "Decades Wasted" Myth

Although, as mentioned above, the "Darwin needed Mendel" myth seems to imply the "decades wasted" myth, only commentators at a much greater distance from Darwin's writings than the likes of Fisher have judged the latter plausible. In his 1930 book, Fisher never raised the question of whether Darwin might have jettisoned his ideas on inheritance had he only read Mendel's paper. In a 1960 essay, the English biologist Julian Huxley did consider it, but only because, over the intervening decades, others had done so. Huxley concluded that almost certainly Darwin would have been unmoved, in part because the form of Mendelism that proved amenable to synthesis with Darwinism required a long time to develop, and in part because Darwin would have regarded Mendel's paper as dealing not with inheritance in general but with a special case. It is worth quoting Huxley at length, beginning with his ringing endorsement of the "Darwin needed Mendel" myth. Note too Huxley's identifying the inheritance of acquired characters as another position that Darwin was forced into supporting to overcome Jenkin's analysis:

Fleeming Jenkin pointed out in 1867 that, on the current theory of blending inheritance, even favourable new variations would tend to be swamped out of effective existence by crossing, if heritable variation in general was rare and infrequent. It was to provide for sources of more abundant variation that Darwin came to ascribe increasing importance to the evolutionary role of "acquired characters." Only when the actual genetic mechanism had been discovered and its particulate (non-blending) nature had been established, could it be shown – notably by R. A. Fisher – that Lamarckian ... theories of evolution were not only unnecessary but inherently incorrect and impossible ... . It has been suggested that Darwin would have avoided falling into these pitfalls if only he had paid attention to Mendel's work, which was published in 1865 [*sic.*], in plenty of time for Darwin to amend his views in later editions of the *Origin*. I do not think this is so. It needed nearly twenty years of intensive research on suitable material such as [the fruit fly] *Drosophila* before the findings of genetics could be fruitfully integrated with evolutionary theory. Before that, most geneticists, obsessed by the obvious mutations with large effects which they naturally first studied, were led to anti-selectionist views and to the idea that evolution would normally take place by discontinuous steps [recall Bateson] ... I suspect that if [Darwin] had known of Mendel's results he would have regarded them as interesting but exceptional and relatively unimportant for evolution, as he had already done for other cases of large mutations and sharp segregation. A premature attempt at generalizing Mendelian principles would merely have weakened the central Darwinian principle of gradual slow change.[12]

---

[11] Morris, "Fleeming Jenkin," p. 317.
[12] Julian Huxley, "The Emergence of Darwinism," in *Essays of a Humanist* (London: Pelican, 1969), pp. 13–38, quotation on 30; first published in 1960.

During the last half century, the scholarly consensus has sided with Huxley.[13] It is nevertheless valuable to look in a little more detail at quite why reading Mendel's paper would probably not have prompted any big changes of mind in Darwin. Here are three considerations that stand out to me.

   1. Darwin had done experiments like Mendel's, and even, sometimes, got results like Mendel's – but never regarded them as clues to some larger new truth about the nature of inheritance. At its most elementary, Mendel's method was to cross two pure-bred varieties, examine the character of the hybrid offspring, then examine the character of the offspring of the offspring. That method led him to his discovery that, for example, when a yellow-seeded variety of garden pea was crossed with a green-seeded variety, all the offspring seeds were yellow, but that, in the next generation, green seeds come back in the famous ratio of three yellow seeds to one green seed. (For Mendel, "dominant" just meant visible in the hybrid generation and "recessive" invisible in that generation.)[14] But in the first chapter of the *Origin of Species*, Darwin reported similar experiments with pigeons – and the results did not much resemble Mendel's. Crossing "some uniformly white fantails" and "some uniformly black barbs" had produced birds that were neither all-white nor all-black but "mottled brown and black." Among the offspring of those crossbreds was a bird with the blue color and black-and-white markings of a wild rock pigeon – something Darwin interpreted as a "reversion" to the wild ancestral form from which, he reckoned, all domesticated pigeons derive. In the early 1860s, he undertook a crossing experiment using two varieties of snapdragon, one with normal flowers (the "common" form), the other with abnormally shaped or "peloric"-form flowers. As described in his 1868 book *The Variation of Animals and Plants Under Domestication*, all of the offspring looked like the common-form parent; and when Darwin allowed them to self-fertilize, the 127 seedlings produced grew into 88 common-form snapdragons and 37 peloric snapdragons. So close to the 3-to-1 Mendelian ratio of dominant to recessive! Yet for Darwin, the pattern was but an instance of "prepotency": when, in the offspring of a cross, one parent's character is visible and the other's is not, though the causal ingredients for the latter can nevertheless be transmitted. The snapdragon pattern was just one of the many possible patterns of inheritance, of no special importance except for illustrating the general truth that manifesting a character and transmitting its causal ingredients are separate things. We should note too that, in an un-Mendelian way, Darwin additionally reported

---

[13] For a recent consensus-affirming paper see Pablo Lorenzano, "What Would Have Happened If Darwin Had Known Mendel (or Mendel's Work)?" *History and Philosophy of the Life Sciences* 33 (2011): 3–48.

[14] Gregor Mendel, "Experiments on Plant Hybrids" (1866), English translation with commentary by Staffan Müller-Wille and Kersten Hall. In *Versuche über Pflanzen-Hybriden*, eds. S. Müller-Wille, K. Hall, and O. Dostál (Brno: Masaryk University Press, 2020), pp. 35, 108.

that among the grandchildren snapdragons were two flowers "in an intermediate condition between the peloric and the normal state."[15]

2. *Darwin's "provisional hypothesis of pangenesis," which he used to explain inheritance patterns and much else, was not a late arrival in his theorizing, hastily put together in desperation at Jenkin's review; furthermore, in the year Mendel published, Darwin found new evidence for it – and from someone whose own work on crossbred peas, also published in that year, in no way supported Mendel's conclusions.* The penultimate chapter of the *Variation* set out what Darwin called his "provisional hypothesis of pangenesis." On this hypothesis, all parts of an adult body constantly shed microscopic buds called "gemmules" which, in sexual organisms, collect in the sperm or pollen and eggs and then, after reproduction, cause the parts that they came from to develop in the offspring (or else to remain latent).[16] It is often remembered as Darwin's worst theory, with the embarrassment for Darwin's admirers made the more acute by the thought that it could have been avoided, had Darwin only followed up the references to Mendel in books on the shelf at Down House ...[17] But part of the attraction of pangenesis for Darwin was that, as he saw it, the many and varied patterns of inheritance – including, yes, the inheritance of the effects of the use or disuse of limbs and organs, or so-called "Lamarckian" inheritance, after the French naturalist Jean-Baptiste de Lamarck (see Burkhardt, Chapter 8, this volume) – were thereby explained. Better still, they were explained by the same ideas that also explained a remarkably wide range of other patterns to do with living tissue, from healing and regeneration to the curiously independent lives that parts of the body sometimes seemed to lead. Darwin found it hard to believe that an idea that brought explanatory order to so much diverse evidence could be wrong. Indeed, for all that he had been nurturing pangenesis for decades, new evidence still came in. In 1866, Darwin was delighted to read that a breeder of peas, Thomas Laxton, had found that when he transferred pollen to a female pea plant, the paternal influence was visible not just in the offspring but on the maternal plant: yet another of the patterns that Darwin held pangenesis to explain. And that year, Laxton published a paper on his observations about seed color and seed shape in experimental crosses he had done with his garden peas. Unlike Mendel, Laxton found that pretty much anything could happen – a conclusion that Darwin would have been prepared to accept partly from his own impressions

---

[15] Darwin, *Origin*, p. 25; Charles Darwin, *The Variation of Animals and Plants Under Domestication*, 2 vols (London: John Murray, 1868), vol. 2, pp. 70–1.

[16] Darwin, *Variation*, vol. 2, ch. 27.

[17] See, e.g., Mario Livio, *Brilliant Blunders: From Darwin to Einstein – Colossal Mistakes by Great Scientists That Changed Our Understanding of Life and the Universe* (London: Simon and Schuster, 2013), esp. chs. 2 and 3.

of how unruly inheritance could be, and partly from his positive regard for Laxton's abilities. (He was one of the most successful breeders of the Victorian era.)[18]

3. *On reading Mendel's paper, Darwin would have found himself the unnamed target of Mendel's criticisms of the belief – dear to Darwin – that under domestication, plants and animals become far more variable than they are in a natural state: a line of argument not calculated to make Darwin embrace Mendel.* It has long been known that, although Darwin did not read Mendel, Mendel read Darwin. A close analysis by the biologist-historian Daniel Fairbanks comparing Mendel's annotations in his copy of a German translation of the *Origin* with Mendel's 1866 paper shows that Mendel's language becomes most strikingly Darwinian when, in the conclusion of the paper, he takes up the question of whether cultivated plants should be thought of as so variable as to be beyond the scope of natural law. Darwin of course, in stating that organisms became more variable when conditions change, and that domestication, involving the imposition of maximally changed conditions, brought on maximal variability, never meant thereby to suggest that cultivated plants were lawless. But that seems to be how Mendel understood him. And since Mendel's entire project in the paper concerns natural law governing the fate of hybrid characters in a certain class of cultivated plants, Mendel responded as if that project's possibility was under threat. The result is the one part of the paper where Mendel is almost sarcastic. It is hard to imagine Darwin reading it and feeling overjoyed.[19]

18.4   Conclusions

At the John Innes Centre in Norwich, England, there is a copy of Bateson's *Defence* with annotations from the Austrian breeder Armin von Tschermak, older brother of one of the 1900 "re-discoverers" of Mendel's work, Erich von Tschermak. Next to Bateson's speculation about the history-altering consequences of Darwin reading Mendel, Armin scribbled: "Ich glaube nicht" – "I don't think so."[20] As we have seen, subsequent historical scholarship bears out Armin's skepticism. But so what? Does it really matter that old myths about Darwin, Mendel, and the former needing the latter live on? How does the continued circulation of these myths leave us worse off? I want to suggest in closing that it does matter – that their anachronism is not merely false but impoverishing, and in two directions.

---

[18] For Darwin and Laxton on pangenesis and peas, see Radick, *Disputed Inheritance*, ch. 1.

[19] Daniel J. Fairbanks, "Mendel and Darwin: Untangling a Persistent Enigma," *Heredity* 124 (2020): 263–73.

[20] Thanks to Kersten Hall for this information.

When the myths inflect our thinking about Darwin, they encourage us to be incurious about his perspective on his theorizing, not least his much-derided pangenesis hypothesis. That incuriosity in turn deprives us not just of deeper understanding of a thinker that so many of us (including readers of this volume) profess to admire, but of the pleasure that can come from inhabiting an alien point of view and, at least temporarily, finding oneself at home in it. Turn yourself into a half-decent applier and defender of pangenesis, and your relationship with it, and with Darwin, will be forever different.

When, in the other direction, the myths inflect our thinking about Mendel, we potentially lose out in even more significant ways. Nowhere in our culture is the mythic treatment of Mendel as the be-all and end-all on inheritance more pronounced than in education. As many commentators have noted, in the standard genetics curriculum, elementary Mendelian examples typically have a prominence that, from the standpoint of twenty-first-century biology, looks downright misleading. Some years ago, I led a project to teach introductory genetics in a more "Weldonian" way, frontloading multifactorial causation and the variability it brings about. What my colleagues and I found was that, where students taking a traditional Mendelian course were on average as determinist about genes at the end of teaching as they were from the start, students on our Weldonian course were on average less determinist.[21] Helping present-day students understand inheritance in an up-to-date way may thus depend in part on liberating ourselves far more completely from the grip of a historical myth about what would have happened had Darwin read Mendel.

---

[21] Annie Jamieson and Gregory Radick, "Genetic Determinism in the Genetics Curriculum: An Exploratory Study of the Effects of Mendelian and Weldonian Emphases," *Science & Education* 26 (2017): 1261–90.

Myth 19   That Darwin Faced a Conspiracy of Silence in Lamarck's Country

Liv Grjebine

19.1   Introduction

Charles Darwin was concerned about the reception of his theory in France. In a 1863 letter to Camille Dareste, he wrote: "Several naturalists in England, North America and Germany have declared that their opinions on the subject have in some degree been modified, but as far as I know my book has produced no effect whatever in France and this makes me the more gratified by your very kind expression of approbation."[1] Thomas Huxley knew the cause of this:

In France, the influence of Elie de Beaumont and of Flourens – the former of whom is said to have "damned himself to everlasting fame" by inventing the nickname of "la science moussante" for Evolutionism . . . to say nothing of the ill-will of other powerful members of the Institut, produced for a long time the effect of a conspiracy of silence.[2]

Why should we care about how Darwin was received in France? We should care because it's a fascinating case study on how science can sometimes take an unexpected path as it circulates in a society. The myth of "a conspiracy of silence" was born in the 1860s from what appeared to be a paradox. Although France was one of the first countries to endorse transformist theories, it was one of the last in Europe to recognize Darwin's theory. The myth spread in the 1870s, when Darwin experienced his first rejection at the Academy of Sciences in Paris. According to this myth, not only did the French remain silent regarding Darwin's thesis, but there was also a plot intended to nip in the bud any attempt to discuss the theory of evolution. For some commentators, this silence was also the result of the nationalism of the French, who could not stand that a British tried to recycle Lamarck's ideas. Darwin's close friend Thomas

---

[1] Darwin to Camille Dareste, February 16, 1863, Darwin Correspondence Project, "Letter no. 3992," (accessed September 30, 2022), www.darwinproject.ac.uk/letter/?docId=letters/DCP-LETT-3992.xml.

[2] C. Darwin, 1887. *The Life and Letters of Charles Darwin, Including an Autobiographical Chapter.* Edited by Francis Darwin. 3 vols. London: Murray.

Huxley popularized the expression in 1884, when he wrote about the French reception of the *Origin of Species* that there was a sort of "conspiracy of silence" fomented by influential members of the Academy.

French scientists were among the first to point out the attitude of their peers toward Darwin. In the 1860s, one of Darwin's correspondents, the geologist and paleontologist Albert Gaudry denounced a new conspiracy of silence that erupted against the British scientist.[3] In 1867, the eminent zoologist Henri Milne-Edwards regretted how, on the subject of the transformist theories and in particular of the Darwinian theory, "the too timid, or perhaps too proud reserve of the few masters prevented them from not even wanting to examine the hazardous ideas."[4] In 1870, the transformist physician and physiologist Eugène Dally complained about the indifference of French scientists for a doctrine which was successfully touring the world.[5] Darwin's foreign acquaintances also expressed surprise at the behavior of French scientists. In August 1871, as he was traveling to Paris, the Russian Paleontologist Vladimir Kovalevsky warned Darwin: "the French are dead against You and I must really mitigate my *Darwinisme* not to irritate them. Still you have here a great friend of Your views, this is Mr Gaudry but even he dares not to do it openly as 'Darwinism is extremely unpopular in the botanic garden'."[6]

Pietro Corsi has shown how transformism had been a subject of debate for decades in Europe before the publication of Darwin's theory (see Corsi, Chapter 2, this volume). This is why Darwin expected a lively debate when the *Origin of Species* was published. In 1870, he still lamented the lack of interest in his theory on the part of French scholars: "It is curious how nationality influences opinion: a week hardly passes without my hearing of some naturalist in Germany who supports my views, and often puts an exaggerated value on my works; whilst in France I have not heard of a single zoologist except M. Gaudry (and he only partially) who supports my views."[7]

But Darwin did not seem to be well informed about the evolution of power relations within the French scholarly community. In the 1830s, there was a controversy between two famous scientists: Georges Cuvier, the leader of fixism, and Etienne Geoffroy Saint-Hilaire, a supporter of transformism

---

[3] Cédric Grimoult, *Evolutionnisme et fixisme en France: Histoire d'un combat, 1800–1882* (Paris: CNRS Editions, 1998), p. 124.

[4] Henri Milne-Edwards, *Rapport sur les progrès récents des sciences zoologiques en France*, 1 vol. grand in-8° (Paris: Hachette, 1867), pp. 430–2.

[5] Eugène Dally, "Bulletin scientifique. Société d'anthropologie de Paris, séances des 3 et 17 mars 1870," *Revue des cours scientifiques de la France et de l'étranger*, April 9, 1870.

[6] Kovalevsky to Darwin, August 19, 1871, Darwin Correspondence Project, "Letter no. 7911," accessed on September 30, 2022, www.darwinproject.ac.uk/letter/?docId=letters/DCP-LETT-7911.xml.

[7] Darwin to Quatrefages, 28 May [1870], Darwin Correspondence Project, "Letter no.7204," accessed on September 30, 2022, www.darwinproject.ac.uk/letter/?docId=letters/DCP-LETT-7204.xml.

(a)   (b)

Figure 19.1 (Left) Etienne Geoffroy Saint-Hilaire. Lithograph.
Credit: Wellcome Collection. Attribution 4.0 International (CC BY 4.0); (Right)
Georges-Léopold-Chrétien-Frédéric-Dagobert, Baron Cuvier. Line engraving
by W. H. Lizars, 1840. Credit: Wellcome Collection. Public Domain Mark

(Figure 19.1). Contrary to the transformists who considered that species
evolve, the fixists defended the idea of the immutability of species. This debate
brought two conceptions of the history of life face to face. In fact, this controversy was not really about the transformist question, but many commentators
have interpreted it as a clash between transformism and fixism. Neither of them
appeared as the winner. For want of a better alternative theory, and upon the
deaths of Lamarck in 1829 and Geoffroy Saint-Hilaire in 1844, fixism, though
greatly weakened, retained its status as an established theory. Many scientists
adhered in part to transformism. But the fixist hypothesis remained influential
among the old guard who controlled the most prestigious scientific institutions.
This old guard was challenged by the transformist hypothesis, of which Darwin
was seen as the leading proponent. Thus, while the conspiracy of silence was
often associated with his name, Darwin often represented the transformist
cause as a whole.

Many researchers have endorsed the view of a conspiracy of silence against Darwin.[8] In their opinion, this silence in France originated from two main factors. First, the scientific community did not understand the specificity of Darwin's theory of natural selection. Second, the fixists strategically blocked Darwin's theory by ignoring it. The problem with these arguments, as we shall see, is that they are correct, but only when one considers a handful of academics.

In many aspects, this myth was a continuation of the myth on Lamarck: some supporters of transformism argued that both scientists had to face a conspiracy of silence. But, Lamarck was portrayed as the defeated hero of transformism, fighting alone against a scientific milieu that was committed to the fixist Georges Cuvier, while Darwin was described as the triumphant champion who finally managed to break the silence. For a long time, the history of science was perceived as a linear sequence of progress, with a clear genealogy. After the discovery of a theory, a long line of scientists would follow, each one contributing a stone to the edifice of knowledge. Hence, to establish its legitimacy, the transformist movement needed founding heroes: "This natural filiation of beings was established by Lamarck at the beginning of this century, fought by the official scientists, buried in the conspiracy of silence, and taken up again by Darwin, who victoriously and definitively demonstrated it."[9]

Denunciations of a conspiracy of silence became part of a mainstream rhetoric in the twentieth century, as evidenced by the use of the term by several institutional figures. During the 1909 inauguration of the statue of Lamarck in the *Jardin des Plantes*, Gaston Doumergue, Minister of Public Education, gave a speech recalling Lamarck's isolated old age, and pointed to the conspiracy of silence organized by top scientific institutions. Léon Guignard, an academician, then declared that since the death of Lamarck, silence had won over the scientific community and that it was Darwin who "came to pull transformism out of the oblivion into which it had unjustly fallen for the last fifty years."[10] Once again, Darwin appeared as the savior of the transformist school and other leading scientists who played a decisive role in the circulation of evolutionary theses, such as Ernst Haeckel or Herbert Spencer, were put aside.

In this chapter, I will argue that the story behind the conspiracy of silence was intended to provide an explanation for a situation that goes beyond both Darwin and his followers. A myth seeks precisely to explain situations by

---

[8] Toby Appel, *The Cuvier-Geoffroy Debate: French biology in the decades before Darwin* (Oxford: Oxford University Press, 1987). Robert Stebbins, "France," in Thomas Glick (dir.), *The Comparative Reception of Darwinism* (Chicago: The University of Chicago Press, 1988).

[9] Camille Flammarion, *Le monde avant la creation de l'homme; origines de la terre, origines de la vie, origines de l'humanité* (Paris: C. Marpon et E. Flammarion, 1886), pp. 107–8.

[10] "Le monument de Lamarck," *Journal des débats politiques et littéraires*, June 14, 1909, Retronews.

mixing elements of reality with fictional components. I will start by analyzing the facts that gave some legitimacy to this myth. I will then show what this myth fails to grasp is the circulation of Darwin's theory in France. As for the "silence," it has never really existed either in scientific circles or in the public. In fact, if we consider French society as a whole, the debate on Darwinism was popularized and spread in extremely varied social and geographical circles, which in turn helped to accelerate the debate on Darwinism within the scientific community. Contrary to what the myth suggests, the dynamics of the diffusion of Darwinism in France has thus taken various paths.

19.2   Origins of the Myth

Several factors combined to explain the late acceptance of Darwin's ideas in the French scientific community. From the start, it took Darwin a lot of effort to find both a translator and a publisher in France, which gave the impression that the French scientific elite ignored the British scientist. As the failures followed one another, Darwin turned to circles increasingly outside the French scholarly world. In December, Darwin asked Quatrefages de Bréau, Chair of Anthropology at the Museum of Natural History in Paris (Figure 19.2), to recommend a translator among the eminent naturalists of his acquaintance.[11] None of them answered positively. In 1860, a translation project was considered with a French professor exiled in England, without scientific reputation, Pierre Talandier. The project was aborted for lack of a publisher. Discouraged, Darwin wrote to Quatrefages: "I have failed to find a French publisher for *Origin*. The gentleman who wished to translate my *Origin of Species* has failed in getting a publisher: Bailliere, Masson and Hachette all rejected it with contempt."[12]

A woman, without scientific training and moving in the liberal economists' circles of Switzerland finally became Darwin's translator. The fact that Clémence Royer was in charge of the French edition of the *Origin of Species* contributed to the idea that the French scientific elite was ignoring the value of Darwin's research. Her translation, published in 1862, was also accused of making it more difficult for scientists to publicly support his theory. Indeed, in an inflammatory preface of about fifty pages which aroused intense polemics, she developed a pamphlet against religious obscurantism.[13] She traced the

---

[11] Darwin to Quatrefages, 5 December [1859], Darwin Correspondence Project, "Letter no.2571," accessed on September 30, 2022, www.darwinproject.ac.uk/letter/?docId=letters/DCP-LETT-2571.xml.

[12] Darwin to Quatrefages, 30 March [1860], Darwin Correspondence Project, "Letter no.2736," accessed on September 30, 2022, www.darwinproject.ac.uk/letter/?docId=letters/DCP-LETT-2736.xml.

[13] Clémence Royer, "Préface," in Charles Darwin, *De l'Origine des espèces, ou Des lois du progrès chez les êtres organisés* (Paris: Guillaumin et Cie, Victor Masson et Fils, 1862).

JEAN-LOUIS-ARMAND
DE QUATREFAGES DE BRÉAU,
MEMBRE DE L'ACADÉMIE DES SCIENCES.

Figure 19.2 Jean-Louis-Armand de Quatrefages de Bréau. Wood
engraving. Credit: Wellcome Collection. Public Domain Mark

evolution of humanity under the effect of "the vital competition," by distinguishing between those who were selected to survive and the others. For some
historians, Royer's translation was also aimed at adapting Darwin's ideas to the
French context, making them more acceptable.[14] But for Darwin himself and
for many of his contemporaries, it was through a distorted translation that his
theory spread in France. Disagreements with Royer over the translation convinced Darwin to look for alternatives, and in 1873 a new translation by Jean-
Jacques Moulinié appeared.

After this publication, Darwin faced an even greater challenge to be recognized by the Academy of Sciences. When his first candidacy was presented
before the Academy in 1870, several academicians spoke harshly against his
theory. In the following decades, their remarks were repeated by his opponents:
"that's a bunch of baloney" said the prominent geologist Elie de Beaumont; "a

[14] Joy Harvey, "Darwin in a French Dress: Translating, Publishing and Supporting Darwin in
Nineteenth-Century France," in Eve-Marie Engels, Thomas Glick (eds.), *The Reception of
Darwin in Europe*, 2 vols., Vol 2: *The Twentieth Century* (London: Continuum International,
2008), pp. 354–74.

fairytale" for the well-known botanist Adolphe Brongniart.[15] The academician Emile Blanchard, Chair of Zoology at the Museum, did not consider Darwin a real scientist and claimed that "It would be a shame for Science, if the Academy were to open its door to Darwin."[16] From 1870 to 1878, Darwin failed on six occasions to be elected to the Academy.

Insults and pressures against those interested in the theory of evolution were somewhat normalized by the attacks of academicians against Darwin. The scientific periodical, *La Revue des cours scientifiques*, explained that this rejection was meant to "definitively ostracize Darwin because of the content of his theory."[17] These criticisms weighed heavily on the scientific debate since the political context was hostile to materialism. Indeed, in the 1860s and 1870s, French naturalists were under institutional pressure not to discuss Darwinism and to remain silent. The political climate was tense, and scientists were closely monitored. The fall of the Empire in 1870 and the advent of the Republic did not allow a liberation of the debate on evolutionism in scientific circles. After the Empire's fall, a coalition of Conservatives governed France from 1873 to 1877, and their goal was to reinstitute the monarchy.

Thus, to study Darwin meant facing criticism from colleagues and the public, risking personal reputation as well as prospects of advancement. Huxley insisted on this point:

> The "effect" of the known repugnance to Mr. Darwin's views of some of the most prominent members of the Institute, to which I refer, is the effect upon the younger generation of French naturalists. Considering the influence of the Institute upon scientific appointments, the chances of a candidate known to be an evolutionist would have been small indeed; and prudence dictated silence.[18]

Darwin's correspondence with French scholars reveals the tense climate in which they evolved. In June 1877, the French botanist Charles Martins evoked, in a letter to Darwin, the threats weighing on scholars associated with his theory of evolution: "All the intelligent young French naturalists are your disciples, but the official professors who know well the degree of truth in your ideas remain reserved, they dare not approach such questions for fear of being accused of materialism, atheism, communism."[19] These pressures had also

---

[15] "Editorial, Darwin à l'Académie des Sciences, Paris, 5 août 1870," *Revue des cours scientifiques de la France et de l'étranger*, August 6, 1870.

[16] Victor Meunier, "La Science et les Savants," *Le Rappel*, April 2, 1872, BnF.

[17] "Editorial, Darwin à l'Académie des Sciences, Paris, 5 août 1870," *Revue des cours scientifiques de la France et de l'étranger*, August 6, 1870, BnF.

[18] Thomas Huxley, "The Duke of Argyll's Charges Against Men of Science," *Nature*, February 9, 1888, p. 342, accessed on September 30, 2022, www.nature.com/articles/037342a0.pdf?origin=ppub.

[19] Charles Martins to Charles Darwin, June 7, 1877, Darwin Correspondence Project, "Letter no.10990," accessed on September 30, 2022, www.darwinproject.ac.uk/letter/?docId=letters/DCP-LETT-10990.xml.

direct consequences on publications. In October 1877, the publisher Reinwald informed Darwin of the deteriorating political climate.[20] The publication of his latest work *Form of Flowers* would have to wait. In November 1877, the "Moral Government" fell and in 1878, the Republic was securely established. The book appeared a year later, in October 1878.

The conservative and catholic press largely contributed to this climate of fear among scholars. In the 1860s and 1870s, a desire to stifle debate with insults spread through the general press like a contagion. Many savants were the targets of this press for their alleged interest in Darwin's theory; among the most famous are Paul Broca, Alphonse de Candolle, Albert Gaudry, and Emile Littré. It was not even necessary to adhere to the Darwinian theory to be criticized by this watchful press. Scientists who were tempted to support Darwin's theory were warned: they would be singled out and mocked in public. In fact, leading catholic journals harshly rejected all forms of evolution. This was particularly true of *L'Univers*, the most influential Catholic newspaper of the nineteenth century. Officially backed by the Pope, it distinguished itself by its vigorous attacks against Darwin and those suspected of being supporters. The repetitive and violent character of these comments contributed to a climate of fear among scholars willing to discuss evolutionary theses.

Darwin published his theory in context with a rejection of modernity by Pope Pius IX. In 1864, the *Quanta cura* encyclical and its appendix, the *Syllabus Errorum* marked this crucial turning point in the history of the contemporary Catholic Church. The *Syllabus* grouped eighty propositions condemned by the Church, including materialism, scientism, and positivism. For the Pope, the priority was to defend the Church in the face of secularization which was weakening its authority. The German translation in 1860 of the *Origin of Species* was immediately condemned by the episcopate at the Council of Cologne: "the doctrine that the human body comes from animal species is in formal contradiction with Scripture and irreconcilable with the Catholic faith."[21] But, in France, the Church had other preoccupations, in particular the debates on the prehistory of man and on spontaneous generation. As a result, while the lower clergy multiplied their accusations and encouraged their followers to scorn the proponents of the theory of evolution, the top ecclesiastical authorities typically temporized on the need for a condemnation.

But their position changed when Darwin published *The Descent of Man* in 1871. Thus, in the 1870s, Pius IX privately expressed the Church's condemnation of Darwinism. In 1877, he congratulated the French polemicist and devout

---

[20] Charles Reinwald to Darwin, October 13, 1877, Darwin Correspondence Project, "Letter no. 11183," accessed on September 30, 2022, www.darwinproject.ac.uk/letter/?docId=letters/DCP-LETT-11183.xml.

[21] Georges Minois, *L'Eglise et la science: Histoire d'un malentendu*, Vol. 2 (Paris: Fayard, 1991), p. 226.

Catholic Constantin James, author of a demonization of the theory of evolution, *Moïse et Darwin. L'homme de la Genèse comparé à l'homme-singe ou l'enseignement religieux opposé à l'enseignement athée*: "We have received with pleasure, my dear Son, the work in which you so well refute the aberrations of Darwinism."[22] In the late 1870s, as the balance of political power swung in reverse between the monarchists and the republicans, to the benefit of the latter, the French clergy, who felt threatened by the secularization of society, multiplied their actions against the theory of evolution. Hence, only a minority of Catholic thinkers took a position in favor of a conciliation between Darwinian theses and Christian dogma. The 1890s were marked by an upsurge in censorship and forced retractions against the theory of evolution emanating from the Vatican, under the pontificate of Leo XIII.

The study of the multiple fronts that stood in the way of Darwinism in France seems to support the idea of a conspiracy of silence. But, when one considers French society as a whole, the debate on Darwinism was very much alive.

19.3   Darwin's Influence in France

After Darwin's death, the botanist Alphonse de Candolle published a book on the causes of the British scientist's success and underlined that: "The conspiracy of silence, which sometimes succeeded, was not possible. The most recalcitrant were obliged to listen, to discuss, and those who at first granted Darwin a small part of the truth soon gave him half or more."[23]

The myth of a conspiracy of silence has perpetuated a false image of the debate on Darwin's theory in France. It has also perpetuated the false idea that only scientists play a role in the circulation of a scientific theory. In fact, it is sometimes easier for the public to be open to a new theory than for scientists who have dedicated their lives to establishing another one.

What made some people believe that there was a conspiracy of silence was the fact that no influential scientist wrote in favor of Darwin's theory. They only wrote negative reviews. Among the leading scholars who contributed to the debate was the fixist Pierre Flourens, the perpetual secretary of the Academy of Sciences, Chair of Comparative Physiology at the Museum. Flourens responded to Darwin's arguments in an attempt to stop its circulation in France. In 1863, he published three critical articles on the *Origin of Species*, then a book in 1864, *Examen du livre de M. Darwin sur L'Origine des espèces*, which met with some success in the following decades. His harsh criticism contributed to the characterization of any debate on Darwin's theory as

---

[22] Ibid.
[23] Alphonse de Candolle, *Darwin considéré au point de vue des causes de son succès et de l'importance de ses travaux* (Genève: H. Georg, 1882), pp. 18–19.

unwelcome in the scientific elite: "But what obscure ideas, what false ideas! What metaphysical jargon thrown badly in the natural history, which falls in the galimatias as soon as it leaves the clear ideas, the right ideas. What pretentious and empty language!"[24] Coming from such a prominent scientist, these remarks had a long-lasting impact in the scientific community. Indeed, the centralization of learning in France made it difficult to discuss or to work on subjects that were deemed irrelevant by the Parisian scientific elite.

Yet, several scholars were working on the theory of evolution in alternative institutional platforms (for instance at the Société d'Anthropologie and at the Société Géologique de France). As the professor of anatomy Mathias Duval summed up in the late 1880s:

In France, if, as Huxley says, the unwillingness of some members of the Institute produced for some time the effect of a conspiracy of silence, and if many years passed before the Academy was sheltered from the reproach that could be made against it for not counting Darwin among its members, at least unofficial science gave an enthusiastic welcome to the new form of the transformist doctrine.[25]

Many transformists were doing research based on Darwin's book in the provinces. The professor of zoology Alfred Giard turned the northern town of Lille into one of the main centers of research and diffusion of the theory of evolution in France. He sought to develop applied research based on Darwinian theory by studying the interactions between marine organisms and their environment. His private laboratory was inaugurated in 1874. By self-financing and relying on private initiatives, he kept control of the orientation of his research and was able to train a young generation of scholars, whose influence would be decisive in French and Belgian scholarly circles at the turn of the century. In the early 1860s, fixists controlled key positions in the most prestigious scientific institutions. But this theory was not as influential in France as they let on.

Not only did several scholars discuss and work with Darwin's theory, but some also publicly defended it – even at the Academy of Sciences. Two leading scholars distinguished themselves in their support of Darwin: Henri Milne-Edwards, Chair of Entomology and then Zoology at the Museum and the powerful Chair of Anthropology at the Museum, Quatrefages. Although he was a prominent fixist, Quatrefages defended Darwin's nomination at the Academy with remarkable tenacity for eight years. Many supporters of fixism resorted to personal attacks against Darwin, but Quatrefages was not one of them. Not only did he seek a debate with Darwin, but he also wished to establish fairness between the participants in the debate. The idea of

---

[24] Pierre Flourens, *Examen du livre de M. Darwin sur l'origine des espèces* (Paris: Garnier Frères, 1864), p. 65.

[25] Mathias Duval, "Le transformiste Lamarck," *Bulletins de la Société d'anthropologie de Paris*, Tome Douzième, Troisième Série, 1889, p. 371.

a conspiracy of silence is largely challenged by the involvement of leading scholars such as Quatrefages and Milne-Edwards. But they represented a minority that failed to convince their peers of the importance of having an open debate on Darwin's theory.

In fact, while the debate was largely thwarted in elite scientific institutions, it was instead growing among the educated public. Public discussions were decisive in spreading Darwin's theory, even before the scientific establishment had embraced him. Indeed, when his theory was debated in the press, some explanations were provided, and it became more widely known. Darwin's theory was seen as an answer to questions that dominated the political and social agenda. Indeed, many contemporaries thought that it provided a stimulating reading grid to understand current issues such as revolutionary violence, the succession of political regimes or the national defeat at the hands of Prussia.

As a result, in the 1870s, Darwinism became a mass cultural phenomenon: the subject of plays, novels, paintings. It was quickly incorporated into popular entertainment.[26] Ape-women and ape-men were particularly successful and attracted the attention of the media, who saw in them the famous Darwinian missing link. In addition, from Victor Hugo to Emile Zola and Jules Verne, many of the great names of French literature discussed Darwin's ideas. The many interpretations of his theory that circulated served to fuel the debates on the future of French identity in the wake of the fall of the Empire. Although not necessarily in favor of Darwin, these authors contributed to a larger debate that included a large variety of geographical and social settings.

In the late 1870s, public curiosity about Darwinism intensified when Darwin became the center of a political clash. Competing factions of the press explicitly linked the debate on Darwinism to advance either conservative or republican political agendas. Many Republican newspapers sided with Darwin. To support Darwin, not only did republican newspapers criticized the lack of scientific arguments of the Conservatives. They also sought to correct the misconceptions that circulated around his thesis. Media at a national, regional, and even local level contributed to this discussion. Several popular scientists protested against the rejections of the Academy of Sciences. Under the pseudonym A. Vernier, the engineer and scientific popularizer Auguste Laugel published many articles in the scientific column of *Le Temps*, one of the most influential newspapers during the Third Republic. He regretted the lack of openness of the official science toward the British scientist.[27]

---

[26] Rae Beth Gordon, *Dances with Darwin, 1875–1910: Vernacular Modernity in France* (Burlington: Ashgate, 2009).

[27] A, Vernier, "Darwin rejeté par l'Académie des sciences," *Le Temps*, August 23, 1870, BnF.

19.4   Conclusions

The study of the reception of a scientific theory should not be limited to scientific circles. Otherwise, the case of the reception of Darwinism in France shows that by adopting this approach, one does not really grasp the channels of its diffusion. In reality, public debate proved to be an essential stage in its diffusion. The introduction of Darwinism in France was not a failure. It simply took unusual paths.

Science must constantly fight against the crystallization of hypotheses – that is, against the regular rebirth of dogmatism within the scientific world. The history of science is often presented as a conflict between the dogmatism of religions and ideologies and the open approach of science. Yet, in fact, there has always been controversy, if not conflict, also within the sciences.

Hence, many scientists have tried to bypass the scientific community by directly addressing a literate audience, even if it means simplifying their way of thinking. In the seventeenth century, Galileo Galilei rallied part of the literate public by writing his works in Italian and not in Latin. But the comprehensible character of his demonstration of the absurdity of the "old physics" outside of scholarly circles earned him the enduring hatred of Aristotelian academics.[28] Similarly, a few decades later, Newton composed the last sections of the *Principia* following a popular method in order to be read by the greatest number. In the nineteenth century, Darwin chose to rely partially on accessible representations and common perceptions to increase his audience.[29] In France, his strategy was particularly effective. Indeed, Darwinism triggered a vast public debate and became a central political issue before being acknowledged by Parisian academic institutions. However, Darwin became more critical of approaches which were intended to reach the public directly to sidestep scholars' reviews. He regretted his own strategy that made his theory so popular, yet so misunderstood.

---

[28] Massimo Bucciantini, Michele Camerota, and Franco Giudice, *Galileo's Telescope: A European Story* (Cambridge MA: Harvard University Press, 2015).

[29] Janet Browne, *Charles Darwin: Power of Place*, 2 vol. (Princeton: Princeton University Press, 2003).

Myth 20   That Hitler Endorsed and Was Influenced by Darwin's Theory

Robert J. Richards

20.1   Introduction

A presumption exists among some scholars that Hitler endorsed Darwinian evolutionary theory, especially natural selection: "If you open *Mein Kampf* and read it, especially if you can read it in German, the correspondence between Darwinian ideas and Nazi ideas just leaps from the page";[1] as well as that Darwin's conception became a driving force in the Führer's ideology and that of the Nazis more generally: "Darwin's theory thus provided justification, not only for racism, but for racial struggle and even genocide."[2] As Richard Weikart, an historian at California State University (Stanislaus), expressed it: "No matter how crooked the road was from Darwin to Hitler, clearly Darwinism and eugenics smoothed the path for Nazi ideology, especially for the Nazi stress on expansion, war, racial struggle, and racial extermination."[3] In a recent book, *Darwinian* Racism, Weikart – a member of The Discovery Institute, an Intelligent Design operation – introduces his claim of a strong connection between Darwin and Hitler with the observation that on April 20, 1999, Eric Harris and Dylan Klebold celebrated Hitler's birthday by killing thirteen fellow students at Columbine High School, outside of Denver. Harris was wearing a shirt with the emblem "Natural Selection," which indicated to Weikart the strong connection between Hitler and Darwin.[4] Even the astute historian Peter Bowler has suggested that Hitler's malignity was due to Darwin: "By making death a creative force in nature ... Darwin may indeed have

---

This chapter is based on my book, *Was Hitler a Darwinian? Disputed Questions in the History of Evolutionary Theory* (Chicago: University of Chicago Press, 2013).

[1] In "Expelled: No Intelligence Allowed" (Rocky Mountain Pictures, 2008), a documentary film written by Kevin Miller and Ben Stein and directed by Nathan Frankowski. The line by Berlinski comes sixty-four minutes into the film.

[2] Richard Weikart, *Darwinian Racism: How Darwinism Influenced Hitler, Nazism, and White Nationalism* (Seattle: Discovery Institute Press, 2022).

[3] Richard Weikart, *From Darwin to Hitler: Evolutionary Ethics, Eugenics, and Racism in Germany* (New York: Palgrave Macmillan, 2004), 6.

[4] Richard Weikart, *Darwinian Racism: How Darwinism Influenced Hitler, Nazism, and White Nationalism* (Seattle: Discovery Institute Press, 2022), 11.

Figure 20.1 Ernst Haeckel, Photogravure after F. Haack. Credit: Wellcome Collection. Public Domain Mark

unwittingly helped to unleash the whirlwind of hatred that is so often associated with his [Hitler's] name."[5]

Some scholars have focused their obloquy on Darwin's disciple Ernst Haeckel (Figure 20.1). Daniel Gasman claimed that Haeckel was "largely responsible for forging the bonds between academic science and racism in Germany in the later decades of the nineteenth century."[6] Stephen Jay Gould repeated Gasman's charge, partly as a measure to protect Darwin from the indictment by claiming it was not Darwin but Haeckel who delivered evolutionary theory to Hitler's Eagle's Nest. I don't believe this effort to divert the claim of responsibility from Darwin to Haeckel will work, because on the important issues regarding evolutionary theory there is no substantial difference separating the two evolutionary thinkers; they are in essential agreement on matters of: descent of species, struggle for existence, natural selection, inheritance of acquired characters, recapitulation theory, progressivism, and

[5] Peter Bowler, "What Darwin Disturbed: The Biology That Might Have Been," *Isis* 99 (2008): 560–7, quotation at 564–5.

[6] Daniel Gasman, *Haeckel's Monism and the Birth of Fascist Ideology* (New York: Peter Lang, 1998), 26.

hierarchy of races.[7] So if Hitler endorsed Haeckel's version of evolutionary theory, he thereby endorsed Darwin's version as well.

Those scholars of religious disposition (Berlinski and Weikart, for instance) who have attempted to cement a link between Darwin's biological views and Hitler's racial beliefs apparently intend to undermine evolutionary theory and morally indict Darwin and his allies, like Ernst Haeckel. But this effort can be disposed of immediately: even if Hitler celebrated Darwinian theory, this would be irrelevant to the validity of the theory. Hitler supported anti-smoking campaigns among the Nazis, but this has no relevance for judging the validity of such campaigns.[8]

20.2   Neither Darwin Nor Haeckel Were Anti-Semitic

As a point of historical fact, Darwin rarely mentioned Jews. He does remark that Jews and Aryans shared similar features, due, he supposed, to "the Aryan branches having largely crossed during their wide diffusion by various indigenous tribes."[9] Darwin's remark contrasts with Hitler's contention that Jews and Aryans were pure (i.e., unmixed) races.

Haeckel spoke directly to the question of anti-Semitism. He was interviewed, along with some forty other intellectuals and artists, on the phenomenon by Hermann Bahr, a journalist and playwright. Haeckel indicated to Bahr that some of his students were anti-Semitic but he explicitly disavowed that prejudice himself. He acknowledged that some nations, including Germany, were judicious in barring the immigration of Slavic Jews since they would not adopt the customs of their new countries. But he remained adamant in celebrating the "*gebildeten Juden*" of Germany. He told Bahr: "I hold these refined and noble Jews to be important elements in German culture. One should not forget," he continued, "that they have always stood bravely for enlightenment and freedom against the forces of reaction, inexhaustible opponents, as often as needed, against the obscurantists [*Dunkelmänner*]. And now in the dangers of these perilous times, when Papism again rears up mightily everywhere, we cannot do without their tried and true courage."[10] As this quotation suggests, Haeckel's long-tern opponent was the Catholic Church; he had no animus against Jews.

---

[7] I make this argument in Robert J. Richards, *The Tragic Sense of Life: Ernst Haeckel and the Struggle over Evolutionary Thought* (Chicago: University of Chicago Press, 2008), 135–70.

[8] See Robert Proctor, *The Nazi War on Cancer* (Princeton: Princeton University Press, 1999).

[9] Charles Darwin, *The Descent of Man and Selection in Relation to Sex*, 2 vols. (London: Murray, 1871), 1:240.

[10] Haeckel as quoted in Hermann Bahr, "Ernst Haeckel," in *Der Antisenutusnhs: Ein international Interview* (Berlin: S. Fischer, 1894), 69.

So neither Darwin himself nor Haeckel, the leading German Darwinian, can be accused of anti-Semitism, certainly not the kind of racism that fueled Hitler's animus and stoked the fires of the Holocaust. Both Darwin and Haeckel simply adopted the kind of racial hierarchy that had long been standard in European science (see Peterson, Chapter 22, this volume). Both scientists used evolutionary theory to explain these differences; they did not use it to establish or standardize them. Two thinkers, who did have a measured impact on Hitler's racial views, are distinctive because of the strength of that impact and because they explicitly rejected Darwinian theory: Arthur comte, de Gobineau and Houston Stewart Chamberlain.

20.3   Gobineau and Chamberlain: Sources for Hitler's racism

Gobineau's four-volume *Essai sur l'Inégalité des Races Humaines* (*Essay on the Inequality of the Human Races*, 1853–55) was translated into several languages.[11] It went through five German editions from 1895 to 1940 and served as the intellectual rationale for the anti-Semitic Gobineau societies that spread through Germany at the turn of the century. Chamberlain's *Die Grundlagen des neunzehnten Jahrhunderts* (*The Foundations of the Nineteenth Century*) flooded Germany with an amazing thirty editions from 1899 to 1944. Chamberlain was inspired by Gobineau's analysis of race and became a member of the elite Gobineau society, along with other partisans of the cult of Richard Wagner.[12]

Gobineau argued that modern nations had lost the vitality characterizing ancient civilizations and that the European nations, as well as the United States, faced inevitable decline, with the French Revolution being an unmistakable sign of the end. When Gobineau learned of Darwin's evolutionary theory he disdainfully dismissed it, thinking its anemic progressivism a distortion of his own rigorously grounded empirical study; certainly the time was near, he believed, when Haeckel's phantasms of ape-men would evanesce.[13] He was assured of the decline of human societies and proposed a very simple formula to explain it: race mixing.

Gobineau held that though we might have to give notional assent to the Biblical story of a common origin, the fundamental traits of the white, yellow, and black races were manifestly different and their various branches displayed

---

[11] I have used the second German edition in this analysis: Joseph Arthur Grafen Gobineau, *Versuch über die Ungleichheit der Menschenracen*, trans. Ludwig Schemann, 2nd ed., 4 vols. (Stuttgart: Fr. Frommanns Verlag, 1902–4).

[12] Paul Weindling provides a trenchant account of the Gobineau Society, with its elitist and nonscientific membership. See Paul Weindling, *Health, Race and German Politics between National Unification and Nazism, 1870–1945* (Cambridge: Cambridge University Press, 1989), 106–9.

[13] Gobineau, *Versuch über die Ungleichheit der Menschenracen*, 1:xxxi–xxxiii.

intrinsically diverse endowments. To support this contention, he spun out, over four substantial volumes, a conjectural anthropology whose conclusions, he ceaselessly claimed, had the iron grip of natural law. In these volumes, he asserted, unsurprisingly, that the white race harbored members who were the most beautiful, intelligent, orderly, and physically powerful; they were lovers of liberty and aggressively pursued it. They played the dominant role in any civilization that had attained a significant culture. The yellow race was lazy and uninventive, though given to a narrow kind of utility. The black race was intense, willful, and with a dull intellect; no civilization ever arose out of the pure black race.

Gobineau postulated two contrary forces operative on the races of mankind: revulsion for race mixing, especially powerful among the black groups, and a contrary impulse toward intermarriage, which oddly was characteristic of those peoples capable of great development.[14] As a result of the impulse to mate with conquered peoples, the pure strains of the higher stocks had become alloyed with the other strains, the white race being constantly diluted with the blood of inferior peoples, while the latter enjoyed a boost from white blood. Contemporary societies, according to Gobineau, might have more or less strong remnants of the hereditary traits of their forbearers, but they were increasingly washed over as the streams of humanity ebbed and flowed. The modern European nations thus lost their purity, especially as the white component had been sullied in the byways of congress with the yellow and black races. So even the modern Germans, who yet retained the greatest measure of Aryan blood and still carried the fire of modern culture and science, had begun to decline and would continue to do so as the tributaries of hybrid stocks increasingly muddied the swifter currents of pure blood. The Jews were subject to the same degradation. For a while they enjoyed the advantageous of a homogeneous population and then slipped silently down the racial slope into their current mongrel state.[15]

Chamberlain, born in 1855, got married as a nineteen-year-old to a woman ten years his senior. They attended the premier of *Der Ring des Nibelungen* in Munich, which infused him with a passion for the numinous music and deranged doctrines of Richard Wagner. He enrolled as a member of the Wagner Society, formed after the composer's death in 1882. He helped found a new French journal devoted to the art of the master. Chamberlain was so taken with the music and ideology of Wagner that in the period between 1892 and 1900, he wrote dozens of articles and four books on the man and his music. The close engagement with Wagner's work brought Chamberlain closer to Cosima Wagner, second wife of the maestro, daughter of Franz Liszt and titular head of the inner circle of the cult, which fed on the racial theories of Gobineau, now

---

[14] Ibid., 38.   [15] Ibid., 2:92–3.

growing into Teutonic glorification and pernicious anti-Semitism. His passion for Wagner spilled over to the composer's daughter, Eva, whom Chamberlain married in 1908, following an expense divorce from his first wife.

Chamberlain's Wagnerian enthusiasms were incised in a volume he had been working on since the second marriage, *Die Grundlagen des neunzehnten Jahrhunderts*[16] (later published in English as *The Foundations of the Nineteenth Century*, and referred to hereafter as *Foundations*), which would eventually flood Germany with a farrago of Goethean sentiment, Kantian epistemology, Wagnerian mysticism, and Aryan anti-Semitism. The medley would echo through the German reading public for almost half a century. While Gobineau maintained that the races originally were pure but tended to degenerate over time because of miscegenation, Chamberlain contended that purity of race was achieved over long periods of time; once achieved, however, it could be endangered by race mixing.[17] He was vague about the origins of human beings, simply observing that as far as history testified, human beings have always existed.[18] He dismissed as a "pseudo-scientific fantasy" Haeckel's argument that the human races descended from apelike forbearers.[19] For Chamberlain, the two principal races that achieved purity and retained it were the Aryan and the Jewish. The Aryans, which in their more recent incarnation he referred to as Germans, were the bearers of culture, science, and the arts. Their mental accomplishments flowed from blood, he argued (or really, simply stipulated). In a wonderful piece of quasi-idealistic morphology, he described the real German as having an ideal type: "great, heavenly radiant eyes, golden hair, the body of a giant, harmonious musculature, a long skull [and] . . . high countenance."[20] All of this notwithstanding, individual Germans might be dark-haired, brown-eyed, and small of stature.

Against the blond giant stood the threatening Jew. Chamberlain devoted 135 continuous pages to dissecting the Jewish type, its physiology and character. Throughout the *Foundations*, this Anglo-German would vacillate between referring to the Jews as a pure race, meaning relatively permanent, but also of a "mongrel character [*Bastardcharakter*]."[21] The Jews' very "existence is a sin [*Sünde*]; their existence is a transgression against the holy laws of life."[22] Thus any mating between Jew and Aryan could only corrupt the nobility of the latter: the Jewish character "is much too foreign, firm, and strong to be refreshed and ennobled by German blood."[23] This could only mean a struggle between the Aryans and the Jews, "a struggle of life and death [*ein Kampf auf Leben und Tod*]."[24]

---

[16] Houston Stewart Chamberlain, *Die Grundlagen des neunzehnten Jahrhunderts*, 2 vols. (Munich: Bruckmann, 1899).
[17] Chamberlain, Grundlagen, 1:266–7.   [18] Ibid., 1:277.   [19] Ibid., 1:122n.   [20] Ibid., 1:496.
[21] Ibid., 1:372.   [22] Ibid., 1:374.   [23] Ibid., 1:325.   [24] Ibid., 1:531.

Chamberlain used the trope of racial struggle frequently in the *Foundations*. Indeed, the phrase usually identified with Darwinian theory, "struggle for existence" (*Kampf ums Dasein*), appears eight times in the *Foundations*. The single word "struggle" (*Kampf*) turns up 112 times. But these terms were not markers of Darwin's theory of natural selection. Chamberlain rejected Darwin's conception completely, comparing it to the old, discredited "phlogiston theory."[25]

Hitler first met Chamberlain when the young politician was invited by the Wagner family to Bayreuth in late September 1923. The two spoke almost continuously over a two-day period. Chamberlain was so impressed with the young man that he wrote a quite fulsome letter to Hitler: "My faith in Germanness [*Deutschtum*] has never wavered for a moment. But my hopes – I will confess – had ebbed. With one blow, you have transformed the core of my soul. That Germany in the hour of her greatest need has given birth to a Hitler, that shows her vital essence."[26] It was a letter Hitler never forgot. On the occasion of Hitler's thirty-fifth birthday, celebrated in Landsberg Prison after the failed Munich Putsch, Chamberlain published another letter, an open letter that extolled this man, so different from other politicians, a man who "loves his German people with a burning passion." "In this feeling," he continued, "we have the central point of his politics, his economics, his opposition to the Jews, his battle against the corruption of values, etc."[27] After his release from jail, Hitler visited Chamberlain on several occasions and mourned him at his funeral. In the depths of World War II, Hitler recalled with gratitude vising Bayreuth for the first time and meeting Chamberlain. In his "Table Talk" – conversations stenographically recorded – Hitler mentioned that "Chamberlain's letter came while I was in jail. I was on familiar terms with them [Chamberlain and the Wagner family]; I love these people."[28]

20.4   Hitler's *Mein Kampf*

It was while in jail, comforted as he was by Chamberlain's recognition, that Hitler composed the first volume of *Mein Kampf*. He began the composition of *Mein Kampf* in July 1924, and it quickly became inflated into two large volumes by the next year. He initially wanted to title it "A Four and a Half Year Battle against Lies, Stupidity and Cowardice," but finally shortened the

---

[25] Ibid., 2:805.

[26] Chamberlain to Adolf Hitler (October 7, 1923), in Houston Stewart Chamberlain, *Briefe. 1882–1924 und Briefwechsel mit Kaiser Wilhelm. Erster Band*. Verlag: München, F. Bruckmann A.-G., 1928), 2:124–25.

[27] The letter was originally published in *Deutsche Presse*, nos. 65–66 (April 20–21, 1924), 1; reprinted in *Houston Stewart Chamberlain, Auswahl aus seinen Werken*, ed. Hardy Schmidt (Breslau: Ferdinand Hirt, 1935), 66.

[28] Hitler, *Monologe im Führer-Hauptquartier* (Hamburg: A. Knaus, 1980), 224.

title simply to "My Battle" – *Mein Kampf*. The book brewed up a mélange of autobiographical sketches, a theory of race, a declaration of the need to expand the land of the Germans, principally to the east, and foreign policy exhortations to restore the honor and power of the nation. Flavoring the stew throughout was the bitter vitriol of scorn for those who had destroyed the means to win the last war and connived to push the nation into collapse after the war – the Jews, capitalists, and Bolsheviks.

Neither Darwin's nor Haeckel's name is mentioned in *Mein Kampf*, rather strange if Hitler's racial theory comes from Darwinian conceptions, as claimed by the conservative critics I've cited in the first pages of this essay. Indeed, the only name carrying any scientific weight cited in the book is that of Houston Stewart Chamberlain, an avowed anti-Darwinian. Perhaps, however, evolutionary ideas became the backbone of *Mein Kampf*, without naming names. But nowhere does Hitler even use the terms *Evolutionslehre, Abstammungslehre, Deszendenz-Theorie*, or any word that obviously refers to evolutionary theory. If Hitler's racial views stemmed from Darwinian theory, without perhaps naming it, one would at least expect some term in general use for evolutionary theory to be found in the book – but not so.

Darwin argued for the descent of species over time, the crux of his theory in the *Origin of Species*. We don't find anything comparable to that idea in Hitler's works, which rejected any notion of the descent of human beings from lower animals. In Hitler's "Table Talk," the German leader was recorded as positively dismissing any notion of the descent of human beings from lower animals. In the late evening of January 25–26, he remarked that he had read a book about human origins and that he used to think a lot about the question. He was particularly impressed that the ancient Greeks and Egyptians cultivated ideas of beauty comparable to our own, which could not have been the case were these peoples quite different from us. He asked:

Whence have we the right to believe that man was not from the very beginning [*Uranfängen*] what he is today? A glance at nature informs us that in the realm of plants and animals alterations and further formation occur, but nothing indicates that development [*Entwicklung*] within a species [*Gattung*] has occurred of a considerable leap of the sort that man would have to have made to transform him from an apelike condition to his present state.[29]

Could any statement be more explicit? Hitler simply rejected the cardinal feature of Darwin's theory as applied to human beings. How could Darwin's conception have been responsible for Hitler's racial theory regarding human beings when that conception was in fact completely rejected by the latter? Hitler simply dismissed an essential component of Darwinian theory. Hitler's rejection

---

[29] Hitler, *Monologe im Führer-Hauptquartier*, 232 (January 25–26, 1942). Hitler's German is an inelegant tangle, even granted that "Table Talk" records spontaneous conversations.

of human evolution is quite consistent with the main thrust of *Mein Kampf*. As a point of historical fact, a German (really a Moravian) who contributed to the biological analysis of race – though certainly unwittingly – was Gregor Mendel, whose pedigree analysis of species varieties lent itself to the application to human racial varieties.[30]

Hitler's overriding racial concern in *Mein Kampf* was purity. He maintained that a general drive toward racial homogeneity, toward "racial purity" (*allgemein gültigen Triebes zur Rassenreinheit*), characterized all living organisms. This drive was exemplified by the uniformity and stability of species: "The consequence of this racial purity [*Rassenreinheit*], which is characteristic of all animals in nature, is not only a sharp separation of the particular races externally, but also in their uniformity of the essence of the very type itself. The fox is always a fox, the goose a goose, the tiger a tiger, and so on."[31] Of course, for a Darwinian, there is no "essence of the very type"; the fox was not always a fox, the goose not always a goose, and in future they would not remain fixed in their types. Fixity of type is the very antithesis of a theory that contends species are not static but vary and are transformed into other species over time.

Darwin's principle of diversity, which he regarded as important as natural selection, maintains that there is a general tendency of varieties and species to diversify, that is, to become heterogeneous as opposed to maintaining homogeneity. Weikart's claim that Hitler "believed that humans were subject to immutable evolutionary laws" simply cannot be true.[32] Racial purity became endangered by race mixing, especially the sullying of the higher Aryan type with the lower Jewish. Reflecting the warnings of Gobineau and Chamberlain, Hitler specified the extreme danger of miscegenation for the race of higher culture.[33] Hitler was assured that "all great cultures of the past were destroyed because the original, creative race died off through blood poisoning"[34] – the diagnosis of Gobineau and Chamberlain. This aspect of Hitler's argument needs to be emphasized. The Aryans, Hitler maintained, were the original bearers of culture – another verse of the gospel according to Gobineau and Chamberlain – and they propagated art and science to the rest of the world. The pure blood of the Aryans could not be improved upon, only degraded by race mixing. In a line reflecting Chamberlain's assertion that the Jews' very existence was a "sin," Hitler declared that such racial mixing would be "a sin against the Will of the eternal Creator."[35] Not, it must be noted, a sin against the theory of Charles Darwin. "Regeneration" of the primitive German people and an

---

[30] See Amir Teicher, *Social Mendelism: Genetics and the Politics of Race in Germany, 1900–1948* (Cambridge: Cambridge University Press, 2020).

[31] Hitler, *Mein Kampf*, 312.

[32] Richard Weikart, *Hitler's Ethic: The Nazi Pursuit of Evolutionary Progress* (New York: Palgrave Macmillan, 2009), 3.

[33] Hitler, *Mein Kampf*, 313.   [34] Ibid., 316.   [35] Ibid., 314.

elimination of blood poisoning can occur "so long as a fundamental stock of racially pure elements still exists and bastardization ceases."[36] Hitler thus sought a return to an ideal past, not an evolutionary advance to a transformed future.

20.5   Struggle for Existence

Most authors who try to connect Darwin with Hitler focus on Hitler's idea of "struggle," as if this implied Darwin's principle of "struggle for existence," that is, natural selection. The very title of Hitler's book, *My Battle* (or *Struggle*, *War*) hardly resonates of Darwinian usage – especially when one considers the title he originally planned: "A Four and a Half Year Battle [*Kampf*] against Lies, Stupidity and Cowardice." A simple word count indicates that Hitler had a mania for the notion of struggle that no simple acquaintance with the idea in a scientific work could possibly explain. The term appears in one form or another some 266 times in the first 300 pages of the 800-page book: from the simple *Kampf* (struggle) to *Bekämpfung* (a struggle), *ankämpfen* (to fight), *Kampffeld* (field of struggle), *Kampfeslust* (joy of struggle), and so forth.

Darwin's principle of natural selection was, of course, used to explain the transmutation of species. But if someone like Hitler denies the transmutation and descent of species, then no matter what language he employs, the concept behind the language cannot be that of natural selection. Let me set aside for the moment this crucial objection to Hitler's supposed employment of Darwin's device and examine the role of "struggle" in *Mein Kampf*.

The phrase used in the German translation of the *Origin of Species* for "struggle for existence" is "*Kampf um's Dasein*."[37] Hitler uses that phrase, or one close to it, twice in *Mein Kampf*. Those two instances occur in an almost 800-page book in which some form of the word *Kampf* appears on almost every page; by sheer accident such a phrase might spill from the pen of an obsessed individual who seems to know hardly any other word. His fundamental view is that "mankind becomes great through eternal struggle – in eternal peace men come to nothing."[38] Most of these usages come almost verbatim from Chamberlain, not Darwin.

Hitler's general conception that humanity develops culturally through struggle and that racial mixing causes degeneration – these ideas replicate those of

---

[36] Ibid., 443.

[37] Heinrich Georg Bronn was the first translator into German of Darwin's *Origin of Species: Über die Entstehung der Arten im Thier- und Pflanzen-Reich durch natürliche Züchtung, oder Erhaltung der vervollkommneten Rassen im Kampfe um's Daseyn*, trans. H. Bronn (Stuttgart: E. Schweizerbart'sche Verlagshandlung und Druckerei, 1860).

[38] Ibid., 149.

Chamberlain, who likewise signaled to his reader that "the idea of struggle governs my presentation [in *Foundations*]."[39] Chamberlain accepted Gobineau's contention that miscegenation caused cultural decline, but he insisted that such decline was not inevitable; one could struggle against degeneration and keep the Aryan folk, the bearers of culture, pure. The fight, however, had to be constantly renewed. "The struggle in which the weaker human material is eradicated [*zu Grunde geht*]," Chamberlain argued, "steels the stronger; moreover the struggle for life [*Kampf ums Leben*] strengthens the stronger by eliminating the weaker elements."[40]

Hitler clearly echoed Chamberlain's observation that a peaceful land sows only cultural mediocrity; such a land, according to Chamberlain, "knows nothing of the social questions, of the hard struggle for existence [*vom bittern Kampf ums Dasein*]."[41] Compare this phrase with Hitler's "the hard struggle for existence [*unerbittlichen Kampf ums Dasein*],"[42] Hitler is thus not recycling Darwin but rather aping Chamberlain.

20.6   Conclusion

Countless conservative religious and political tracts have attempted to undermine Darwinian evolutionary theory by arguing that it was endorsed by Hitler and led to the biological ideas responsible for the crimes of the Nazis. These dogmatically driven accounts have been abetted by more reputable scholars who have written books with titles such as *From Darwin to Hitler*. Ernst Haeckel, Darwin's great German disciple, is presumed to have virtually packed his sidecar with Darwinian theory and monistic philosophy and delivered their toxic message directly to Berchtesgaden – or at least, individuals such as Daniel Gasman, and Stephen Jay Gould, so argued. In this chapter, I have maintained that these assumptions simply cannot be sustained after a careful examination of the evidence.

There is no doubt that one can find racists who stretched Darwinian theory to promote white-nationalist views. Weikart, in his new book, digs up a few, obscure though they may be: Kevin Strom, Richard McCulloch, Revilo Oliver, and Mary Pennington[43] writes Weikart. Then there is the example of Harris and Klebold, the Columbine killers, cited by Weikart in the introduction to his *Darwinian Racism*. But I believe the evidence to have shown that Hitler could not be listed among these; he was not an advocate of Darwinian theory, one who turned the theory to support the kind of racism with which Führer is rightly associated. His main support in that regard was Houston Stewart Chamberlain, an avowed anti-evolutionist. The myth of Darwin's complicity in the crimes of Hitler, has, I believe, been thoroughly dispatched.

---

[39] Chamberlain, *Grundlagen*, 2:536.   [40] Ibid., 1:277–8.   [41] Ibid., 1:44.
[42] Hitler, *Mein Kampf*, 149.   [43] Weikart, *Darwinian Racism*, 140–8.

Myth 21 That Sexual Selection Was Darwin's Afterthought to Natural Selection

Kimberly A. Hamlin

21.1   Introduction

One way to explore the function and meaning of myths – scientific or otherwise – is to look closely at the celebrations and rituals during which myths are created and enacted. As Pnia Abir-Am wrote in a special issue of *Osiris* on the importance of commemorative events in science, such practices often function to "produc[e] and dissementat[e] a collective memory" of a heroic figure.[1] Students of Darwin, undoubtedly one of the most heroic figures in modern science, can look, most recently, to the global hoopla that greeted the Darwin bicentennial in 2009. What themes emerged at these global celebrations? Which aspects of Darwin's vast writings were celebrated and which were obscured? Who was invited to give the keynote addresses and whose research was overlooked?

One core theme that emerged from the 2009 bicentennial was the scientific and historical communities' hyper-focus on *Origin of Species* (published in 1859) and its cornerstone theory, natural selection. The University of Chicago, for example, professional home to many of the world's most esteemed Darwin scholars and a leading university press in the history of science, hosted one of the most prestigious 2009 commemorations. Scholars spoke about the portraiture of Darwin, Darwin's finches, religious challenges to evolution, speciation, and modern scientific approaches such as evo-eco and evo-devo. Not one of the more than 25 session titles[2] mentions, or even alludes to, sexual selection or *The Descent of Man and Selection in Relation to Sex*, published in 1871 (hereafter *Descent of Man*).[3] Similarly, the Darwin Online Project compiled a list of 2009 commemorations[4] and hardly any – save for the launch of Adrian

---

[1] P. G. Abir-Am, "Commemorative Practices in Science: Historical Perspectives on the Politics of Collective Memory Commemorative Practices in Science," *Osiris* 14, (1999), p. 13.

[2] http://darwin-chicago.uchicago.edu/participants.html (accessed December 15, 2022).

[3] The full list of sessions and participants can be found here: http://darwin-chicago.uchicago.edu/participants.html (accessed December 15, 2022).

[4] http://darwin-online.org.uk/2009.html (accessed December 15, 2022).

Desmond and James Moore's 2009 book *Darwin's Sacred Cause: How a Hatred of Slavery Shaped Darwin's Views on Human Evolution* – mention sexual selection or *The Descent of Man*. In part, the bicentennial's focus on natural selection is because 2009 also marked the sesquicentennial of the publication of *Origin*, but this coincidence does not fully explain the near total exclusion of sexual selection and *The Descent of Man* from the 2009 events, especially considering how central the theory was to Darwin's overall thinking.

What about the 2021 sesquicentennial of *The Descent of Man*? Perhaps that anniversary witnessed a flurry of international exhibitions, conferences, and publications on sexual selection similar in scope to those that marked the 2009 sesquicentennial of the *Origin of Species*? Not so much. Several magazines and journals published essays about sexual selection theory, and the *British Journal of the History of Science* published a special "themes" issue on *The Descent of Man* at 150, edited by Erika Lorraine Milam and Suman Seth.[5] And at least two conferences held plenary sessions reflecting on the 150th anniversary of *The Descent of Man* at their annual meetings.[6] But this pales in comparison to the scholarly and popular attention bestowed upon natural selection and the *Origin of Species* by the 2009 events, begging the question: How and why has sexual selection been relegated to second-tier status in the world of Darwin studies and beyond?

In part, this is due to the persistent and pervasive myth that sexual selection was Darwin's afterthought to the "real" agent of evolution: natural selection. Darwin did refer to sexual selection as the "second" evolutionary mechanism, but by no means did he think of it as secondary in importance or effectiveness. To the contrary, over time he became only more convinced of the importance of mate choice as a driver of evolutionary change. In a letter read before the Zoological Society in 1882, just hours before his death, Darwin once again affirmed his commitment to sexual selection, writing: "I may perhaps be here permitted to say that, after having carefully weighed, to the best of my ability, the various arguments

---

[5] Erika Lorraine Milam and Suman Seth, eds, *British Journal for the History of Science*, Summer 2021. *Themes* 6 (2021): Descent of Darwin: Race, Sex, and Human Nature. www.cambridge.org/ core/journals/bjhs-themes/article/charles-darwin-sexual-selection-and-the-evolution-of-otherre garding-ethics/9914E7B322270C98C1F4D2635904FAB2. See also the special issue of *Evolutionary Human Sciences*, www.cambridge.org/core/journals/evolutionary-human-sci ences/article/celebrating-the-150th-anniversary-of-the-descent-of-man/657B01DE3EF0C 4089CA7812D78B03CB6.

[6] Conference plenary "Revisiting Darwin's *Descent of Man* 1871–2021," International Society for the History, Philosophy and Social Study of Biology, Cold Spring Harbor Laboratory, MA, July 15, 2021 (virtual), full program here https://ishpssb.org/meetings/ishpssb2021; and The Descent of Man and Selection in Relation to Sex: 150 Years of Feminist Engagement with Science, Anniversary Symposium, University of Stockholm, December 3, 2021 (virtual).

which have been advanced against the principle of sexual selection, I remain firmly convinced of its truth."[7]

21.2   The Causes of the Second-Tier Status of Sexual Selection

Why, then, have subsequent generations of scientists and historians more or less accepted the myth that sexual selection was an afterthought, subordinate in importance to natural selection? There are several factors. For starters, Darwin's contemporaries, most famously Alfred Russel Wallace who also came up with the idea of natural selection, did not accept key elements of sexual selection theory, namely that animals could exercise choice-based behavior in their selection of mates, leading generations of naturalists to downplay the theory's overall importance in the grand Darwinian narrative. (Wallace also endorsed group selection, which made it hard for him to accept the function of individual selection so foundational to mate choice; see Ruse, Chapter 11, this volume.) In the twentieth century, as Erika Lorraine Milam has established, internal debates between biologists, fueled by the rise of molecular biology, relegated organismal biology to the margins and, eventually, to fields such as anthropology.[8] Compounding official science's skepticism of sexual selection was the fact that feminists, socialists, novelists, and sex radicals vocally championed the theory, especially its motive agent – female choice of sexual partners – at the turn of the twentieth century, thus making the theory appear even less scientific in the eyes of many scientific leaders and gatekeepers.[9]

In the second half of the twentieth century, feminist historians and science studies scholars engaged with Darwin and confronted his many statements about the "natural" inferiority of women, most of which appear in *The Descent of Man*. More often than not, these scholars concluded that evolutionary theory was inherently sexist, further obscuring sexual selection theory and perpetuating a second myth about this vital evolutionary mechanism.[10] As Fiona Erskine

[7] George Romanes, *Darwin and after Darwin* vol. 1 (Chicago: The Open Court, 1901), 400. Also quoted in Helena Cronin, *The Ant and the Peacock: Altruism and Sexual Selection from Darwin to Today* (New York: Cambridge University Press, 1991), 249.

[8] Erika Lorraine Milam, *Looking For a Few Good Males: Female Choice in Evolutionary Biology* (Baltimore, MD: Johns Hopkins University Press, 2010).

[9] This argument is more fully developed in Kimberly A. Hamlin, "Darwin's Bawdy: The Popular, Gendered, and Radical Reception of the *Descent of Man* in the U.S., 1871–1910," special issue marking the 150th Anniversary of Darwin's *Descent of Man*, *British Journal for the History of Science*, Summer 2021. *Themes* 6 (2021): Descent of Darwin: Race, Sex, and Human Nature eds. Erika Lorraine Milam and Suman Seth.

[10] For work arguing that evolutionary theory and/or sexual selection are fundamentally sexist, see, e.g., Flavia Alaya, "Victorian Science and the 'Genius' of Woman," *Journal of the History of Ideas* 38 (1977); Susan Sleeth Mosedale, "Science Corrupted: Victorian Biologists Consider 'The Woman Question,'" *Journal of the History of Biology* 11 (Spring 1978): 1–55; Janice Law Trecker, "Sex, Science and Education," *American Quarterly* 26 (October 1974): 352–66; Ruth Hubbard, Mary Sue Henifin, and Barbara Fried, eds., *Women Look at Biology Looking*

concluded in 1995 "Sexual selection is intrinsically anti-feminist: outlined in the Origin as the means by which males become stronger, fitter, more beautiful or more talented, it can only be a mechanism for the disproportionate development of male prowess. The Descent gives voice to Darwin's deeply-rooted beliefs."[11] In the twenty-first century, the combined, snowballing effects of 150 years of these two myths – that sexual selection was an afterthought to natural selection and that it is fundamentally sexist – continue to keep sexual selection, and in some cases those who research it, in a secondary position to natural selection.

To deconstruct the myth that sexual selection was and remains secondary in importance to natural selection, we must pay close attention to the theory's intellectual history and to the key role that contextual concerns about race, sex, and gender have played in the theory's popular and scientific reception. In her comprehensive history of *The Descent of Man*, Evelleen Richards argues that "the history of sexual selection was never one of straightforward, unadulterated science, but from its beginnings was intertwined with cultural and social beliefs and shaped by professional and institutional power plays and the larger issues of the day."[12] Only our collective discomfort with a book granting females reproductive agency and containing the word "sexual" in the subtitle has prevented so many of us from understanding the centrality of sexual selection in the Darwinian program.

21.3   Why Sexual Selection Was Not an Afterthought to Natural Selection

Far from being an afterthought to natural selection, sexual selection developed in concert with it. The key elements of sexual selection – male battle, female

at Women: A Collection of Feminist Critiques (Cambridge: Schenkman Publishing Company, 1979); Ruth Hubbard, Mary Sue Henifin, and Barbara Fried, ed., *Biological Woman – The Convenient Myth: A Collection of Feminist Essays and a Comprehensive Bibliography* (Cambridge: Schenkman Publishing Company, 1982); Marian Lowe and Ruth Hubbard, eds., *Woman's Nature: Rationalizations of Inequality*, The Athene Series (New York: Pergamon Press, 1983); Sue Rosser and Charlotte Hogsett, "Darwin and Sexism: Victorian Causes, Contemporary Effects," in *Feminist Visions: Toward a Transformation of the Liberal Arts Curriculum*, ed. Diane Fowlkes and Charlotte McClure (Tuscaloosa: University of Alabama Press, 1984): 42–52. In particular, see Hubbard's essay "Have Only Men Evolved?" in both *Women Look at Biology* and *Biological Woman*.

[11] Fiona Erskine, "The *Origin of Species* and the Science of Female Inferiority," in *Charles Darwin's The Origin of Species: New Interdisciplinary Essays*, ed. David Amigoni and Jeff Wallace, Texts in Culture Series (New York: Manchester University Press, 1995), p. 100.

[12] Evelleen Richards, *Darwin and the Making of Sexual Selection Theory* (Chicago: University of Chicago Press, 2017), p. xviii. For other work on the history of sexual selection, see Milam, *Looking for a Few Good Males*, and Kimberly A. Hamlin, *From Eve to Evolution: Darwin, Science, and Women's Rights in Gilded Age America* (Chicago: University of Chicago Press, 2014).

choice, and the evolutionary importance of beauty/attraction – are evident in Darwin's species notebooks as far back as the 1830s and even earlier when we look to the notes he took while reading the writings of his iconoclastic grandfather, Erasmus Darwin (Fara, Chapter 3, this volume).[13] Young Charles' encounters with the Fuegians while aboard the *HMS Beagle* convinced him that different races had different standards of beauty; this is the intellectual link that enabled him to believe that all humans were part of the same species even as he remained convinced of racial hierarchies (Peterson, Chapter 22, this volume).[14]

Decades later, when he received word that young upstart Alfred Russel Wallace was set to announce his own theory of evolution, Darwin hastened to compile all his notes about sex differences, racial divergence, and mate selection. He was confident in his ability to explain natural selection but felt he needed more evidence for sexual selection because he believed that, in order for natural selection to be compelling, sexual selection had to account for what natural selection could not, namely sexual dimorphism, racial differences, and beauty (Figure 21.1).[15] This realization seemed to have surprised even Darwin. "It could never have been anticipated," he subsequently confessed, "that the power to charm the female has sometimes been more important than the power to conquer other males in battle."[16] Until Darwin had worked out these puzzles of transmutation – and, vitally, how he could explain them without the intercession of a divine creator – he did not feel comfortable publishing his "big species book." As Evelleen Richards demonstrates in *Darwin and the Making of Sexual Selection*, Darwin formulated the main tenets of sexual selection theory between 1856 and 1858 in his rush to publish the *Origin of Species*.[17] Far from being an afterthought, then, Darwin considered sexual selection interdependent with natural selection.

Though he only devoted two pages of the *Origin of Species* to sexual selection, Darwin thoroughly outlined the core elements of the mechanism there because, in fact, he had more or less worked out the entire theory by the summer of 1858. According to Richards, Darwin began to differentiate the two evolutionary mechanisms between 1842 and 1844. Previous scholars, including me, have written that Darwin defined the contours of sexual selection in between the publication of *Origin*, in 1859, and the publication of *The Descent of Man*, in 1871. But Richards clearly establishes the error of this long-held truism. She demonstrates that Darwin mapped both natural and sexual selection simultaneously between 1844 and 1858. In 1855, Darwin

---

[13] Richards, *Darwin and the Making*, p. 63.   [14] Ibid., p. 90.   [15] Ibid., p. 291.

[16] Charles Darwin, *The Descent of Man, and Selection in Relation to Sex* (1871), introduction by John Tyler Bonner and Robert M. May (Princeton: Princeton University Press, 1981), pp. 262–3.

[17] Richards, *Darwin and the Making*, p. 106.

Fig. 49.   Spathura underwoodi, male and female (from Brehm).

Figure 21.1 Highly decorated male hummingbirds provided evidence of sexual selection at work. Credit: Reproduced with permission from John van Wyhe ed. 2002–. The Complete Work of Charles Darwin Online. (http://dar win-online.org.uk/)

wrote down his idea that different races resulted from different standards of beauty, and he used the term "sexual selection" for the first time in 1856.[18] In the course of her exhaustive research, Richards also found two notebook pages long believed to have been lost. These pages show that "Darwin had decided on female choice as the efficacious agency in sexual selection among birds" by the summer of 1858, which prompted him to first articulate the process which came to be known as "female choice."[19] Richards argues that,

[18] Ibid., p. 293.   [19] Ibid., pp. 259–61.

in fact, having done so was what enabled Darwin to feel comfortable publishing the *Origin of Species* in 1859.

21.4   Why the Importance of Sexual Selection Was Hard to Accept

Prior to 1858, the main obstacle to Darwin's ability to see female choice as a motive force was his ingrained masculine bias. Or, in Richards' words, "his acculturated presumption of the predominance of male sexual preference in sexual selection."[20] How could female animals be the sexual selectors when Victorian women exercised so little autonomy, sexual or otherwise? Ingrained male bias also prevented most of Darwin's scientific colleagues from accepting sexual selection as a powerful evolutionary mechanism. After all, natural selection described strong males – males displaying huge antlers, beautiful plumage, and fierce martial skills – battling for survival. This process made intuitive sense and appealed to upper-class men deeply invested in their own success and survival. But a process predicated on the reproductive agency, rational or otherwise, of female animals seemed much more far-fetched. Whereas organized naturalists in the U.S., Britain, and Europe had greeted the *Origin of Species* with praise and almost immediate acceptance, the major scientific practitioners, institutions, and publications remained skeptical, silent, or hostile toward sexual selection theory well into the twentieth century.[21] Wallace, who became one of the most outspoken critics of sexual selection, wrote in 1877:

The explanation of almost all the ornaments and colours of birds and insects as having been produced by the perceptions and choice of the females has, I believe, staggered many evolutionists, but has been provisionally accepted because it was the only theory that even attempted to explain the facts. It may perhaps be a relief to some of them, as it has been to myself, to find that the phenomena can be shown to depend on the general laws of development, and on the action of "natural selection," which theory will, I venture to think, be relieved from an abnormal excrescence, and gain additional vitality by the adoption of my view of the subject.[22]

Eager to establish the evolutionary necessity of a divine creator, Wallace argued that seemingly miraculous traits, such as the peacock's stunning plumage, could surely not be the result of natural processes. His insistence on divine intervention is what prompted Darwin to enumerate so precisely just how exactly a natural process – sexual selection – could and did account for all

---

[20] Ibid., p. 332.
[21] For examples of the scientific reception of the *Descent of Man*, see Hamlin's introduction to *From Eve to Evolution*.
[22] Wallace, RECORD: S272.1. Wallace, A. R. 1877. The colours of animals and plants, I. The colours of animals. Macmillan's Magazine 36 (215), pg. 408, John van Wyhe, ed. 2012–. *Wallace Online* (http://wallace-online.org/).

seemingly miraculous features of the natural world, including even a woman's beauty (hairlessness in women became a surprisingly heated sticking point of the debate).[23] Eventually, Wallace, a fervent socialist, championed female choice among humans for political purposes but his scientific colleagues remained skeptical.[24]

While sexual selection theory in no way disappeared (another myth), the majority of naturalists in the USA, the UK, and Europe acted as if sexual selection was not part of the Darwinian evolutionary system, especially after Darwin's death in 1882.[25] The countless events organized for the 1909 Darwin centennial, for example, contained no mention of *The Descent of Man* or sexual selection, its cornerstone theory.[26]

At the same time, countless socialists, novelists, feminists, and political radicals – far beyond Wallace – invoked female choice as integral to their political programs, gendering the theory "female" and unscientific in the eyes of university-trained scientists.[27] In fact, sexual selection's reach into the era's popular culture dwarfed that of natural selection. Sexual selection featured prominently in all sorts of turn-of-the-century novels and short stories (including those penned by the era's most popular writers Edith Wharton and Harriet Beecher Stowe), cartoons, satires, and even parlor games. In *Darwin, Literature, and Respectability,* Gowan Dawson argues that a major stumbling block for the naturalistic worldview presented in the *Descent of Man* was "Darwin's surprisingly recurrent connection with sexual immorality," a connection popularized by the political radicals and literary figures who made frequent reference to *The Descent of Man.*[28] Many of Darwin's critics in England went so far as to claim that *The Descent of Man* not only "transgressed Victorian standards of respectability" but also the "acceptable boundaries of nineteenth-century scientific publishing" because the work dealt so frankly with matters of sexual attraction and reproduction.[29] Surely then, many naturalists believed, this book

---

[23] This debate and the importance of hair/hairlessness to it is elaborated in Kimberly A. Hamlin, "'The Case of a Bearded Woman': Hypertrichosis and the Construction of Gender in the Age of Darwin," *American Quarterly* 63 (December 2011), pp. 955–81.

[24] Wallace's views are further elaborated in Hamlin *From Eve to Evolution* and "Darwin's Bawdy."

[25] Erika Lorraine Milan, "Myth 14. That after Darwin (1871), Sexual Selection Was Largely Ignored until Robert Trivers (1972)," in *Newton's Apple and Other Myths about Science,* edited by Ronald L. Numbers and Kostas Kampourakis (Cambridge, MA: Harvard University Press, 2015).

[26] For a list of the 1909 events see http://darwin-online.org.uk/EditorialIntroductions/vanWyh e_1909.html, John van Wyhe ed. 2002–. *The Complete Work of Charles Darwin Online* (http:// darwin-online.org.uk/).

[27] This argument is elaborated in Hamlin, "Darwin's Bawdy."

[28] Gowan Dawson, *Darwin, Literature and Victorian Respectability* (Cambridge: Cambridge University Press, 2007), p. 4.

[29] Ibid., p. 28. See especially chapter 2 "Charles Darwin, Algernon Charles Swinburne and Sexualized Responses to Evolution."

and the evolutionary mechanism it elucidated could not be considered integral to the Darwinian cosmology, much less scientifically valid.

Lack of more recent engagement with sexual selection theory can also be traced to Darwin's pronouncements about women's "natural" inferiority. For example, in *The Descent of Man*, Darwin argued that the male "has been the more modified" due to the males having "stronger passion than the females," which tend to retain "a closer resemblance" to the young. Among humans, Darwin believed that "owing to her maternal instincts" woman differs from man chiefly in her "greater tenderness and less selfishness" and lack of intellectual attainments.[30] According to Darwin, "The greater intellectual vigour and power of invention in man is probably due to natural selection combined with the inherited effects of habit, for the most able men will have succeeded best in defending and providing for themselves, their wives and offspring."[31] Darwin's catalogue of "natural" sex differences convinced generations of naturalists that female inferiority was natural and permanent, and suggested to subsequent feminists that evolutionary science stood in opposition to the equality of women.[32]

21.5   The Unanticipated Influence of *The Descent of Man*

Yet, *The Descent of Man* was a multivalent text that also appealed to those intent on challenging the idea that patriarchy was "natural." Generations of freethinking feminists, socialists, and pioneering sexologists looked to Darwin for evidence of the diversity of sexual and reproductive practices found in the animal kingdom, and many counted Darwin as an intellectual ancestor in the struggle for equality (despite his refusal to join their ranks). Frank discussions of sex and an endorsement of female reproductive agency may have made sexual selection repellant to most naturalists, but left-leaning reformers seized on Darwin's insights, giving lie to the myth that sexual selection was inherently sexist (an argument I chart in my book *From Eve to Evolution: Darwin, Science, and Women's Rights in Gilded Age America*, published in 2014).

In the final decades of the nineteenth century, in fact, no scientific publication inspired more broad-based conversations about sex and gender than did *The Descent of Man*.[33] Freethought, sex reform, and feminist periodicals (including *Lucifer the Light Bearer*, the *Woman's Tribune*, and *Socialist Woman*) all

---

[30] Darwin, *Descent of Man*, 629.   [31] Ibid., 674–5.
[32] Cynthia Eagle Russett, *Sexual Science: The Victorian Construction of Womanhood* (Cambridge, MA: Harvard University Press, 1989).
[33] Hamlin, "Darwin's Bawdy," and Kimberly A. Hamlin, "The Birds and the Bees: Darwin's Evolutionary Approach to Human Sexuality," in *Darwin in Atlantic Cultures: Evolutionary Visions of Race, Gender and Sexuality*, eds. Jeannette Eileen Jones and Patrick Sharp, Research in Atlantic Studies Series, ed. William Boelhower, Stephen Fender, and William O'Reilly (New York: Routledge Press, 2010), pp. 53–72.

recommended readers mine *The Descent of Man* for scientific evidence that patriarchy was not, as generally believed, natural, and inevitable. Creative readers of Darwin looked to the animal kingdom for examples of autonomous females, males who performed domestic labor, and a wide variety of sexual behavior. Female readers took heart from the diverse array of sexual practices and gender norms found in the animal kingdom, as well as from the fact that *The Descent of Man* forever upended what is perhaps the most firmly held myth in Western culture: the myth that woman was made from man's rib to be his helper.

At the same time, the most popular and influential reformist fiction of the era (namely Edward Bellamy's bestselling *Looking Backward*, published in 1888, and the work of Charlotte Perkins Gilman) predicted that female choice of sexual partners would liberate women from subservience and reform capitalist excess.[34] To socialists, including eventually Wallace, female choice redefined the role of women in capitalistic, patriarchal society and seemed, somehow, "natural." The very first issue of the *Socialist Woman* (June 1907) encouraged subscribers to "Read Darwin's 'Descent of Man.' It will give you a pretty good idea of the part the feminine principle has played in the animal kingdom."[35] As Sara Kingsbury observed in "The Lady-like Woman: Her Place in Nature," the modern woman "violates the habit of every other female in the animal kingdom ... She is the only female in the animal kingdom who seeks to charm the male."[36] Kingsbury concluded on an optimistic note, observing that feminists could count Darwin, by which she meant the processes articulated in *The Descent of Man*, as an ally.

21.6   Sexual Selection Today

On the once controversial point that sexual selection might be enlisted for feminist, or at least egalitarian, purposes an increasingly vocal and prominent group of scientists agree. Not only did Evelleen Richards' redefine the place of sexual selection within the history of science in her 2017 book, another book aimed at a more popular audience also made similar claims that same year: *The Evolution of Beauty: How Darwin's Forgotten Theory of Mate Choice Shapes the Animal World – and Us*, by the Brown University biologist Richard

[34] Kimberly A. Hamlin, "Sexual Selection and the Economics of Marriage: 'Female Choice' in the Writings of Edward Bellamy and Charlotte Perkins Gilman," in *America's Darwin: Darwinian Theory and U.S. Culture, 1859-present*, eds. Lydia Fisher and Tina Gianquitto (Georgia: University of Georgia Press, 2014), pp. 151–80.
[35] "Notes," *Socialist Woman*, 1.1, June 1907, pg. 4. *The Descent of Man* was included on another reading list published later that year, "Books on the Woman Question," *Socialist Woman*, October 1907, 1.5, p. 6.
[36] Sara Kingsbury, "The Lady-like Woman: Her Place in Nature," *The Socialist Woman*, August 1908, pg. 9.

O. Plum. In this widely reviewed book, Plum urges us to give sexual selection its due, finally, as a primary agent of evolutionary change (though many historians of science might quibble with his use of the word "forgotten" in the subtitle).[37] In 2022, two new books encourage us to look with fresh eyes at the animal kingdom to better appreciate the wonders of the mammalian world, and to better appreciate the female's role within it: popular British zoologist Lucy Cooke's *Bitch: The Female of the Species* and science journalist Ed Yong's *An Immense World: How Animal Senses Reveal the World Around Us*.[38] While these exciting new scientific works cannot, in and of themselves, dispel 150 years' of mythmaking about sexual selection, these new books – and the mainstream media attention that they are receiving – portend, at least, that the historical and scientific contexts in which we build myths has evolved and is now open to the importance of sexual selection – and female agency – in the animal world.

---

[37] Richard O. Plum, *The Evolution of Beauty: How's Darwin's Forgotten Theory of Mate Choice Shapes the Animal World – and Us* (New York: Penguin Random House, 2017).

[38] Lucy Cooke, *Bitch: The Female of the Species* (New York: Basic Books, 2022); Ed Yong, *An Immense World: How Animal Senses Reveal the World Around Us* (New York: Penguin Random House, 2022).

Myth 22   That Darwin's Hatred of Slavery Reflected His Beliefs in Racial Equality

Erik L. Peterson

22.1   Introduction

In May 2020, the United States erupted in protests against police violence, racism, and the legal system that scaffolded them. Academia, too, felt the impact. In June 2020, social media collectives such as "Black In The Ivory" and activist awareness events like "ShutdownSTEM," challenged purportedly politically left university faculty and administrators to address the deep historical wounds of white supremacy manifested visibly in campus monuments and less visibly in institutional histories and power structures. Statues fell; trustees renamed buildings; faculty created new courses.

As academics researched their disciplines' own checkered racial legacies, scrutiny fell upon the champions of science. Some – Jim Watson of double-helix fame perhaps most glaringly – made for easy targets. But Charles Darwin seemed an unlikely one. For starters, Darwin was a Wedgwood through his mother's side. Josiah Wedgwood cast the iconic "Am I Not a Man and a Brother?" medallion, the unforgettable symbol of the eighteenth-century British abolitionist movement. And in *The Descent of Man; and Selection in Relation to Sex*, published in 1871, Charles Darwin, son of and husband to Wedgwoods, attacked the justifications for polygenism – the scientific theory *de jour* that insisted Blacks and Whites had separate ancestors. Polygenism undercut support for abolition.

Nevertheless, in 2021, on the 150th anniversary of the publication of *Descent of Man*, anthropologist Agustín Fuentes excoriated the book in *Science*, calling it a text that "offer[ed] justification of empire and colonialism, and genocide." Darwin's racism blinded him to both "data" and to positive assessments of nonwhites.[1] Aghast, a collective of British, European, and American scholars lashed out at Fuentes. Far from being a racist, they claimed, Darwin maimed

[1] Agustin Fuentes, "'The Descent of Man,' 150 Years On," *Science* 372(6544), May 21, 2021: 769. doi:10.1126/science.abj4606.

slavery-justifying polygenism.[2] "Darwin was frequently and notably more modern in his thinking than most Victorians," they claimed. The *Descent of Man*, the second of Darwin's trilogy extending his "abstract" of *Origin of Species*, Darwin "demolished the slavery-justifying view of different races as separate species." Beyond merely reforming Victorian biology, this group claimed *Descent* inspired later "anti-racist perspectives" of later anthropologists – even the vaunted "Boaz [*sic*]." Any insinuation to the contrary, the Whiten et al. group worried, opened the proverbial microphones to "anti-evolution voices" and dissuaded the "more gender and ethnically diverse" evolutionary scientists of the future from pursuing their scientific passions. The stakes adjudicating this battle, then, seem high.

Charles Darwin hated slavery just like his Wedgwood family, or so it is taught in many biology classrooms. But why? Was it because he was ahead of his time, a startlingly modern thinker on issues of race, indeed a proto-Boasian anthropologist? Unfortunately, no – this is the myth that I'll attempt to debunk in this chapter. Perhaps it is a helpful myth, however, one that keeps the anti-evolution serpents in their holes and swings wide the door to diverse future scientists – who, if Whiten et al. are to be believed, need to see Darwin as a modern role-model, if not a secular saint, in order to join the ranks of practicing evolutionary scientists themselves.

If we *need* to uphold a historical myth, however kindly, something must be seriously wrong in today's life sciences. Yet, the reaction to Fuentes' letter clearly indicates that to delve into Darwin's own views on human race is to enter the proverbial briar patch. What's a *mere historian* to do if even a prominent anthropologist like Fuentes suffers "friendly fire" from other evolutionists? Probably the best method is the one Darwin himself purported to follow: induce from individual observations.[3]

22.2   Young Darwin's Antipathy Toward Cruelty

Aside from the general Wedgwood family tradition, Darwin explicitly expressed distaste at cruelty to humans and animals. He displayed this family aversion at his first confrontation with the totalizing enslavement system in Salvador, Brazil in 1833:

Near Rio de Janeiro I lived opposite to an old lady, who kept screws to crush the fingers of her female slaves. I have stayed in a house where a young household mulatto, daily

[2] Andrew Whiten et al., "RE: 'The Descent of Man,' 150 years on," June 6, 2021 comment to Fuentes 2021. *Science* 372(6544), May 21, 2021: 769. doi:10.1126/science.abj4606.

[3] "I worked on true Baconian principles, and without any theory collected facts on a wholesale scale, more especially . . . by extensive reading" (Charles Darwin, *Charles Darwin: His Life Told in an Autobiographical Chapter, and in a Selected Series of His Published Letters*, ed. by F. Darwin. London: John Murray, 1902, 40).

and hourly, was reviled, beaten, and persecuted enough to break the spirit of the lowest animal ... I was present when a kind-hearted man was on the point of separating forever the men, women, and children of a large number of families who had long lived together. I will not even allude to the many heart-sickening atrocities ... nor would I have mentioned the above revolting details, had I not met with several people ... [who] speak of slavery as a tolerable evil.[4]

Without question, young Charles Darwin continued to carry the Wedgwood torch against the violence needed to uphold the international economic institution of slavery. When Great Britain moved to finally eradicate it in 1833 (having done so only incompletely just before his birth), Darwin wrote to his sister:

I have watched how steadily the general feeling, as shown at elections, has been rising against Slavery. – What a proud thing for England, if she is the first Europæan nation which utterly abolishes it. – I was told before leaving England, that after living in Slave countries: all my opinions would be altered; the only alteration I am aware of is forming a much higher estimate of the Negros character. – it is impossible to see a negro & not feel kindly towards him; such cheerful, open honest expressions & such fine muscular bodies.[5]

In this widely quoted passage, Darwin revealed both his warm feelings for the enslaved persons' physique and cheerfulness. We may now read it as a cringe-worthy repetition of the "happy slave" anecdote ever after baked into the pernicious American Lost Cause myth. But importantly, this letter suggests that Darwin had no particular *antipathy* toward the Black body. Emotional sentiments like these anchor this particular Darwin myth. As one influential historical account summarized it: "Darwin was never a *racist* in the conventional sense of the word. . . . [H]e expressed his dislike of slavery and his admiration for the [Brazilian] black population."[6]

Partly, defenses such as these are accurate. In his published works, Darwin repeatedly argued for the shared ancestry of all organisms, including humans. This is probably the most well-known of Darwin's contributions, even if others expressed it earlier, including his own grandfather (see Fara, Chapter 3, this volume). So convinced was Darwin of this singular point that he ended the final book in his evolutionary trilogy, *Expression of the Emotions in Man and Animals*, published in 1872, with another appeal to common humanity as shown through shared expressions of joy and sorrow,

---

[4] Charles Darwin, *The Voyage of the Beagle*, 2nd ed., in *Darwin, The Indelible Stamp: The Evolution of an Idea.* J. Watson, ed. Philadelphia, PA: Running Press Books, 2005 [1845], p. 333.

[5] CD to Catherine Darwin (sister), May 14 to July 22, 1833. Darwin Correspondence Project, "Letter no. 206," www.darwinproject.ac.uk/letter/?docId=letters/DCP-LETT-206.xml.

[6] Nancy Stepan, *The Idea of Race in Science: Great Britain, 1800–1960.* London: Macmillan, 1982, 49–50.

which apes also shared.[7] Yet there is more to the story, visible when we dig below the passages reprinted by today's biologists.

22.3   Darwin's Expressed Thoughts on Race

In the weeks preceding the publication of *The Origin of Species* in 1859, Darwin sent printer's proofs to his old mentor, geologist Charles Lyell. The two then corresponded regarding Lyell's objections. In October, Darwin identified what he thought to be the core of their dispute. Though he explicitly avoided discussing humans in *Origin of Species*, Darwin thought an example using human race would drive home the book's message to Lyell:

> [Y]ou doubt the possibility of gradations of intellectual powers. Now it seems to me looking to existing animals alone, that we have a very fine gradation in the intellectual powers of the Vertebrata, with one rather wide gap (not half so wide as in many cases of corporeal structure) between say a Hottentot & an Ourang, even if civilised as much mentally as dog has been from wolf. – I suppose that you do not doubt that the intellectual powers are as important for the welfare of each being, as corporeal structure: if so, I can see no difficulty in the most intellectual individuals of a species being continually selected; & the intellect of the new species thus improved, aided probably by effects of inherited mental exercise. I look at this process as now going on with the races of man; the less intellectual races being exterminated.[8]

Evident is their shared prejudice that southern African Khoikoi and Khoi-san ("Hottentots") people and orangutans remained distinguishable by a wide-but-not-*that*-wide intellectual gap between them, "even if civilized." The implications of that comment are only overshadowed by Darwin's admission that nonwhites were just then being exterminated across the globe because they were "less intellectual."

Darwin and Lyell continued their friendly sparring uninterrupted through 1860, returning repeatedly to issues of human race. They both referred to the much grander and more impressive (in comparison to the *Origin of Species*) collaboration between Louis Agassiz and Alabama physician and race scientist Josiah Clark Nott, *Types of Mankind* published in 1854, in which the authors defended polygenism. Darwin found the scientific arguments for polygenism weaker than did Lyell but did not challenge the book's hierarchy of races:

> ... I sh[d] look at all races of man as having certainly descended from single parent. – I should look at it as probable that the races of man were less numerous & less divergent formerly than now; unless indeed some lower & more aberrant race, even than the Hottentot, has become extinct. ... Agassiz & Co. think the Negro & Caucasian are now

---

[7] Gregory Radick, "How and Why Darwin Got Emotional About Race." In *Historicizing Humans: Deep Time, Evolution, and Race in Nineteenth-Century British Sciences*, ed. Efram Sera-Shriar. Pittsburgh: University of Pittsburgh Press, 2018.

[8] CD to Charles Lyell, October 11, 1859. Darwin Correspondence Project, "Letter no. 2503," www.darwinproject.ac.uk/letter/?docId=letters/DCP-LETT-2503.xml.

distinct species; & it is a mere vain discussion, whether when they were rather less distinct they would, on this standard of specific value, deserve to be called species. ... I agree with your answer which you give to yourself on these points; & the simile of man now keeping down any new man which might be developed strikes me as good & new. White man is "improving off the face of the earth" even races nearly his equals.[9]

Here again, we might find jarring Darwin's cavalier references to white supremacy and genocide as supportive of his overall theory.

Similar themes appear in letters to some of Darwin's other prominent correspondents. Reverend Charles Kingsley, social reformer, historian, and popular author of *Water-Babies*, published in 1863, speculated that myths of dwarves and fairies emerged from real "missing links" between humans and apes: "That they should have died out, by simple natural selection, before the superior white race, you & I can easily understand."[10]

While skeptical about these "missing links," Darwin agreed with Kingsley about whites exterminating all other races: "In 500 years how the Anglo-saxon race will have spread & exterminated whole nations." His next sentence seems still more shocking: "in consequence how much the Human race, viewed as a unit, will have risen in rank."[11] To different correspondents many years later, Darwin reaffirmed this view: "When I look to the future of the world hardly any event seems to me of such great importance as the settling of Australia, New Zealand, &c by the so called Anglo Saxons."[12] And, again:

Lastly I could show fight on natural selection having done and doing more for the progress of civilisation than you seem inclined to admit. Remember what risk the nations of Europe ran, not so many centuries ago of being overwhelmed by the Turks, and how ridiculous such an idea now is ... Caucasian races have beaten the Turkish hollow in the struggle for existence. Looking to the world at no very distant date, what an endless number of the lower races will have been eliminated by the higher civilised races throughout the world.[13]

So much for Darwin's privately expressed sentiments. Before attempting to make sense of them, let's turn to his published works.

In his book the *Origin of Species* Darwin avoided humans but described intergroup variation and competition, by which he also meant inter*racial* variation and competition. We often discuss competition between *individuals*

---

[9] CD to Charles Lyell, September 3, 1860. Darwin Correspondence Project, "Letter no. 2925," www.darwinproject.ac.uk/letter/?docId=letters/DCP-LETT-2925.xml.

[10] Charles Kingsley to CD, January 31, 1862. Darwin Correspondence Project, "Letter no. 3426," www.darwinproject.ac.uk/letter/?docId=letters/DCP-LETT-3426.xml.

[11] CD to Charles Kingsley, February 6, 1862. Darwin Correspondence Project, "Letter no. 3439," www.darwinproject.ac.uk/letter/?docId=letters/DCP-LETT-3439.xml.

[12] CD to E. H. O'Callaughan, July 14, 1879. Darwin Correspondence Project, "Letter no. 12158," www.darwinproject.ac.uk/letter/?docId=letters/DCP-LETT-12158.xml.

[13] CD to William Graham, July 3, 1881. Darwin Correspondence Project, "Letter no. 13230," www.darwinproject.ac.uk/letter/?docId=letters/DCP-LETT-13230.xml.

as the heart of Darwin's theory (so-called "survival of the fittest"). However, the rarely acknowledged subtitle of the *Origin of Species* indicates just how important the notion of racial competition was in Darwin's scientific works: *the Preservation of Favoured Races in the Struggle for Life*. Here, he tells us precisely how important the struggle for existence between *races* is to his work.

In the second chapter of the *Origin of Species*, for instance, Darwin brought home the notion that "dominance leads to greater dominance." Groups already broad and diversified, able to dominate in different environments, will continue to vary and displace rival groups in more and more habitats.[14] He repeated this group variation and competition theme in the first book of his evolutionary "trilogy," *Variation in Animals and Plants Under Domestication* published in 1868. In the animal and plant kingdoms, on average and over time, the rich get richer.

Unsurprisingly, this theme of interracial difference emerges where Darwin explicitly addressed humanity in *The Descent of Man, and Selection in Relation to Sex* published in 1871. He leaned heavily on the painstaking work of Dr. John Beddoe, whose *Races of Britain* published in 1862 became the first "big data" project of the era comparing human skin, hair, and eye color, and that of Dr. Joseph Barnard Davis comparing human cranial capacities. While Darwin denied that quantitative comparisons between two *individuals* yielded much of worth, he upheld the belief that brain size and intellectual capability did correlate when it came to big groups like human *races* with whites on the high end and dark-skinned Australians on the bottom.[15] *The Descent of Man* also attempted to account for racial extinction by appeals to these differences in racial intelligence, cultural flexibility, or "grade of civilization."[16] When Europeans contacted any aboriginal population, the white group always won the struggle. It appeared to be just another aspect of natural selection. For those of us accustomed to thinking of Darwin as primarily interested in *individual* variation and competition, statements such as these might cause us to reevaluate our assumptions.

A broader scope of Darwin's perspective is illustrated by some substantive mentions of race in the first 240-page segment of *The Descent of Man*, most of which are presented in Table 22.1. It helps demonstrate just how mythological is the notion that Darwin's hatred of slavery had roots in any belief in racial equality.

22.4   Clearly a Man: But a Brother?

At best, we might underscore that Darwin openly abhorred *torture and cruelty* to humans and animals and possessed a clear-eyed vision regarding how much

---

[14] Charles Darwin, *The Origin of Species, or the Preservation of Favoured Races in the Struggle for Life*. London: J. Murray, 1859.

[15] Charles Darwin, *The Descent of Man, and Selection in Relation to Sex*. 2nd ed. London: Penguin, 2004 [1879], 74–5.

[16] Darwin, *Descent*, 212.

Table 22.1 *Human race in part 1 of* Descent of Man

Page[*]	Content of passage	Racial similarity?	Racial hierarchy?
35	Greater sense of smell in "dark-coloured races" (mental structure closer to that of lower animals)	No	Yes
74	Cranial capacity ranking from 92.3 ci (Europeans) to 81.0 ci (aboriginal Australians)	No	Yes
134	Savages will only risk life to save own community; civilized will risk life for strangers.	No	Yes
141–42	Slavery "a great crime." Savages treat women like slaves, torture and abuse animals. Show kindness to kin but not to strangers, except Mungo Park. "[C]ommon experience justifies the maxim of the Spaniard, 'Never, never trust an Indian'"	Uncertain (cruelty/torture universal; less among whites)	Yes
143–44	Low morality of savages because: (a) only sympathetic to own tribe; (b) low reasoning skills to see broader consequences of actions; (c) weak "power of self-command," which in the civilized comes through habit "perhaps inherited," instruction, and religion.	Uncertain (because also possible among whites)	Yes
147	Civilized men extend sympathies to all in nation, then all humans, then beyond humans. "Unfelt by savages, except toward their pets."	Uncertain (because also possible among whites)	Yes
153	Civilized nations supplanting barbarous ones through "products of the intellect" perfected through natural selection.	No	Yes
164	Quotes W. R. Greg (1868) and cites Galton (1865) re: the increase of the poor because they reproduce more frequently and earlier; but confident that child mortality rate is double among the poor in cities, so it balances out.	Yes	No (doesn't connect poverty to race)

Pages		Col1	Col2
183–84	Civilized races "will almost certainly exterminate, and replace, the savage races throughout the world." The gap between man and ape will grow wider than "as now between the negro or Australian and the gorilla."	No	Yes
194–210	Ch. 7, pt. 1: monogenism vs. polygenism. Support for monogenism; "subspecies" seems the most appropriate designation for human races.	Yes	No
211–222	Ch. 7, part 2: extinction of races. "Extinction follows chiefly from the competition of tribe with tribe, and race with race."	No	Yes
222–230	Ch. 7, part 3: formation of races. Racial characteristics, except for intelligence, cannot be accounted for by natural selection, Lamarckian use/disuse, or the direct action of environment alone; introduces sexual selection.	Partially	Yes

* Pages are from Penguin's 2004 reissue of the 1879 second edition.

torture proved necessary to uphold chattel slavery. Indeed, this serves as the center of a telling spat he had with Charles Lyell. After a trip down the coasts and riverways of the southeastern United States, Lyell remained impressed by the accounts by white slave owners of their kind treatment of their slaves. His *Travels in North America* published in 1845 sounded all too credulous to Darwin. In the context of a much longer letter covering the heat reflectivity of snow, traveling to tropical islands, and volcanoes, Darwin inserted:

> How could you relate so placidly that atrocious sentiment about separating children from their parents; & in the next page, speak of being distressed at the Whites not having prospered; I assure you the contrast made me exclaim out. – But I have broken my intention, & so no more on this odious deadly subject.[17]

For some, this is a key confrontation revealing the naturalist's righteous fire. But examine the statement closer: it's the cruelty of family separation that sparked Darwin. And even his frustration with Lyell was not enough to compel him to cut off correspondence or even raise the issue again. For this descendent of Wedgwoods, the dark-skinned slave was clearly a *man* – but a *brother*?

Still, we need not take a firm position on whether Darwin's beliefs constitute racism – whether Darwin was a racist. That's just not helpful. It would be helpful, however, to adjudicate the 2021 dispute on the pages of *Science* between Fuentes' charge that *The Descent of Man* is a racist text and the Whiten et al. countercharge that Darwin inspired the anti-racism of cultural anthropology's scion, Franz Boas.

Philosophers helpfully distinguish between two components of racism: antipathy and inferiorization.[18] The former is well illustrated by famed Swiss geologist and biologist Louis Agassiz's first encounter with an African American hotel employee in Philadelphia in 1847:

> It is impossible for me to repress the feeling that they are not of the same blood as us. In seeing their black faces with their thick lips and grimacing teeth, the wool on their head, their bent knees, their elongated hands, their large, curved nails, and especially the livid color of the palms of their hands, I could not take my eyes off their face in order to tell them to stay away.[19]

Darwin never openly demonstrated this sort of emotionally driven bigotry in *The Descent of Man* – with one exception when considering his own relatedness to "savage" Fuegians.[20] Perhaps, as influential historian Nancy Stepan intimated in

---

[17] CD to Charles Lyell, August 25, 1845. Darwin Correspondence Project, "Letter no. 905," www.darwinproject.ac.uk/letter/?docId=letters/DCP-LETT-905.xml.

[18] Lawrence Blum, *"I'm Not a Racist, But . . . ": the Moral Quandary of Race*. Ithaca, NY: Cornell University Press, 2002.

[19] Quoted in Saima S. Iqbal, "Louis Agassiz under a Microscope," *The Harvard Crimson* (March 18, 2021), www.thecrimson.com/article/2021/3/18/louis-agassiz-scrut/.

[20] Darwin, *Descent*, 689.

The Idea of Race in Science, that avoidance keeps us from classifying Darwin and his works as "conventionally" racist.[21]

Yet, we cannot say the same about inferiorization. Several of the above passages I tabulated from *The Descent of Man* very clearly express Darwin's belief that whites led a global racial struggle for existence. Even with the "sin of Brazilian [and American and British] slavery" washed away, he believed, all "uncivilized" people would be exterminated, the gap between humans and apes widened. Private conversations, such as those quoted above, reveal that Darwin accepted and even welcomed this result. This certainly appears to constitute Darwin's acceptance of inferiorization.

Scholars, including Whiten et al., offer two common defenses of Darwin's works against charges of racism, including *Descent*: that (1) even hinting that Darwin would have supported wrong or malicious ideas means biblical creationists will win, and that (2) he was simply a man "of his time" in his views on white superiority (while simultaneously lauded as a man "ahead of his time" when arguing for the science of common descent with modification, which sometimes meant intergroup conflict).

The first defense turns out to be an insidious bit of "ends justify the means" history, unbecoming of individuals who would not permit such shoddy reasoning in their own professional work. As anthropologist Jonathan Marks responded in the Fuentes 2021 forum, "discussing the ideas that are toxic in Darwin's work in addition to those that are prescient and revolutionary helps to make Darwin seem more real, and his followers seem less cult-like."[22]

On its face, the second defense seems more credible. After all, I just quoted Darwin's peers and mentors expressing similar sentiments. But as Zitzer has carefully compiled, many prominent figures across the Empire, as well as figures that Darwin actually cited in *The Descent of Man*, highlighted the immorality of British colonial policy – acts of commission, not the mere working out of natural law – leading to the near extermination of so many aboriginal populations, even after the 1808 and 1833 acts ending slavery.[23] These included philanthropists, members of Parliament, the founders of the Aborigines' Protection Society, and even Darwin's evolutionist grandfather, Erasmus. In other words, to say Darwin was a man "of his time" means to imagine that no prominent voices, or at least none familiar to him, would be crying out against Darwin's own attitude toward the destruction of aboriginal populations. But there were such voices, and Darwin knew about them. Darwin may have been anticruelty, but that's not the same thing as saying he believed in

---

[21] Stepan, *Idea of Race*, 49.

[22] Jonathan Marks, "RE: 'The Descent of Man,' 150 years on," May 26, 2021 comment to Fuentes 2021. doi: 10.1126/science.abj4606.

[23] Leon Zitzer, *Darwin's Racism: the Definitive Case, Along With a Close Look at Some of the Forgotten, Genuine Humanitarians of That Time*. Bloomington, IN: iUniverse, 2016.

tempering colonial expansion to avoid genocide. Racial equality stood even further away.

22.5   Conclusion

It is a shame Darwin could not have wielded that original Wedgwood abolitionist flame to argue for racial equality. Instead, any scientific argument for racial equality lost credence for at least a half-century after Darwin.[24] *Descent* became not an egregious example of scientific racism but all too typical. Scientific texts adopted Darwinian language to justify their racist assumptions with astonishing rapidity following *The Descent of Man* (Figure 22.1). Only

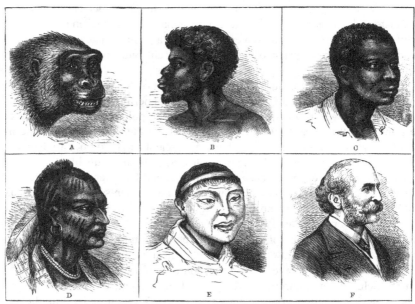

PROGRESSIVE DEVELOPMENT OF MAN.—(2) EVOLUTION ILLUSTRATED WITH THE SIX CORRESPONDING LIVING FORMS.

Figure 22.1 In the wake of Darwin, old racism received a new scientific justification and lexicon. Credit: John C. Ridpath, *History of the World; Comprising the Evolution of Mankind and the and the Story of All Races.* New York: Merrill & Baker, 1897, p. 233. Source: mikroman6/Moment/Getty Images

[24] John S. Haller, Jr., *Outcasts from Evolution: Scientific Attitudes of Racial Inferiority, 1859–1900.* 2nd ed. Carbondale, IL: Southern Illinois University Press, 1995.

someone with Darwin's stature could have trailblazed a different course. Instead, by *breaking* with Darwin (not drawing from him as Whiten et al. suggested), Franz Boas, W. E. B. DuBois, Anténor Firmin, and other social scientists finally did begin to fracture the racist, imperialist foundations of nineteenth-century bioscience.

Myth 23   That the Discovery of *Australopithecus* in 1925 Belatedly Confirmed Darwin's 1871 Scientific Prediction of African Human Origins

Emily M. Kern

23.1   Introduction

Here is a familiar story you may have seen in museums,[1] or encyclopedias:[2] in the nineteenth century, visionary scientist Charles Darwin proposed his theory of evolution. Although it was controversial, he also included humans in that theory, proposing that our species' ancestors had mostly evolved somewhere on the African continent. In the 1920s, South African scientist Raymond Dart (Figure 23.1) discovered the fossil *Australopithecus africanus*, which he identified as a transitional species that fulfilled the terms of Darwin's original proposal and the history of evolution was forever changed. Over the course of the twentieth century, many subsequent excavations across Africa have substantiated Darwin's prediction and validated Dart's evidence.

In broad strokes, this history is not wrong. Darwin *did* propose the idea of African origins – although in his 1871 *Descent of Man*, rather than the 1859 *Origin* as many people assume.[3] Anatomist Raymond Dart *did* announce the discovery of a fossil he had named *Australopithecus*, the "southern man-ape," in Johannesburg in 1925 and mentioned Darwin in his original report.[4] Over the

---

[1] "The strong similarities between humans and the African great apes led Charles Darwin in 1871 to predict that Africa was the likely place where the human lineage branched off from other animals – that is, the place where the common ancestor of chimpanzees, humans, and gorillas once lived. ... Hardly ever has a scientific prediction so bold, so 'out there' for its time, been upheld as the one made in 1871 – that human evolution began in Africa." Smithsonian National Museum of Natural History, "Genetic Evidence" https://humanorigins.si.edu/evidence/genetics.

[2] "In 1924, at a time when Asia was believed to have been the cradle of mankind, Dart's recognition of the humanlike features of the Taung skull, recovered in South Africa near the great Kalahari desert, substantiated Charles Darwin's prediction that such ancestral hominin forms would be found in Africa." The Editors of Encyclopaedia Britannica, "Raymond A. Dart South African anthropologist" www.britannica.com/biography/Raymond-A-Dart.

[3] Charles Darwin, *The Descent of Man and Selection in Relation to Sex*, 2 vols. (London: John Murray, 1871), vol. 1, 199–201.

[4] Raymond A. Dart, "*Australopithecus africanus*: The Man-Ape of South Africa," *Nature* 115, no. 2884 (7 February 1925), 195–9.

Prof. Raymond Dart with the Taungs skull.

Figure 23.1 Raymond Dart with the Taungs skull in 1925. Reprinted with permission from Current Topics and Events. *Nature* 115, 469–73 (1925)

intervening century, researchers working in many different parts of the African continent have discovered a wealth of ancient hominin species, filling out the ancestral family tree while adding new layers of evolutionary and chronological complexity and demolishing the bases for racial discrimination – for, as one prominent paleoanthropologist wrote, it is now known that "under our skins, we are all Africans."[5]

This is a clean and schematic history, with clearly identified (and generally positively regarded) main characters, telling a story of how science and scientists got it right. It's very satisfying, and by this point, you are probably feeling more than a little suspicious. And rightly so – this is a case of myth-making in action, one that hides a vast number of intellectual skirmishes, complex interplay between evidence and theory, and paths toward knowledge production that were, at best, uneven and highly contingent – hardly the smooth progression from Darwin's proposal to Dart's proof that one might assume. The gaps in the myth are also all too easy to fill in – that Darwin's idea was too ahead of its time, not taken seriously by his contemporaries for reasons of

[5] Chris Stringer and Robin McKie, *African Exodus: The Modern Origins of Humanity* (New York: Henry Holt & Co., 1996), 16.

dogmatism, racism, or both, or that once Dart provided the required fossil proof, it was broadly accepted that human evolution began in Africa and that the only resistance came from anti-evolutionists and other contrarians. Neither circumstance is actually true.

This mythological history also quietly obscures the role of race and racism in the history of anthropology, despite its central and inescapable presence as a major research topic before Darwin, between Darwin and Dart, and long after. It trades on the idea of Darwin as an anti-racist thinker (discussed by Erik Peterson in the previous chapter) and ignores the South African context of Dart's original find – and indeed Dart's contemporaneous research on racial "types" among people in the Kalahari Desert. This history is increasingly well-known – there has been significant recent publication by historians and anthropologists, especially in South Africa, on paleoanthropology's implication in colonial and apartheid politics.[6] But very little of this contextual complexity has yet seeped through to museum displays, encyclopedia entries, or classroom textbooks (especially in the United States, where discussion of human evolution at all remains somewhat on thin ice).

Was Darwin's evolutionary proposal of African origins really so "out there"? If so, why and how was Dart's *Australopithecus* accepted as valid evidence? What other stories are being obscured by this clean Darwin-to-Dart progressive narrative? And, last of all, if this is a myth, then when and where did it emerge?

23.2   Darwin on African Origins

To begin, what did Darwin actually have to say about the evolutionary origins of the human species? Famously, he did not discuss human evolution outright in the *Origin of Species*. Even in his self-described "Man-book," *The Descent of Man and Selection in Relation to Sex*, the topics that we would expect to see in an evolutionary history of *Homo sapiens* don't really come up at great length. (In other words, *The Descent of Man* is a lot more about *Selection in Relation to Sex* than it is about the origins of *Man*.) This is not to say that sexual selection was not a critical part of how Darwin understood human evolution. As Evelleen Richards, Erika Lorraine Milam, and, in this volume, Kimberly Hamlin have all shown, sexual selection was neither an adjunct to natural selection nor an afterthought.[7] Some of the conclusions that necessarily followed from the

---

[6] Alan G. Morris, *Bones and Bodies: How South African Scientists Studied Race* (Johannesburg: Wits University Press, 2022); Martin Porr and Jacqueline M. Matthews, *Interrogating Human Origins: Decolonisation and the Deep Human Past* (New York: Routledge, 2020); Christa Kuljian, *Darwin's Hunch: Science, Race, and Human Origins* (Johannesburg: Jacana Media, 2016).

[7] Evelleen Richards, *Darwin and the Making of Sexual Selection* (Chicago: University of Chicago Press, 2017); Erika Lorraine Milam, *Looking for a Few Good Males: Female Choice in Evolutionary Biology* (Baltimore: Johns Hopkins University Press, 2010).

concept, about female choice and sexuality especially, had uncomfortable implications for a Victorian scientific gentleman like Darwin. But, sexual selection was significantly easier for Darwin to speak about in another way, because there was ample evidence to investigate, from the mating calls of lyrebirds to the vagaries of women's contemporary fashion.

Humans were trickier – leaving the political and spiritual dimensions to one side, Darwin simply had very little physical evidence on which to draw. Given that humans were anatomically most similar to the great apes and that the great apes were only found in the Old World, the Americas were unlikely to be humanity's first home – ditto Australia. From the remaining candidate continents – Europe, Asia, and Africa – Darwin ultimately landed on Africa:

> In each great region of the world the living mammals are closely related to the extinct species of the same region. It is, therefore, probable that Africa was formerly inhabited by extinct apes closely allied to the gorilla and chimpanzee; and as these two species are now man's nearest allies, it is somewhat more probable that our early progenitors lived on the African continent than elsewhere.[8]

"But it is useless to speculate on the subject," continued Darwin, because one of the few ancient primate fossils that had been found was an ape "nearly as large as a man" called *Dryopithecus*, which had been unearthed by the French prehistorian and paleontologist Edouard Lartet in 1856.[9] *Dryopithecus* had been found in Miocene deposits in the region of the French Pyrenees – definitely a region devoid of apes in the present day. This suggested to Darwin that the contemporary distribution of apes was not necessarily a guide to the prehistoric distribution of the same species, since "there has been ample time for migration on the largest scale."

In context, this was hardly a strongly stated scientific hypothesis on Darwin's part. In the very next paragraph he referred to the era and place where humans evolved to be less simian, starting with a loss of our all-over hairy coating, "whenever and wherever it may have been." Darwin was uncertain and the point itself seems to have been fairly unimportant to him – it garnered a mere two-and-a-half pages of discussion in a more than 900-page-long book.

23.3   How "Out There" Was Darwin?

Was Darwin's Africa hypothesis all that radical? Darwin was not the first person to suggest an ancestral affinity between humans and apes – his own grandfather Erasmus Darwin had proposed as much several decades earlier – and his argument from morphological similarity was very much within the

---

[8] Darwin, *The Descent of Man*, vol. 1: 199.   [9] Darwin, *The Descent of Man*, vol. 1: 199.

bounds of accepted biological practice.[10] He was also not alone in speculating about the evolutionary geography of humankind and relating that geography to questions of both ancestry and descent. This was a vastly complicated intellectual terrain for anthropology in the nineteenth century and one that consumed the attention of many men of science before 1859, between 1859 and 1871, and after 1871.

Arguably the key question for early nineteenth century anthropologists dealt with the nature of race difference and whether the different human "races" – believed to range from three to more than fifty in number – had been singly or multiply created.[11] Were humans one species sharing a common origin (either secular and natural in nature or endowed by a divine creator), or were the different races in fact not the same species at all? If humanity did share one origin, what forces could explain the vast differentiation that could be observed in physical appearance and cultural behavior across the globe, and were those forces unique to man? This question, of monogenesis versus polygenesis, was freighted with theological as well as political and economic significance, given its implications for debates over the validity of the Genesis account and the continuation or abolition of the transatlantic slave trade.

Darwin was a staunch monogenist – his theory of natural selection and descent with modification could hardly make sense otherwise – and this informed his approach to the problem of humanity's geographic origins. However, plenty of other supporters of monogenesis arrived at other geographic conclusions. The British anthropologist James Cowles Prichard, for instance, was also a strong believer in monogenesis and also held a number of proto-evolutionary beliefs that resembled ideas later proposed by Darwin.[12] In his 1813 *Researches into the Physical History of Man*, Prichard proposed that humans became lighter in color as they became more "civilized"; correspondingly, the first humans would have dark skin and Prichard identified them as "Negroes." However, Prichard did not exclusively identify Blackness with Africa, nor did he locate the origins of the supposed Adam and Eve on the African continent – instead, he argued for Asia, postulating a cradleland somewhere in the wide area between the Nile and the Ganges rivers.[13] He reconciled this seeming incongruity by arguing that there

---

[10] Erasmus Darwin, *The Temple of Nature; or, The Origin of Society*, American edition (Baltimore, MD: Butler and Bonsal & Niles 1804), 68, fn. My thanks to Erik Peterson for this citation.

[11] See discussion in George W. Stocking Jr., *Race, Culture, and Evolution: Essays in the History of Anthropology* (New York: The Free Press, 1968).

[12] George W. Stocking, Jr., "From Chronology to Ethnology: James Cowles Prichard and British Anthropology, 1800–1850," in James Cowles Prichard, *Researches into the Physical History of Man*, edited and with an introductory essay by George W. Stocking Jr. (Chicago: University of Chicago Press, 1973): xi, especially fn. 9. See also discussion in T. K. Penniman, *A Hundred Years of Anthropology* (London: Duckworth, 1935): 79, and T. Hodgkin, "Obituary of Dr. Prichard," *Journal of the Ethnological Society of London (1848–1856)* 2 (1850): 189.

[13] James Cowles Prichard, *Researches into the Physical History of Man* (London: W. Phillips, 1813), 554–5.

was clear evidence that the "Negro" race had been formerly much more widespread, using the example of the dark-skinned and curly-haired people of the Andaman Islands to prove his point.

For Darwin, sexual selection provided the necessary explanation for racial differentiation; for Prichard, it was the selective pressures produced by successive rounds of "civilizing" forces. To identify the first homeland of the human species, Darwin turned to comparative anatomy, while Prichard preferred comparative philology and ancient history. In context, Darwin's approach stands out, although again the argument that human anatomy had meaningful similarities to the anatomy of other primates was hardly new. But for the nineteenth century, his explanation was not especially unusual or radical – or at least, not much more radical than many others. The only minor oddity was that, on the balance, Darwin preferred Africa as the "somewhat more probable" choice.

Darwin's Africa point did not convince many of his contemporary supporters, even the ones who were self-professed enthusiastic Darwinians. Thomas Henry Huxley shared Darwin's preference for anatomical analysis, but identified chimpanzees and *orangutans* (rather than gorillas) as having brains most structurally similar to those of modern humans and therefore as likely equally closely related to modern man. This split the difference between Africa (chimpanzees) and Asia (orangutans) as *Homo sapiens'* first home, although Huxley declined to weigh in on the question outright. Alfred Russel Wallace approached human origins from the standpoint of biogeography, proposing that there was no need to assume that humanity had first originated in the tropics. Africa, Wallace wrote, was "known to have been separated from the northern continent in early tertiary times," only to reconnect with the Eurasian landmass at a much later period, when humans could be assumed to have already evolved.[14] Central Asia, by way of comparison, had the right combination of varied geography and climate to both challenge and sustain human evolutionary development.

German evolutionist and ardent Darwinian Ernst Haeckel also preferred Asia, arguing in 1870 that the ancestors of modern humanity had first appeared on the continent of Lemuria, now lost in the depths of the Indian Ocean, and had then spread to the rest of the world through India, differentiating into racial types along the way.[15] (Famously, or infamously, the "Indo-Germanic" or "Aryan" race was placed at the hierarchy's top.) This was not unusual; Alfred Russel

---

[14] Alfred Russel Wallace, *Darwinism: An Exposition of the Theory of Natural Selection, with Some of Its Applications* (London: Macmillan and Co., 1889), p. 458.

[15] Ernst Haeckel, *The History of Creation: Or the Development of the Earth and Its Inhabitants by the Action of Natural Causes. A Popular Exposition of the Doctrine of Evolution in General and of that of Darwin, Goethe, and Lamarck in Particular*, trans. E. Ray Lancaster, 2 vol. (London: Henry S. King & Co., 1876).

Wallace also felt that an Asian origin point provided a good explanation for the process of race formation and differentiation, although he followed a slightly different logic than Haeckel. Similarly, when Dutch physician and amateur paleoanthropologist Eugène Dubois proposed undertaking research into the prehistoric fauna of Southeast Asia, he quoted Darwin directly from *Descent of Man*, but argued that the discovery of the remains of fossilized chimpanzees in northern India would have likely changed Darwin's mind had he known of their existence. In 1891, Dubois' position was partially validated when he announced the discovery of a fragmented hominin fossil which he named *Pithecanthropus erectus*, the "upright ape-man."[16]

In short, Darwin's Africa proposition was only one of many theories put forward about the geography of the human species in the nineteenth century. Like many others, it was essentially untestable, and even the researchers who did take Darwin extremely seriously had other pieces of theory and evidence that they felt made Asia a better fit. After the discovery of *Pithecanthropus erectus* in 1891, that body of theory looked like it was on better and better footing. And, more to the point and with the benefit of twenty-first century hindsight, we know that both Darwin *and* his critics were partially right. While Africa would prove to be the most likely evolutionary homeland for the human species, elements like environmental factors and the finer points of the primate family tree have also proven to be critically important for understanding how human evolution works. Darwin was correct about *where*, but he was far from complete in understanding the *how* and *why*.

23.4   Dart in the Teeth of the Evidence

Over the fifty years between Darwin and Dart, plenty of people took the idea of human evolution very seriously indeed and spent a great deal of money, time, and intellectual effort investigating the evolutionary fossil record. At least five major expeditions headed into the field to search for ancestral human and transitional hominin fossils in Asia between 1900 and 1940.[17] Plenty of

---

[16] Bert Theunissen, *Eugène Dubois and the Ape-Man from Java: The History of the First "Missing Link" and Its Discoverer* (Dordrecht: Kluwer Academic Publishers, 1989).

[17] Sigrid Schmalzer, *The People's Peking Man: Popular Science and Human Identity in Twentieth-Century China* (Chicago: University of Chicago Press, 2008); Marianne Sommer, *History Within: The Science, Culture, and Politics of Bones, Organisms, and Molecules* (Chicago: University of Chicago Press, 2016); Brian Regal, *Henry Fairfield Osborn: Race and the Search for the Origins of Man* (New York: Routledge, 2002); Pat Shipman, *The Man Who Found the Missing Link: Eugène Dubois and His Lifelong Quest to Prove Darwin Right* (New York: Simon and Schuster, 2001); Chris Manias, "Jesuit Scientists and Mongolian Fossils: The French Paleontological Missions in China, 1923–1928," *Isis* 108, no. 2 (June 2017): 307–32.

human evolutionary research work in the mode of Darwin was carried out between 1871 and 1925 – just not in Africa.

It was not that sub-Saharan Africa was unknown to European science – there had been a certain number of geological, geographical, and even paleontological surveys conducted on the small scale from the end of the nineteenth century. The French nobleman de Bourg de Bozas led an expedition through the fossiliferous deserts of the Omo Valley in southern Ethiopia at the turn of the twentieth century, and identified the remains of prehistoric giant pigs before dying of a tropical fever during an expedition in 1902. In 1913, German anthropologist Hans Reck made a descent into Olduvai Gorge in what was then German East Africa and discovered prehistoric human remains preserved in the deep strata of the gorge wall. Reck planned to return for a second and more extensive expedition the next year, but was stopped by the outbreak of World War I. At no point, though, does it seem that Darwin's Africa suggestion alone was sufficient to justify the practical and financial outlays required. Instead, accident and serendipity provided the context in which discovery took place.

This was certainly true in the case of *Australopithecus*. At the end of 1924, a small hominin fossil skull was found in the wall of a limestone mine in Taungs, South Africa, and brought to the attention of local professor of anatomy Raymond Dart. In February 1925, Dart published his report in *Nature*, under the title "*Australopithecus africanus*: The Man-Ape of South Africa." In it, he described the anatomical features of the tiny, fragmented, and incomplete skull, proposing – with great confidence – that it represented "an extinct link between man and his simian ancestor."[18] He also explicitly invoked Darwin, claiming that the discovery vindicated "the Darwinian claim that Africa would prove to be the cradle of mankind."[19]

And yet, for all Dart's confidence, his claims were poorly received by the wider scientific community. A four-author reply was published in *Nature* a week later, objecting to Dart's confidence in extrapolating from such weak evidence: a partial juvenile skull missing several key portions of the fossil that might actually help substantiate some of Dart's claims about its relationship to other primates, including humans.[20] It was also impossible to work out the fossil's relative age – it had no geological context, since it was found in an active mine, and was accompanied by no other animal or plant fossils that would help index the find to an approximate geological era. As such, there was little evidence to support Dart's claims to have found something of extraordinary evolutionary importance – instead, perhaps, he had just found a somewhat

---

[18] Dart, "*Australopithecus africanus*," 198.   [19] Dart, "*Australopithecus africanus*," 198.
[20] Arthur Keith, G. Elliot Smith, W. L. H. Duckworth, and Arthur Smith Woodward, "The Fossil Anthropoid Ape from Taungs," *Nature* 115.2885 (February 14, 1925), 234–6.

interesting sort of primitive primate. Darwin was irrelevant – or, at the very least, Darwin's hypothesis could not compensate for Dart's evidentiary weaknesses.

Over the following decade and a half, however, *Australopithecus* was joined by a selection of other, similar fossils from excavations and cave sites around South Africa, found by other researchers like anatomist and physician Robert Broom. Dart, in the meantime, took a break from paleoanthropology to conduct research on living "racial types" among people in the Kalahari Desert – "race science" still as much a preoccupation in the 1930s as it had been for researchers fifty years earlier.[21]

In December 1946, the full collection of *Australopithecus* fossils were examined by British anatomist Wilfrid Le Gros Clark, who then flew north and east to Kenya to attend the first Pan-African Congress on Prehistory and Quaternary Geology, organized by Mary and Louis Leakey. At the conference, Le Gros Clark presented a strong case for interpreting *Australopithecus* as an extremely early forerunner of the human species. Increasingly, it looked like Dart had found something interesting in 1925 – but this validation only occurred two decades and several specimens later.

23.5   Making a Darwinian Myth

Over the course of the 1950s – buoyed by both new fossils and the increasing acceptance of African origins by his professional peers – Dart also began to craft his own narrative about his relationship with Darwin's Africa idea. While he had explicitly referenced and relied on Darwin in the 1920s, Dart in the 1950s was interested in narrating and dramatizing his discoveries for wider audiences. In 1953 Dart spoke in Johannesburg as part of a radio series, "Africa's Place in the Human Story," for the South African Broadcasting Corporation. He referenced Darwin's Africa hypothesis, but proffered a novel explanation for why he was ignored: "Darwin's idea of close relationship with apes was too new and people were much too distracted about business affairs to worry about the novelties Africa might have bearing upon this aspect of the human story."[22]

"People did not have too much time to spend on oddities amidst the Industrial Revolution," Dart continued, but since his fortuitous encounter with *Australopithecus* in 1925 it had been his "inescapable fate" to hunt for fossils and pick up the threads dropped by Darwin's distracted capitalist contemporaries. This was a careful framing on Dart's part: above all, he needed

---

[21] Kuljian, *Darwin's Hunch*, 60–75; and Saul Dubow, "Human Origins, Race Typology and the Other Raymond Dart," *African Studies* 55, no. 1 (1996), 1–30.

[22] "Introduction," in Raymond Dart, ed. *Africa's Place in the Human Story* (Cape Town: South African Broadcasting Corporation, 1953), 3.

to avoid stating or implying that anti-evolutionist religious orthodoxy was to blame for a scientific near-miss, since this would have drawn negative attention from both the ruling National Party and their close ally, the South African Dutch Reformed Church. Evolution was a dangerous topic in South Africa in 1953, both because it could be used to support arguments for racial equality, counter to the doctrines of apartheid, and because it was viewed by the Reformed Church as an anti-Christian heresy.[23]

In 1959, however, Dart published his autobiography, *Adventures with the Missing Link*, and told the story of his discovery in a slightly different way.[24] The first half of the book was a rousing scientific adventure story about *Australopithecus*, while the second half was a bid for Dart's controversial theory of *Australopithecus* making tools and weapons and the intrinsic place of violence in human nature. The biggest problem was that no one else thought *Australopithecus* had the brain capacity for this kind of work; theories on the natural origins of violence, however, proliferated in the middle of the twentieth century.[25] In telling the story of the discovery of *Australopithecus*, Dart referenced Darwin and his theory, but he emphasized Darwin as a scientific visionary who was brave and bold enough to face down public and scientific controversy – not over Africa, but over evolution in general. Dart drew a deliberate parallel: science had vindicated Darwin on both evolution and African origins. Surely, it was not a stretch to believe that science might do the same for Dart and his theories on the fundamental role of violence in shaping humankind.

This framing also had the effect of narrowing down the history of the science of paleoanthropology to a very small cast of characters, with Darwin and his self-designated heirs at the center. Certainly, Dart knew there were many contemporary researchers in his field broadly interested in similar questions about anatomy, evolution, anthropology, and origins – he was an active part of a highly cosmopolitan and wide-ranging network of scientific correspondents, connected especially through his former professor, the Australian-born, London-based Grafton Elliot Smith.[26] Very little of that world – intellectually or socially – comes through in the pages of *Adventures with the Missing Link* or in Dart's speeches, while at the same time *Adventures*

[23] Saul Dubow, *Scientific Racism in Modern South Africa* (Cambridge: Cambridge University Press, 1995), 46–7; and Saul Dubow, *A Commonwealth of Knowledge: Science, Sensibility, and White South Africa, 1820–2000* (Oxford: Oxford University Press, 2006), 256–7.

[24] Raymond Dart with Dennis Craig, *Adventures with the Missing Link* (New York: Harper & Brothers, 1959).

[25] Erika Lorraine Milam, *Creatures of Cain: The Hunt for Human Nature in Cold War America* (Princeton: Princeton University Press, 2019).

[26] Ross L. Jones and Warwick Anderson, "Wandering Anatomists and Itinerant Anthropologists: The Antipodean Sciences of Race in Britain between the Wars," *British Journal for the History of Science* 48, no. 1 (March 2015), 1–16.

was increasingly treated as a canonical source in popular media coverage of the history of paleoanthropology.[27] In its simplest form, the narrative became: Darwin proposes, Dart proves, and science wins the day.

23.6   Conclusions

What would the history of paleoanthropology look like without Darwin in it? Let's imagine that Darwin and his evolutionary theories still exist, and he still extends the implications of his theory directly to human beings, but there is no tentative suggestion of Africa as the place where humankind first evolved. What changes? How important is Darwin's intuition about African origins in the development of paleoanthropology?

My suspicion is: not terribly much. To review, Darwin's Africa idea was known to many of his contemporaries, including those who directly headed into the Asian field to look for transitional and ancestral fossils, and it had a minimal determining impact on where those researchers chose to dig. Of the researchers publishing on African fossil discoveries in the early twentieth century, Darwin's idea seems to have played little-to-no role in those researchers going to fossil-hunt in Africa in the first place. Raymond Dart pointed to Darwin's work directly in his publications; but, in 1925, invoking Darwin did not protect Dart from criticism and disbelief, even from sympathetic readers familiar with Darwin's work. Darwin's African origins idea had no actionable content. It was not a hypothesis that could be proven except by going out and finding fossil material that more or less fit paleontological expectations, and those expectations themselves were in the process of being determined – and contested – at the very same time. Ultimately, Darwin's salience to the problem of human origins lies more in his original evolutionary theory than in his geographic intuitions.

The narrowing of the narrative also reflects another change: the increasing de-legitimation of racial thinking in paleoanthropology. In focusing on Darwin and Dart, Africa and *Australopithecus*, we find familiar heroes – and a story of anthropological research neatly severed from the field's overwhelming historical preoccupation with race. Partly, the embrace of this myth reflects an admirable goal: to emphasize how science can be a force for justice and equality, to (rightly) condemn "race science" to the dustbin of history. And yet, to neglect these connections and contexts and to believe that Darwin and Dart's work could be pared down or separated from its racial implications and concerns is, as this and other chapters have shown, a myth in and of itself.

---

[27] See Martin Meredith, *Born in Africa: The Quest for the Origins of Human Life* (New York: Public Affairs, 2012), ch. 2 as an example of this kind of use of *Adventures* as documentary source.

Myth 24 That Darwin's Theory Brought an Instant and Immediate Revolution in the Life Sciences

Shruti Santosh and Anya Plutynski

24.1 Introduction

It has been commonly held that Darwin's *Origin of Species* led to a "revolution" in science.[1,2] Our aim here is to consider one particular variant of this claim: namely, that Darwin's theory brought about an *immediate* change in the life sciences. We will argue here that this claim is false – a myth. Unfortunately, such claims can be found in sources that are aimed to inform the public, popular science writing, and even in work by scientists, educators, and philosophers.

That said, "myths" often contain some truths. Myths are not simply "falsehoods," but may perform the function of simplifying and explaining complex situations. The reception of Darwin's theory was one such complex situation. Some of Darwin's ideas were in wide circulation before the publication of the *Origin of Species*, and thus not revolutionary in the sense of "novel." Moreover, Darwin's view that natural selection was a – if not "the" – major cause of evolutionary change, was not immediately and widely adopted, at least among many natural historians.[3] Rather than an *immediate* "revolutionary" change, then, the gradual acceptance of natural selection as a significant cause of the adaptation and diversification of life was part of a gradual transformation of the life sciences over a span of some fifty years, during which objections and competing theories

---

[1] Wallace wrote, "[Y]our book has revolutionized the study of Natural History, & carried away captive the best men of the present Age." Darwin Correspondence Project, "Letter no. 4514," accessed on June 16, 2023, www.darwinproject.ac.uk/letter/?docId=letters/DCP-LETT-4514.xml.

[2] "Charles Robert Darwin (1809–1882) transformed the way we understand the natural world with ideas that, in his day, were nothing short of revolutionary." Kerry Lotzof, "Charles Darwin: history's most famous biologist," Natural History Museum website, www.nhm.ac.uk/discover/charles-darwin-most-famous-biologist.html.

[3] There were a few whose perspective was quite suddenly changed (or so they claimed). Haeckel claims that after reading Darwin, his perspective on the relationships among radiolaria was quite dramatically altered. On Haeckel's transformation, see, R. J. Richards (2008) *The Tragic Sense of Life*. Chicago: University of Chicago Press.

were widely debated. That said, broadly "Darwinian" ideas were widely discussed and engaged with in literature, politics, agriculture, and the public after the publication of the *Origin of Species*, and thus might be said to have "revolutionized" *popular thought* on a range of theological, political, and social questions.[4]

24.2   What Is a Scientific Revolution?

To say whether Darwin's ideas caused an immediate revolution in the life sciences, we need to first understand what it means to refer to scientific change as "revolutionary." The source of much of the talk of "revolution" is political history. A vivid example of sudden or dramatic changes in the organization or governance of a state is the French Revolution. Contrast this with the "Industrial Revolution," which (arguably) took nearly a century. When one looks to purported instances of such "revolutionary" change, it's clear that there's a great deal of variation in both the rapidity, and the nature of such change. Perhaps not surprisingly then, there's a great deal of disagreement among historians about common features, causes or consequences of "revolutions."

The same is true for historians of science, and for (purported) "scientific" revolutions. Are all revolutions "sudden"? If so, what count as "sudden" changes? Do revolutionary changes require "radical" breaks with the past? If so, how radical need such breaks be? Which sorts of changes in science are "revolutionary"? Need there be change in conceptual frameworks, institutional organizations, theoretical commitments, methodological standards, technological tools, or all of the above? Given that genuinely sudden shifts in thought are rare in science, historians of science have been led to question "revolution" as an appropriate label for scientific change, or at very least, to argue that whatever features such revolutions have, rapidity is not one of them.[5]

Granting this ambiguity around the concept of "revolution," given the question posed in our title, we must at least provide a specification of some

---

[4] What it meant to be a "Darwinian" was itself hotly debated. See, e.g., M. Ruse (2019) *The Darwinian Revolution*. Cambridge: Cambridge University Press, and K. Hamlin (2014) *From Eve to Evolution: Darwin, Science, and Women's Rights in Gilded Age America*. Chicago: University of Chicago Press. Darwin's ideas on sexual selection (more or less aptly understood, or transformed via his interpreters), transformed literature and social thought (on the matter of marriage, sexual selection and free will), and the import for his theory shaped social and political thinkers, from Pierce to Oliver Wendell Holms, Jr. Thanks to Michael Ruse and Kimberly Hamlin for drawing our attention to this point.

[5] See, e.g., S. Shapin (1996) *The Scientific Revolution*. Chicago: University of Chicago Press. For discussion of the Darwinian "revolution," it's worth noting that the phrase "Darwinian revolution" was not coined by historians, but by scientists. Hodge argues that scientists' promotion of the idea of a "Darwinian revolution" had as much to do with promoting a particular historical narrative as it had to do with any univocal agreement on a set of scientific ideas. (See, e.g., M. J. Hodge (2005) Against "Revolution" and "Evolution." *Journal of the History of Biology*, 38(1), 101–21.)

features of "revolutionary" change in science. There are (roughly) two ways of talking of scientific change as "revolutionary": it may be "revolutionary" in the sense of "sudden," or "substantial." With respect to the first, as we've mentioned, what counts as "sudden" change is disputed. With respect to the second, there's a great deal of disagreement about what counts as "substantial" change in the history of science. Here, turning to Kuhn's *Structure of Scientific Revolutions*, and in particular, his ideas about "paradigm change" may be fruitful. Kuhn was neither the first nor the last to write about the nature of scientific change.[6] However, he was one of the most influential. So, in assessing whether Darwin's ideas were "revolutionary," it will be helpful to consider how Kuhn understood "paradigm shifts," if only as a useful foil.

A "paradigm" is, for Kuhn, not only a conceptual and theoretical framework defining the domain of inquiry, but also a set of methods, techniques, and tools of inquiry for studying these objects, and proscriptions regarding which sort of problems we ought to go out and investigate. A paradigm is thus not only a "theory" in the narrow sense of a set of general claims about the natural world and how it's organized or "lawfully" governed; it also involves a commitment to what there is, and how good scientists ought to study such things.

Kuhn believed that it becomes *necessary* for scientists to shift from one paradigm to another when anomalies began to accumulate. An anomaly is some pattern, entity, event, or process that is unexpected, or which violates the predictions and expectations of a given paradigm. These are only recognized as genuine challenges to the received view when anomalies have been accumulating. In response to this moment of "crisis," a "revolution" or "paradigm shift" can come about.

According to Kuhn, such "paradigm shifts," or, scientific revolutions, are "non-cumulative developmental episodes in which an older paradigm is replaced in whole or part by an *incompatible* new one." The key word here is "incompatible." There is no independent standard of evaluation for competing scientific paradigms. Thus, Kuhn controversially argued, it is not simply a matter of new evidence or theory that leads to a shift of paradigm, but also community assent, or mass persuasion: "As in political revolutions, so in paradigm choice – there is no standard higher than the assent of the relevant community."[7]

This picture of scientific change contrasts with a gradual, cumulative view of scientific change, according to which, science proceeds by collecting facts about the world, and gradually building theories from them. New theories may simply be theories which extend the scope of the previous domains. On Kuhn's view,

---

[6] I. B. Cohen's classic (1985) *Revolutions in Science* (Boston: Harvard University Press) is a wide-ranging history of the idea of "revolution" in science, tracing back to the eighteenth century.

[7] T. Kuhn (1970) *The Structure of Scientific Revolutions*. Chicago: University of Chicago Press, p. 94.

revolutionary science involves a dramatic shift in worldview – including, but not limited to, the acceptance of novel theories, classifications, descriptions of the problems worth solving, and how best to solve them. Successive paradigms are altogether different views about the things that populate the universe, and about that population's behavior. Is anything like this story of scientific change a fair characterization of Darwin's "revolution"?

24.3   Did Darwin's Theory Bring About a Scientific Revolution?

Darwin's *Origin of Species* is typically characterized as one "long argument" for two central theses: common descent, and natural selection as one – though not the only – process by which organisms diversify and adapt over time. As Love (Chapter 9, this volume) argues persuasively, this typical characterization of Darwin's "theory" is an idealized account of a dynamic, shifting set of observations, claims, and hypotheses, all of which he defended over the course of his career. "Common descent" is the thesis that all life is descended from common ancestors; this is in contradiction to the fixity of species, and of course, "special creation." Natural selection is the means by which populations change over time; given that individuals vary in traits that better enable them to survive and reproduce in a given environment, relative to their peers, over time, individuals with traits that are more "fit" will (by and large) increase in representation in future generations. Darwin's long argument, however, did not consist only in a defense of natural selection as a cause of change. He also took natural selection to be one of several causes of evolutionary change. He endorsed "use and disuse" (sometimes glossed as the inheritance of "acquired" characters; see Burkhardt, Chapter 8, this volume),[8] and he considered the "principle of divergence," "sexual selection," and an analogue of what is today called "drift," the "random sampling" of variant types, via geographical isolation of small subpopulations, or chance elimination of some subpopulations due to various ecological conditions, to all be important factors, as well.

Was Darwin's theory revolutionary in Kuhn's sense? In several ways, the answer is clearly no. The idea of evolution was not novel to Darwin.[9] Several prominent naturalists such as Owen, Goethe, Erasmus Darwin (Charles Darwin's grandfather), Lamarck, and Chambers all accepted a "theory of descent" – indeed, it was a widespread view among many naturalists that

---

[8] R. Burkhardt and W. Richard (2011) "Lamarck, Cuvier, and Darwin on Animal Behavior and Acquired Characters." In S. B. Gissis, S. Gissis and E. Jablonka (eds.), *The Transformations of Lamarckism*. Cambridge, MA: MIT Press, 33–44. See also chapters in this book by Corsi, Chapter 2, Fara, Chapter 3, and Rupke, Chapter 14.

[9] Though how Darwin understood and articulated his reasoning on behalf of common descent was genuinely novel (see e.g., Hodge, Against "Revolution").

species in existence today were not all that had ever existed, but that patterns of extinction were evident in the fossil record. Fossils were thus not some special anomaly but were well accommodated by a variety of competing views, some of which attempted also to accommodate a Biblical flood, and others of which set aside the problem of consistency with Biblical narrative. Indeed, Darwin drew upon some of these theories in developing his own views. Thus, the field cannot be said to have been in a state of "crisis" in the Kuhnian sense, at least insofar as the problem of accumulating anomalous observations was concerned.[10]

That said, there were a variety of features of Darwin's methods, mode of justification, conceptualization of patterns and process of common descent, and of course, his account of natural selection, that were original to him.[11]

24.4   Responses to the Publication of the *Origin of Species*

Many responses to the *Origin of Species* noted that Darwin's ideas were not "revolutionary" in the sense of "novel" – remarking on the fact that Darwin was not the first to posit the idea of evolution. In their reviews of the book, both Bishop Samuel Wilberforce and the comparative anatomist Richard Owen referred to Erasmus Darwin's ideas of evolution and common descent.[12] Moreover, Darwin's theory regarding the primary process of descent – natural selection – was not immediately accepted. Some reviews of the *Origin of Species*, such as that of the anatomist Thomas H. Huxley, who would become an advocate of "Darwinism"[13] were positive. Nonetheless, in his 1860 review of the *Origin of Species*, Huxley argued that it did not provide decisive evidence that natural selection could in fact create new species.

Religious concerns motivated other objections.[14] Though, even those who offered religious objections did not necessarily treat such objections as central.

---

[10] That said, one could argue that the theory of "special creation" was in a state of crisis, around 1850, insofar as there were debates concerning whether there was one or several special creations, and certainly Wallace's observations regarding the emergence of novel species by and large in the vicinity of similar variants seemed to challenge a single special creation event. (Thanks to Michael Ruse and David Depew for drawing our attention to this perspective.)

[11] D. Hull (1973/1989) "Charles Darwin and Nineteenth-Century Philosophies of Science." In *The Metaphysics of Evolution*. Albany, NY: SUNY Press, 27–42. Originally published in R. N. Giere and R. Westfall (eds.), (1973) *Foundations of Scientific Method: The Nineteenth Century*. Bloomington: Indiana University Press; P. Kitcher (1981) "Explanatory Unification." *Philosophy of Science*, 48(4), 507–31.

[12] Though Darwin disavowed his grandfather's influence, Erasmus Darwin's ideas clearly had some influence on Darwin's ideas. See Fara, Chapter 3 this volume.

[13] What was meant by "Darwinian" or "Darwinist" changed over time. Even figures as central to Darwinism's dissemination such as Huxley did not accept all elements of Darwin's view (see Bowler, Chapter 12, this volume).

[14] Though, Darwin himself left room for the possibility of God's influence, insofar as the "laws" of variation, growth, etc., that formed the Malthusian foundation for natural selection themselves

Bishop Wilberforce, for instance, is famous in the popular imagination for his heated debate with Huxley in Oxford, where he purportedly asked Huxley whether it was through his grandmother or his grandfather that Huxley considered himself descended from a monkey. According to one witness (Vivian Green), Huxley is said to have replied that he would not be ashamed to have an ape for his ancestor. However, he would be ashamed to share ancestry with someone who was using his gifts of reason to obscure the truth (presumably, Wilberforce, see Brooke, Chapter 13, this volume). However, in his 1860 review of the *Origin of Species*, Wilberforce's religious objections were secondary to his concerns with evidence for natural selection as a mechanism of descent. Likewise, the distinguished botanist Asa Gray, in his own review of the *Origin of Species*, focused on empirical justifications for the theory. Owen accepted evolution but did not accept the view that natural selection was sufficient to drive species transmutation.

It should not of course be a surprise that the publication of the *Origin of Species* attracted the attention of cartoonists. The magazine *Punch* was well established at the time, and on May 25, 1861, it included "The Lion of the Season" (Figure 24.1), showing a gorilla in evening dress being introduced at a grand function by an alarmed servant who announces him as "Mr. G-g-g-o-o-o-rilla!" According to art historian David Bindman, the focus on the gorilla rather than another kind of ape was due to the great publicity of the presence of the explorer Paul Du Chaillu in London in 1860.[15]

24.5   Regression to the Mean and the Problem of Inheritance

Another objection to Darwin's theory came from Fleeming Jenkin (Figure 24.2), who cast doubt on the power of natural selection to act as the main driving force in the origin of species, given the problem of "blending inheritance." Issues surrounding heredity were one central factor which delayed the acceptance of Darwin's hypothesis that natural selection was a major cause of diversification and adaptation, and hence, a Darwinian revolution. An engineer by trade, Jenkin published an anonymous review of the *Origin of Species* in *the North British Review* in 1867. This review articulated

could have been legislated by God. That is, his theological commitments seemed to permit a "clockwork" divinity, who set the laws in place and nature in motion. For discussion, see, B. F. Brown (1986) The Evolution of Darwin's Theism. *Journal of the History of Biology*, 19, 1–45; R. Richards (2009) "Natural Selection and Its Moral Purpose." In M. Ruse and R. Richards (eds.), *The Cambridge Companion to the "Origin of Species."* Cambridge: Cambridge University Press, 47–66; and, J. H. Brooke (2010) Darwin and Religion: Correcting the Caricatures. *Science & Education*, 19(4), 391–405.

[15] D. Bindman (2009) "Mankind after Darwin and Nineteenth-Century Art." In Diana Donald and Jane Munro (eds.), *Endless Forms: Charles Darwin, Natural Science and the Visual Arts*. New Haven, CT: Yale University Press, 146–7.

Figure 24.1 *Punch*, May 25 1861, p.213, Satirical cartoon "The Lion of the Season". Credit: Wellcome Collection, Attribution 4.0 International (CC BY 4.0)

several concerns about the process of species transmutation. Jenkin's first concern was that variability within species was limited. The second concern, known as the "Swamping Argument," is that rare novel variation is lost in each generation. A third concern was with the enormous timescales required for natural selection to work on species. Such timescales were incompatible with then current scientific estimates of the age of the earth.

Of these criticisms, it was the Swamping Argument that was potentially most threatening. Jenkin argued that natural selection would not be effective in selecting for a unique "sport" (an adaptive anomaly) because it would be "swamped" by the variants typical of the "original" population. Effectively, Jenkin was drawing attention to how it would be difficult for an advantageous trait to spread across the whole population, given both the assumed "blended"

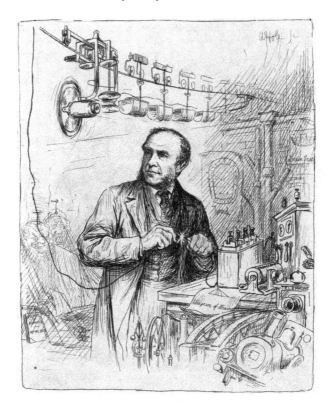

Figure 24.2 Fleeming Jenkin. Source: Science & Society Picture Library/
Contributor/SSPL/Getty Images

model of inheritance (when offspring inherit an approximate mean of their
parent's characteristics), and the rarity of the novel variant. In large populations, advantageous novel traits would eventually be diluted through interbreeding. However advantageous, chance alone would likely lead to the
elimination of rare variants.

Many have argued that this line of argument was a central reason for why
Darwin's hypothesis was not immediately adopted.[16] Jenkin's objection was
only addressed after the emergence and wide acceptance of a Mendelian or

[16] cf., P. J. Bowler (2005) "Variation from Darwin to the Modern Synthesis." In B. Hallgrímsson
and B. K. Hall (eds.), *Variation and Variability: Central Concepts in Biology.* Cambridge, MA:
Academic Press, 9–27.

particulate theory of inheritance, arguably, a process that was not completely resolved until the "modern synthesis" – the unification of Darwin's theory of natural selection with Mendelian genetics.[17]

The force of the swamping argument was tied to several open questions about the nature of hereditary variation. On the one hand, most of the observable variation in biological traits was continuous – height, weight, or milk yield for instance, vary continuously. The (yet invisible) causes of this variation were not well understood. If such causes were "discrete," however, such variations might be preserved across generations. In this way, advantageous variations would *not* be "swamped away" and so, natural selection could make headway in ensuring that individual variations could be preserved across generations. But the fact of continuous variation seemed to argue for the view that inheritance was blending, which would thus prevent natural selection from having lasting effects. Darwin accepted that inheritance was blending. He thus felt the force of Jenkin's objection: selection should not be able to move populations beyond a limited range. Without some means by which hereditary variation might be preserved, proponents of natural selection did not have an answer to the Swamping Argument.

Francis Galton, Darwin's cousin, was one of many critics of Darwin who struggled with this challenge to the theory. Galton's key observation was that children tend to resemble the typical means of their "race" rather than their parents. He explained this "stability of racial types" by arguing that "latent" characteristics in parents are later inherited by offspring.[18] Galton believed that while natural selection eliminated 'inferior' types, it could not change the essential character of a race. That is, while natural selection could select between races, it could not *modify* them.[19] This apparent incompatibility of natural selection with the observed facts of heredity was a major problem for Darwinism and prevented its widespread acceptance.

The research program of the biometricians from 1893–7 centered around finding a solution to the issues raised by Galton. These projects also aimed to demonstrate empirical cases of natural selection. What would count as an empirical demonstration of natural selection? This has been a subject of debate

[17] The modern synthesis refers to a formulation of evolutionary theory which married Darwinian natural selection with a population-oriented view of Mendelian genetics. What counts as the start of the Modern Synthesis is subject to debate, but 1920–50 is an approximate time frame, although some narrow the dates to 1936–47.

[18] cf. J. Gayon (1998) *Darwinism's Struggle for Survival: Heredity and the Hypothesis of Natural Selection*. Cambridge: Cambridge University Press.

[19] The concept of "race" of course is a historically contentious one, and Galton's ideas are no longer widely accepted (cf. J. P. Jackson and D. J. Depew (2017) *Darwinism, Democracy, and Race: American Anthropology and Evolutionary Biology in the Twentieth Century*. London: Routledge.

among evolutionary biologists for many decades. While there are widely agreed upon methodological standards for such demonstrations today,[20] methodological disputes over what count as sufficient tests of the effects of selection, or how to distinguish the role of selection and drift, continued well into the twentieth century.[21] Moreover, concerns about over-emphasis on "adaptationist" hypotheses, particularly with respect to human behavior, continue to this day.[22]

Starting in the 1860s, however, the hope was to demonstrate that subpopulations heterogenous for a particular trait, given divergent selective conditions, could be shown to diverge over time. In an 1893 paper, Raphael Weldon appeared to demonstrate exactly that through an analysis of the morphological characteristics of the common littoral crab, *Carcinus mænas*. While Weldon's crab study seemed to demonstrate the possibility of genuinely stable "divergence of character," it was still not an empirical demonstration of the *formation* of a new species.

Starting roughly in 1900, there was a confrontation between "Mendelian" or "mutationist" schools, and "biometrical" schools of thought. While the historiography of this controversy and the confrontation has been contentious,[23] there was broad disagreement surrounding the nature of hereditary variation, and its role in evolution. Recall that for Darwin, "natural selection acts solely by accumulating slight, successive, favourable variations; it produces no great or sudden modifications."[24] This was at odds with a view that came to be associated with the "Mendelians" or "Mutationists" – namely, that continuous and "fluctuating variation" was not terribly significant in evolution. For instance, Hugo de Vries argued that major "mutations," leading to novel phenotypes, were the main driving force of species change. While he granted natural selection some role in species change, by culling maladapted individuals, natural selection could not by itself affect positive change.

[20] M. R. Rose and G. V. Lauder (1996) *Adaptation*. San Diego: Academic Press; E. Sober (2008) *Evidence and Evolution: The Logic Behind the Science*. Cambridge: Cambridge University Press.

[21] Gayon, *Darwinism's Struggle*; R. L. Millstein (2008) Distinguishing Drift and Selection Empirically: "The Great Snail Debate" of the 1950s. *Journal of the History of Biology*, 41(2), 339–67.

[22] D. J. Buller (2006) *Adapting Minds: Evolutionary Psychology and the Persistent Quest for Human Nature*. Cambridge, MA: MIT Press.

[23] Y. Shan (2021) Beyond Mendelism and Biometry. *Studies in History and Philosophy of Science Part A*, 89, 155–63; G. Radick (2005) Other Histories, Other Biologies. *Royal Institute of Philosophy Supplements*, 56, 21–47; G. Radick (2023) *Disputed Inheritance*. Chicago: University of Chicago Press.

[24] C. Darwin (1859) *Origin of Species*, Chapter XIV. Recapitulation and Conclusion.

24.6   Darwinism in the Twentieth Century

This conflict over how and whether natural selection should be accepted as a major cause of evolution led to a situation which has been described as the "Eclipse of Darwinism."[25] This historical claim has been contentious. Historian Mark Largent points out that the idea of an "eclipse" was promoted during the modern synthesis by Huxley, in his 1942 book, *Evolution: The Modern Synthesis*. Largent argues that talk of the "eclipse of Darwinism" is a deliberate attempt on the part of the architects of the modern synthesis to portray themselves as the saviors of Darwin's legacy. A section from the opening chapters of Vernon Kellogg's popular 1907 textbook *Darwinism Today, "On the Deathbed of Darwinism"* is often hailed as providing evidence for this "eclipse." There, Kellogg documents a variety of criticisms of Darwin's theory, and a variety of alternatives. However, Kellogg's actual position is more conciliatory toward Darwin than this chapter title would suggest.[26]

All that said, there were serious criticisms of "Darwinism" circulating around this period which Kellogg gives voice to. Neo-Lamarckian, orthogenesist, and saltationist accounts of evolution were all widely in play and regarded as either alternatives or supplements to natural selection. The neo-Lamarckians held that individuals could acquire evolutionary significant variations over the course of their lifetime which could be inherited by their offspring. The defenders of orthogenesis held that evolution could be directed, and the saltationists, like the mutationists and Mendelians, believed that new species could be generated suddenly through discontinuous "leaps."

Many critics of Darwin argued that while natural selection did play a role in evolution, it was not the sole driver of the transmutation of species. They thus viewed Darwinism as a solely negative process.[27] Another common feature these theories shared, and which set them apart from Darwin's theory, was the

---

[25] Bowler, Variation, see also: P. J. Bowler (2017) "Alternatives to Darwinism in the Early Twentieth Century." In R. G. Delisle (ed.), *The Darwinian Tradition in Context: Research Programs in Evolutionary Biology*. Cham: Springer, 195–217.

[26] M. A. Largent (2009) The So-Called Eclipse of Darwinism. *Transactions of the American Philosophical Society*, 99(1), 3–21. Likewise, also, Richmond's (2006) critical analysis of the 1909 Darwin celebration at Cambridge suggests that even among boosters of the "revolution," there was a great deal of disagreement regarding which aspects or features of "Darwinism" were worth retaining, and which to take issue with. See, e.g., M. L. Richmond (2006) The 1909 Darwin Celebration: Reexamining Evolution in the Light of Mendel, Mutation, and Meiosis. *Isis*, 97(3), 447–84.

[27] Some proponents of Darwinism such as August Weismann took issue with the notion that selection could not create and could only reject, pointing out that natural selection, "in rejecting one thing preserves another, intensifies it, combines it, and in this way *creates* what is new." (*Darwin in Modern Science*, ed. A. C. Seward. (1909). *Darwin and Modern Science: Essays in Commemoration of the Centenary of the Birth of Charles Darwin and of the Fiftieth Anniversary of the Publication of the Origin of Species*. Cambridge: Cambridge University Press.)

idea that evolutionary change was directed and determined along specific channels. Kuhn himself notes that the Darwinian denial of such teleological evolution was major point of contention among its detractors. For example, defenders of orthogenesis took there to be "internal" laws of form that meant variation followed predetermined channels separate from any environmental constraints. This was in contrast to Darwin, who argued that mutations or "sports" arose "by chance"; accidents of natural variation in his view were by and large not "directed" or necessarily progressive.

The above picture can be analyzed through a Kuhnian lens: Darwinism and its alternatives were paradigms in their own right, with different theoretical and methodological commitments. However, the differences between "Darwinism" and its "alternatives" were not always so clear cut. During the years of Darwinism's "eclipse," various more or less devout proponents of "Darwinism" placed greater or lesser emphasis on natural selection as a process of change. Even proponents of orthogenic views held that natural selection did play some role in the formation of new species. This historical ambiguity about what it means to be a "Darwinian" thus rebuts any simple Kuhnian picture of the "Darwinian revolution" as a case of scientists committed to "incommensurable" world views. Both the "Darwinians" and their critics agreed on many matters of both fact and theory.

The Modern Synthesis also challenges a simple Kuhnian interpretation of the Darwinian "revolution," at least with respect to Kuhn's ideas regarding incommensurability. The integration of Mendelism with evolutionary theory provided a means to address the swamping objection, and effectively synthesized two arguably competing theories of heredity and evolution.[28] In other words, the purported divide between "Mendelism" and "biometry" was not an instance of incompatible paradigms, but of greater (or lesser) emphasis on various mechanisms of change and modes of inheritance. This further complicates a simplistic Kuhnian reading of a Darwinian revolution.

24.7   Conclusion

The above evidence suggests that Darwin's theory did not bring an immediate revolution to the life sciences. While aspects of Darwin's theory were (already) widely accepted, natural selection faced consistent opposition from life scientists well into the twentieth century. This opposition was in part due to theological concerns, but largely due to concerns over whether natural selection was sufficient to explain the origins of novelty or the diversity of life. Moreover, there were a variety of competing and complementary explanatory

---

[28] Gayon, *Darwinism's Struggle*.

accounts of evolution, which were only gradually challenged, rejected, or integrated into a more comprehensive "Darwinian" worldview.

There were a shifting array of views over whether (or how) Darwin's theory could accommodate novel discoveries about heredity and variation in natural populations. Jenkin's famous objection was only the first of several challenges growing out of the empirical study of heredity, and the rise of Mendelism. Thus, it took some time for Darwin's theory to be adopted. Nonetheless, at least some naturalists at the time saw his ideas as "revolutionary." The botanist Hewett Cottrell Watson wrote to Darwin to convey is admiration for the presentation copy of the *Origin of Species* he had just received: "Your leading idea will assuredly become recognized as an established truth in science, i.e., 'natural selection.' – (It has the characteristics of all great natural truths, clarifying what was obscure, simplifying what was intricate, adding greatly to previous knowledge.) You are the greatest Revolutionist in natural history of this century, if not of all centuries."[29]

So although Darwin did not initiate an *immediate* revolution in science, at least some of his readers perceived and commended his contributions as "revolutionary," in the sense of original and insightful.[30] There is no question, at least, that Darwin's naturalistic and eclectic approach to the biological world – drawing upon a diversity of evidence, ranging from fossil records, to biogeography and developmental homologies – founded a research program, and became a model for the biological sciences.

[29] Darwin Correspondence Project, "Letter no. 2540," accessed on December 28, 2022, www.darwinproject.ac.uk/letter/?docId=letters/DCP-LETT-2540.xml.

[30] Thanks to Frank Sulloway for the following astute observation that "the longer it takes for a set of ideas to triumph, owing to ideological opposition and/or to technical issues that take many years to resolve, the more revolutionary (in the sense of boldly prescient) these ideas really were at the time they were first set forth . . . " Regarding the distinction between scientific revolutions, which tend to take a long time to resolve, and more technical revolutions that are over relatively quickly, see F. J. Sulloway (2014) "Openness to Scientific Innovation." In D. K. Simonton (ed.), *The Wiley Handbook of Genius*. John Wiley & Sons, 546–63.

Conclusion: What Inferences About Science Can We Draw from Charles Darwin's Life and Work?

Kostas Kampourakis

C.1 Introduction

In this concluding chapter, I benefit from the scholarship, the historical evidence and the arguments in the twenty-four chapters of the present book in order to arrive at some broad inferences about the methods of science and about science as a social process. Most importantly, I want to draw on these chapters in order to humanize Darwin and to show that he was one of us: an everyday person who used his brilliance to achieve a goal and who achieved that goal after decades of diligent work and meticulous study. Darwin was not a super-genius who excelled in his school or university studies; he was not a gifted mathematical mind; he was not working against the thought of his time; he was not ahead of his time. Rather, he tried to answer questions of his time, being a man of his time who was lucky to be given the means and the opportunity to do something important, and made excellent use of that! He read widely, he experimented intensively, and he collected data from a vast network of correspondents all over the world. Let us then consider some aspects of his work in some detail.

C.2 The Brilliance of the Individual

Stating that Darwin was not a super-genius does not in any way diminish the brilliance of his ideas. Darwin was unquestionably the first to come up with a plausible and empirically grounded theory about the transmutation of species. He was also brilliant enough to put together ideas that were widely known at the time, but which had nevertheless not been considered together. In some sense, Darwin found different pieces of a jigsaw puzzle and realized that they could nicely fit to one another.

Perhaps the most important aspect of his work was the three creative analogies upon which he built his theory. These analogies were those between:

1. Artificial selection (based on his reading about and experience with animal breeding) and natural selection (see Ruse, Chapter 11, this volume).

2. The struggle for existence in human societies (an idea by political economist Thomas Malthus) and the struggle for existence in nature (see Sulloway, Chapter 6, this volume).

3. The (physiological) division of labor (and idea that goes back to political economist Adam Smith) and ecological specialization, which led him to the principle of divergence (see Love, Chapter 9, this volume).

After reading the pamphlets written by animal breeders, such as John Sebright and John Wilkinson, who were explicit about the nature and power of artificial selection, Darwin thought that sustained selection for small changes could be taking place in nature. He even joined several pigeon-breeding clubs to see for himself how far selective breeding could go in producing new varieties. Darwin's analogy between artificial and natural selection was based on the assumption that breeders' selection resulted in modifications in the domesticated organisms that were permanent and that had not existed in their wild ancestors. However, these two processes differ in a very important aspect. Artificial selection requires an intelligent external selector who picks variants according to particular aims or goals. No such selector exists in the process of natural selection, which is the outcome of an unintentional, natural process of struggle among individuals. Therefore, the analogy between artificial and natural selection initially seems weak. This is where the idea of struggle for existence comes in, by analogy with Malthus. According to Darwin, it is the competition between individuals of the same species that takes place simultaneously with competition between individuals of different species that takes the place of the intelligent selector of artificial selection and that makes natural selection strongly analogous to artificial selection. And when Darwin considered the developmental concepts and the generalizations of Karl Ernst von Baer and Henri Milne-Edwards, he came up with the idea of the principle of divergence. The core concept of the principle of divergence was the ecological division of labor, which is based on the analogy with the physiological division of labor by Milne-Edwards or even the division of labor in human societies by Adam Smith.[1] Darwin was the first to see the connection between these ideas, which form the basis for the main arguments in the *Origin of Species* (Figure C.1).

No matter if others came up with some of these ideas independently, it was Darwin who constructed the theory, creatively putting its elements together. The theory of evolution by natural selection was not waiting out there to be discovered. It was constructed by Darwin, who was explicit about this in his correspondence. In a letter written to Baden Powell on January 18, 1860, he wrote:

No educated person, not even the most ignorant, could suppose that I meant to arrogate to myself the origination of the doctrine that species had not been independently created.

---

[1] See chapter 4 of K. Kampourakis (2020) *Understanding Evolution*. Cambridge: Cambridge University Press.

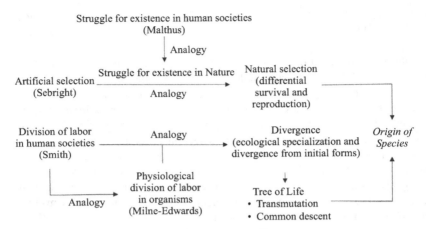

Figure C.1 The conceptual foundations of Darwin's theory, which was based on three analogies, and the arguments in the *Origin of Species*

> The only novelty in my work is the attempt to explain how species become modified, & to a certain extent how the theory of descent explains certain large classes of facts; & in these respects I received no assistance from my predecessors.[2]

As this quotation shows, Darwin acknowledged that others long before him had questioned the stability of species (but see Corsi, Chapter 2, this volume). As he wrote in the historical sketch in the third edition of the *Origin of Species*:

> It is curious how largely my grandfather, Dr. Erasmus Darwin, anticipated the errone-ous grounds of opinion, and the views of Lamarck, in his "Zoonomia" (vol. i. p. 500–510), published in 1794. According to Isid. Geoffroy there is no doubt that Goethe was an extreme partisan of similar views, as shown in the Introduction to a work written in 1794 and 1795, but not published till long afterwards. It is rather a singular instance of the manner in which similar views arise at about the same period, that Goethe in Germany, Dr. Darwin in England, and Geoffroy Saint Hilaire (as we shall immediately see) in France, came to the same conclusion on the origin of species, in the years 1794–5.[3]

Darwin's theory was also the outcome of very hard work (see Pearn, Chapter 7, this volume). Whereas Darwin came up with the idea of natural selection (stemming from points 1 and 2 above) in 1839, it took him almost another twenty years to come up with the complete theory (see Love, Chapter 9, this volume).

---

[2] Darwin Correspondence Project, "Letter no. 2655," accessed on October 19, 2022, www.darwin project.ac.uk/letter/?docId=letters/DCP-LETT-2655.xml.

[3] C. R. Darwin (1861) *On the Origin of Species by Means of Natural Selection, or the Preservation of Favoured Races in the Struggle for Life.* London: John Murray. 3rd ed., p. xiv (note).

And there were also questions he did not manage to answer such as the basis of heredity – and it would not have made any difference had he read Mendel's famous paper (see Radick, Chapter 18, this volume). Darwin, following the standards of the philosophy of science of his time, wanted to establish in a solid manner that natural selection was a *vera causa* (true cause). In a letter to his son Horace written in 1871, he wrote:

I have been speculating last night what makes a man a discoverer of undiscovered things, & a most perplexing problem it is. – Many men who are very clever, – much cleverer than discoverers, – never originate anything. As far as I can conjecture, the art consists in habitually searching for the causes or meaning of everything which occurs. This implies sharp observation & requires as much knowledge as possible of the subjects investigated.[4]

Darwin's statement stems from his experience, as he had indeed done what according to him a scientist ought to do.

At the end of his autobiography, he recounted his work in science. Here is his personal account of his own brilliance:

Therefore, my success as a man of science, whatever this may have amounted to, has been determined, as far as I can judge, by complex & diversified mental qualities & conditions. Of these the most important have been – the love of science – unbounded patience in long reflecting over any subject – industry in observing & collecting facts – and a fair share of invention as well as of common-sense. With such moderate abilities as I possess, it is truly surprising that thus I should have influenced to a considerable extent the beliefs of scientific men on some important points.[5]

C.3   The Influence of (Historical, Social, Cultural, Scientific) Milieu on the Individual

At the same time, it is important to note that in spite of Darwin's brilliance and hard work, he relied on intellectual and practical resources that were characteristic of the society in which he lived: natural theology was a tradition of the Anglican Church that put emphasis on the idea of adaptation; animal and plant breeding was a very successful Victorian technology; Thomas Malthus and Adam Smith were political economists who developed their theories in the English context; the British Empire was at the time expanded almost everywhere in the world, and thus not only Darwin traveled around the world for five years aboard the HMS *Beagle*, he also developed a vast network of correspondents, thanks to the British ships traveling around the world. These

---

[4] Darwin Correspondence Project, "Letter no. 8107," accessed on October 19, 2022, www.darwin project.ac.uk/letter/?docId=letters/DCP-LETT-8107.xml.

[5] N. Barlow (Ed.) (2005) [1958]. *The Autobiography of Charles Darwin 1809–1882*. New York: W.W. Norton, pp. 117–18.

and other features of Darwin's cultural, social, and historical milieu were crucial for the insights he had and for the development of his theory. Would it have been possible for Darwin to come up with this theory without these resources? We cannot really tell in hindsight, but it seems doubtful as their influence is evident.

How about the influence by his predecessors and contemporaries, both intellectually and practically? The names mentioned in the twenty-four chapters of the present book are too many to list here, but it is unquestionable that Darwin was influenced by many people. First there were those before him who developed evolutionary theories, such as his grandfather Erasmus (see Fara, Chapter 3, this volume) and Jean Lamarck (see Burkhardt, Chapter 8, this volume), even if he did not agree with them. Others like Aristotle had an influence that is sometimes hard to discern (see Lennox, Chapter 17, this volume). To say the least, their theories prepared the grounds for the reception of an evolutionary theory (with oftentimes exaggerations about its impact or nonimpact; see Richards, Chapter 20, Kern, Chapter 23, Grjebine, Chapter 19, this volume), and also convinced him about the importance of establishing the empirical foundations of his theory. As he wrote in the third edition of the *Origin of Species*:

Lamarck was the first man whose conclusions on this subject excited much attention. This justly-celebrated naturalist first published his views in 1801, and he much enlarged them in 1809 in his "Philosophie Zoologique," and subsequently, in 1815, in his Introduction to his "Hist, Nat. des Animaux sans Vertèbres." In these works he upholds the doctrine that all species, including man, are descended from other species. He first did the eminent service of arousing attention to the probability of all change in the organic as well as in the inorganic world being the result of law, and not of miraculous interposition.[6]

There were also those of his contemporaries, such as Herbert Spencer (see Depew, Chapter 15, this volume), Asa Gray (see Lennox, Chapter 16, this volume) and Joseph Dalton Hooker (see van Wyhe, Chapter 10, this volume). His exchanges with them certainly helped him clarify his own views, and convince him that he had come up with something important. As he wrote in a letter to Hooker on October 23, 1859:

I remember thinking above a year ago; that if ever I lived to see Lyell, yourself & Huxley come round, partly by my Book & partly by their own reflexions, I sh[d]. feel that the subject was safe; & all the world might rail, but that ultimately the theory of Natural Selection (though no doubt imperfect in its present condition, & embracing many errors) would pre-vail. Nothing will ever convince me that three such men, with so much diversified knowledge, & so well accustomed to search for truth, could err greatly.[7]

---

[6] Darwin, *On the Origin of Species*, p. xiii.
[7] Darwin Correspondence Project, "Letter no. 2509," accessed on October 19, 2022, www.darwin project.ac.uk/letter/?docId=letters/DCP-LETT-2509.xml.

Those of his contemporaries who disagreed with him, such as Richard Owen (see Rupke, Chapter 14, this volume) and Fleeming Jenkin (see Santosh and Plutynski, Chapter 24, this volume), forced him to continuously try to improve his theory, and to even include responses to potential criticisms in the *Origin of Species*. In a letter to Hooker on January 16, 1869, Darwin wrote:

Fleeming Jenkins [Fleeming Jenkin] has given me much trouble, but has been of more real use to me, than any other Essay or Review.[8]

Finally, there were those of his contemporaries who practically supported him in a variety of ways, from Henslow to Lyell (see Sulloway, Chapter 5, this volume), Huxley (see Bowler, Chapter 12, this volume) and Hooker. As he wrote to Huxley on July 3, 1860, with respect to the Oxford debate (see Brooke, Chapter 13, this volume): "As I am never weary of saying I sh^d. have been **utterly** smashed had it not been for you & three others. . . . I fancy from what Hooker says he must have answered the Bishop well. – God knows, I honour & thank you both."[9] In his autobiography he also wrote: "I have not as yet mentioned a circumstance which influenced my whole career more than any other. This was my friendship with Prof. Henslow."[10] The point is that, without diminishing Darwin's brilliance and work, we should keep in mind that his social, cultural, and scientific milieu mattered.

Last, but not least, there is another important factor in his milieu. Thanks to his father Robert, Charles never really had to work to earn his living. This allowed him to devote all his time and effort to the study of nature, and not to put effort on things he did not like such as studying medicine. As he acknowledged in his autobiography:

As I was doing no good at school, my father wisely took me away at a rather earlier age than usual, and sent me (October 1825) to Edinburgh University with my brother, where I stayed for two years or sessions. My brother was completing his medical studies, though I do not believe he ever really intended to practise, and I was sent there to commence them. But soon after this period I became convinced from various small circumstances that my father would leave me property enough to subsist on with some comfort, though I never imagined that I should be so rich a man as I am; but my belief was sufficient to check any strenuous effort to learn medicine.[11]

This is really admirable, as others in the same situation could have, and actually did, just enjoy their lives. But not Darwin.

---

[8] Darwin Correspondence Project, "Letter no. 6557," accessed on October 19, 2022, www.darwinproject.ac.uk/letter/?docId=letters/DCP-LETT-6557.xml.
[9] Darwin Correspondence Project, "Letter no. 2854," accessed on October 19, 2022, www.darwinproject.ac.uk/letter/?docId=letters/DCP-LETT-2854.xml.
[10] Barlow, The Autobiography, p. 54.   [11] Barlow, The Autobiography, pp. 40–1.

C.4   Humanizing the Individual

In his autobiography Darwin wrote:

In order to pass the B.A. examination, it was, also, necessary to get up Paley's Evidences of Christianity, and his Moral Philosophy. This was done in a thorough manner, and I am convinced that I could have written out the whole of the Evidences with perfect correctness, but not of course in the clear language of Paley. The logic of this book and as I may add of his Natural Theology gave me as much delight as did Euclid. The careful study of these works, without attempting to learn any part by rote, was the only part of the Academical Course which, as I then felt and as I still believe, was of the least use to me in the education of my mind. I did not at that time trouble myself about Paley's premises; and taking these on trust I was charmed and convinced by the long line of argumentation.[12]

As a young student in Cambridge, Darwin was initially convinced by Paley's arguments about the existence of purpose and design in nature (see Ruse, Chapter 4, this volume). But this later changed as he acknowledged in his autobiography:

The old argument of design in nature, as given by Paley, which formerly seemed to me so conclusive, fails, now that the law of natural selection has been discovered. We can no longer argue that, for instance, the beautiful hinge of a bivalve shell must have been made by an intelligent being, like the hinge of a door by man. There seems to be no more design in the variability of organic beings and in the action of natural selection, than in the course which the wind blows. Everything in nature is the result of fixed laws.[13]

It is often said that Darwin's study of nature drove him to atheism. However, this does not seem to be the case. Both in his autobiography, which was not intended to be published and in his personal correspondence, Darwin consistently described himself as an agnostic. It is true that he underwent several fluctuations of belief during his life, especially after the death of his beloved daughter Annie in 1851. However, he never explicitly rejected the existence of God. As historian Dov Ospovat showed many years ago, the main shift that Darwin underwent was conceptual: the shift from natural theology and a view of adaptation as perfect to natural selection and a view of adaptation as relative to the environment.[14] Even though his white beard grown during his fifties (Figure C.2) made him look like a prophet, Darwin was not one. He was a human with existential concerns such as those we all have.

---

[12] Barlow, The Autobiography, p. 59.   [13] Barlow, The Autobiography, p. 87.
[14] D. Ospovat (1981) *The Development of Darwin's Theory: Natural History, Natural Theology and Natural Selection, 1838–1859*. Cambridge: Cambridge University Press.

Figure C.2 Darwin in 1878 photographed by his son Leonard Darwin. This version of the photograph is L. Darwin 1878a.5. in the iconography by John van Wyhe, *Darwin: A Companion*, 2021, p. 176. Credit: Wellcome Collection. Public Domain Mark.

Here is what Darwin wrote about these matters in a letter to John Fordyce, on May 7, 1879:

Dear Sir

It seems to me absurd to doubt that a man may be an ardent Theist & an evolutionist. – You are right about Kingsley. Asa Gray, the eminent botanist, is another case in point – What my own views may be is a question of no consequence to any one except myself. – But as you ask, I may state that my judgment often fluctuates. Moreover whether a man deserves to be called a theist depends on the definition of the term: which is much too large a subject for a note. In my most extreme fluctuations I have never been an atheist in the sense of denying the existence of a God. – I think that generally (& more and more so as I grow older) but not always, that an agnostic would be the most correct description of my state of mind.

Dear Sir | Yours faithfully | Ch. Darwin.[15]

---

[15] Darwin Correspondence Project, "Letter no. 12041," accessed on October 19, 2022, www.darwinproject.ac.uk/letter/?docId=letters/DCP-LETT-12041.xml.

Figure C.3 Henrietta Emma Darwin. CUL DAR 225: 52 Cambridge
University Library (www.darwinproject.ac.uk/file/darwin-h-e-01-01225jpg)

Furthermore, in his autobiography, he wrote:

I cannot pretend to throw the least light on such abstruse problems. The mystery of the
beginning of all things is insoluble by us; and I for one must be content to remain an
Agnostic.[16]

Describing himself as an agnostic is especially important, given the difficulties
he had to perceive a beneficent and omnipotent God:

With respect to the theological view of the question; this is always painful to me. – I am
bewildered. – I had no intention to write atheistically. But I own that I cannot see, as
plainly as others do, & as I sh[d] wish to do, evidence of design & beneficence on all sides
of us. There seems to me too much misery in the world. I cannot persuade myself that
a beneficent & omnipotent God would have designedly created the Ichneumonidæ with

---

[16] Barlow, The Autobiography, p. 87.

the express intention of their feeding within the living bodies of caterpillars, or that a cat should play with mice. Not believing this, I see no necessity in the belief that the eye was expressly designed. On the other hand I cannot anyhow be contented to view this wonderful universe & especially the nature of man, & to conclude that everything is the result of brute force. I am inclined to look at everything as resulting from designed laws, with the details, whether good or bad, left to the working out of what we may call chance. Not that this notion *at all* satisfies me. I feel most deeply that the whole subject is too profound for the human intellect. A dog might as well speculate on the mind of Newton. – Let each man hope & believe what he can. – [17]

One can infer what one wants, but I do not think that Darwin could have been any clearer about his religious views.

C.5   That Was a Man's World

One astonishing feature of the present book is that among the various portraits of the individuals discussed in its chapters that it includes, there is not even one of a woman. This does not reflect a prejudice or bias on behalf of the contributors. The science of the time was considered as the exclusive preoccupation of men, and it was really difficult for any woman to advance in those circles. Therefore, the fact that the major figures portrayed in the present book are exclusively men, rather reflects the prejudices and biases of the nineteenth-century Western society. An example of this kind of prejudice can be found in Darwin's book *The Descent of Man*:

The chief distinction in the intellectual powers of the two sexes is shewn by man attaining to a higher eminence, in whatever he takes up, than woman can attain – whether requiring deep thought, reason, or imagination, or merely the use of the senses and hands. If two lists were made of the most eminent men and women in poetry, painting, sculpture, music, – comprising composition and performance, history, science, and philosophy, with half-a-dozen names under each subject, the two lists would not bear comparison. We may also infer, from the law of the deviation of averages, so well illustrated by Mr. Galton, in his work on "Hereditary Genius," that if men are capable of decided eminence over women in many subjects, the average standard of mental power in man must be above that of woman.[18]

So, what do we have here? Not only a rather racist Darwin (see Peterson, Chapter 22, this volume), but also a sexist one who considers women as inherently inferior to men intellectually? Perhaps. This was a standard view of women at the time, even though not all men considered women as

---

[17] Darwin Correspondence Project, "Letter no. 2814," accessed on October 19, 2022, www.dar winproject.ac.uk/letter/?docId=letters/DCP-LETT-2814.xml.

[18] C. R. Darwin (1871). *The Descent of Man, and Selection in Relation to Sex*. London: John Murray. Vol. 2. 1st ed., pp. 326–7. Available on John van Wyhe, ed. 2002– *The Complete Work of Charles Darwin Online* (http://darwin-online.org.uk).

intellectually inferior. I would refrain from judging Darwin as being sexist or not, either by today's criteria or by the criteria of his time. Anachronism is slippery anyway. Rather, I find it more interesting to consider Darwin's own attitude towards women, and their skills. This reveals, I believe, a more complex picture.

The first woman that comes to mind is of course his wife Emma. Whereas she certainly had an influence on Darwin, she was not as involved in his work as their daughter Henrietta (Figure C.3) was. Consider for instance a letter Darwin wrote to her on February 8, 1870.

My dear H.

Please read the Ch. first right through without a pencil in your hand, that you may judge of general scheme; as, also, I particularly wish to know whether parts are extra tedious; but remember that M.S is always *much*more tedious than print. – The object of Ch. is simply comparison of mind in men & animals: in the next chapt. I discuss progress of morals &c. – Some sentences are at back of Page marked thus @. –

I do not send foot-notes, as I have no copy & they are almost wholly mere author-ities. – After reading once right through, the more time you can give up for deep criticism or corrections of style, the more grateful I shall be. – Please make any long corrections on separate slips of paper, leaving narrow blank edge, & pin them to margin of each sheet, so that I can turn each back, & read whilst still attached to its proper page. – This will save me a world of troubles. Heaven only knows what you will think of the whole, for I cannot conjecture. – You are a very good girl indeed to undertake the job. –

Your affect Father | C. Darwin[19]

It is evident in this letter that Darwin not only trusts Henrietta's judgment, but also that he really values it. Why would the great naturalist entrust his most controversial book, the *Descent of Man* – it would be about the evolution of humans that he had entirely refrained from discuss-ing in the *Origin of Species,* as well as about sex and reproduction (see Hamlin, Chapter 21, this volume) – to a woman, if he considered women as intellectually inferior to men? Well, she was his daughter, you may think, so he made an exception. But Darwin also had several sons, who were younger than Henrietta but who had received a significantly stronger formal education than her. Yet, it was her who he depended on for the editing of his work. It would be interesting of course to have her own thoughts on the passage from the *Descent of Man* quoted above, but we will probably never know.

---

[19] Darwin Correspondence Project, "Letter no. 7124," accessed on December 20, 2022, www.da rwinproject.ac.uk/letter/?docId=letters/DCP-LETT-7124.xml.

Figure C.4 Antoinette Louisa Brown Blackwell (Image from archive.org; https://archive.org/stream/historyofwomansu01stan#page/n465/mode/2up; retrieved from www.darwinproject.ac.uk/file/blackwell-l-b-01-00471jpg)

Darwin took for granted that any good work was likely to come from a man. Thus, when Antoinette Louisa Brown Blackwell (Figure C.4) sent him in 1869 a copy of her book *Studies in General Science*,[20] Darwin responded with the following letter:

Dear Sir

I am much obliged to you for your kindness in sending me your "Studies in General Science", over which, as I observe in the Preface, you have spent so much time. – In turning over the pages I notice that you quote some statements made by me & very little known to public. I received your work only yesterday.

With my best thanks | Dear Sir | Yours faithfully | Ch. Darwin[21]

I imagine that Antoinette Louisa Brown Blackwell was not thrilled when she read that letter . . . but perhaps she was also used to this kind of attitude. Darwin did think that men were more keen to doing science than women. In another letter to Elinor Mary Dicey, he wrote: "I sh^d look at it as a Sin to discourage any boy from studying physiology who had the wish to do so; & I make the distinction between a boy &

---

[20] Antoinette Louisa Brown Blackwell (1869) *Studies in General Science*. New York: G. P. Putnam & Son.
[21] Darwin Correspondence Project, "Letter no. 6976," accessed on December 20, 2022, www.darwinproject.ac.uk/letter/?docId=letters/DCP-LETT-6976.xml.

Figure C.5 Mary Treat, Public Domain (Wikipedia, citing www.darwinpro ject.ac.uk/mary-treat)

a girl, because as yet no woman has advanced the science." Yes, men had done more than women in physiology at the time. But this does not mean that Darwin thought that women did not deserve any education, by stating in the previous sentence of the same letter that "I should regret that any girl who wished to learn physiology sh^d. be checked, because it seems to me that this science is the best or sole one for giving to any person an intelligent view of living beings, & thus to check that credulity on various points which is so common with ordinary men & women."[22]

But what is more important, and certainly more interesting, is that Darwin actively corresponded with women naturalists and also actively supported the publication of their work. He did not hesitate to express his admiration for the good work they had done and to encourage them to publish their work in scientific journals. Here, for instance, is a letter he wrote to US botanist Mary Treat (Figure C.5) on January 5, 1872.

Dear Madam

Your observations & experiments on the sexes of butterflies are by far the best, as far as known to me, which have ever been made. They seem to me so important, that I earnestly hope you will repeat them & record the exact numbers of the larvæ which you tempt to continue feeding & deprive of food, & record the sexes of the mature insect.

Assuredly you ought then to publish the result in some well-known scientific journal. I am glad to hear that your observations on Drosera will be published.

---

[22] Darwin Correspondence Project, "Letter no. 10746," accessed on December 20, 2022, www.darwinproject.ac.uk/letter/?docId=letters/DCP-LETT-10746.xml.

Figure C.6 Lydia Ernestine Becker, *The Graphic*, Jan 1874, p44. CUL NPR.
c.53 Cambridge University Library (www.darwinproject.ac.uk/lydia-becker)

I have attended to this subject during several years, & have almost M.S enough to
make a volume; but have never yet found time to publish it.

I am very much obliged for yr courteous letter & remain dear Madam | yours faithfully
| Charles Darwin[23]

Whether or not she was motivated by this warm and encouraging letter, Treat
repeated her experiments and published the results in the *American Naturalist* in
1873.[24]

Darwin also had a correspondence with Lydia Ernestine Becker (Figure C.6) on
scientific matters, especially during 1863. She was a leading member of the
suffrage movement, perhaps best known for publishing the *Women's Suffrage
Journal*. In 1866, she asked Darwin to send her a paper to read in the first meeting
of *Manchester Ladies' Literary Society*, and he did send three papers (whether or
not he realized that he thus supported a feminist organization is not clear, although
in her letter of December 22, 1866, Becker had attached a pamphlet of that
organization). Becker even asked Darwin for advice as to where she should submit
her work, which probably entails that she was expecting a positive response.[25] In
other cases, Darwin communicated himself the work of women, as in the case of

[23] Darwin Correspondence Project, "Letter no. 8146," accessed on December 21, 2022, www.da
rwinproject.ac.uk/letter/?docId=letters/DCP-LETT-8146.xml.
[24] M. Treat (1873) Controlling Sex in Butterflies. *The American Naturalist*, 7(3), 129–32.
[25] Darwin Correspondence Project, "Letter no. 7037," accessed on December 21, 2022, www.da
rwinproject.ac.uk/letter/?docId=letters/DCP-LETT-7037.xml.

Mary Elizabeth Barber, whose paper Darwin recommended to the President and the Council of the Linnean society in 1869: "Mrs Barber's paper seems to me worth publishing, because the dependence of the fertilization of a plant on one kind of insects alone, though not an unknown case, is very rare. Nor has any instance been recorded, as far as I can remember, of the access of other insects, being prevented by a mechanical obstacle requiring strength to be overcome."[26] In other instances, Darwin communicated or forwarded her work to journals.[27]

C.6   Envoi

Here is then a take-home message about the nature of science from Darwin's life and work. Science is a fascinating and creative endeavor done by communities who arrive at conclusions about the natural world. It is not done by lonely geniuses, but by brilliant, hard-working individuals who are influenced by their milieu. It is scientific communities that validate the advancement of a scientific field, and even if an individual takes most of the credit, that person has certainly depended on the work and the insights of others. This is why we should not focus on one or few individuals. Historian Janet Browne has described the recent trends in the historiography of Darwinism as follows: "historians of science are changing their focus from the individual to the collective, from treating Darwin as a single dominant figure in biology to embracing a more panoramic picture that takes account of wider historical trends and transformations."[28] I hope that this book is a small contribution toward this end, if only by showing how many other people besides Darwin were important. By humanizing and demythologizing Darwin, perhaps we will achieve a more authentic representation of history and of how science was done at the time. By stop talking about "fathers" of disciplines, we may come to better understand the contributions of many, men and women, and not only of the celebrated ones to our understanding of nature.

[26] Darwin Correspondence Project, "Letter no. 6740," accessed on December 21, 2022, www.darwinproject.ac.uk/letter/?docId=letters/DCP-LETT-6740.xml.

[27] Mary Elizabeth (Mrs) Barber (1874) Notes on the peculiar habits and changes which take place in the larva and pupa of Papilio Nireus. Communicated by Charles Darwin. Transactions of the Entomological Society of London (Part IV, Dec.): 519–21; A Lady. 1875. Extinction of the Macartney Rose. (Forwarded by Darwin) Gardeners' Chronicle (17 Jul.): 78 (both available at John van Wyhe, ed. 2002–. *The Complete Work of Charles Darwin Online*. (http://darwin-online.org.uk/)).

[28] J. Browne (2021) Charles Darwin and the Darwinian Tradition. In *Handbook of the Historiography of Biology*, ed. M. Dietrich, M. E. Borrello, and O. Harman, 7–32. Cham: Springer, p. 9.

Further Reading

There are numerous books about Darwin. Janet Browne's *Darwin's "Origin of Species": A Biography* (Atlantic Books, 2006), provides a nice overview of Darwin's life until the *Origin of Species* was published. Browne has also coauthored with Adrian Desmond and James Moore, the other major Darwin biographers, a rather short biography of Darwin (Oxford University Press, 2007). Desmond and Moore are the authors of a major biography, entitled *Darwin: The Life of a Tormented Evolutionist* (W. W. Norton 1994/1991). Janet Browne later published a massive two-volume biography of Darwin; the first volume is entitled *Charles Darwin: Voyaging* (Pimlico, 2003/1995) and the second one *Charles Darwin: The Power of Place* (Pimlico, 2003/2002). Finally, Erik Peterson's *Understanding Charles Darwin* provides a different, more human perspective of Darwin's life (Cambridge University Press, 2023).

If you want to read the *Origin of Species*, then a highly recommended edition is the one edited by Jim Endersby (Cambridge University Press, 2009), which includes a lengthy introduction that nicely sets the context in which the book was written. Another option is Harvard facsimile edition, edited by Ernst Mayr (Harvard University Press, 1964/1859). All of Darwin's books, manuscripts and much more are freely available online at Darwin Online: http://darwin-online.org.uk, under the direction of John van Wyhe. Similarly, most of his correspondence is freely available by the Darwin Correspondence Project online at www.darwinproject.ac.uk. Dov Ospovat's book, *The Development of Darwin's Theory: Natural History, Natural Theology and Natural Selection, 1838–1859* (Cambridge University Press, 1981), provides a detailed account of Darwin's writings from his manuscripts until the publication of the *Origin of Species*. Perhaps the best history of evolutionary thought is *Evolution: The History of an Idea* by Peter J. Bowler (The University of California Press, 2009). Quite interesting is also the book *Darwin Deleted: Imagining a World without Darwin* (University of Chicago Press, 2013), by the same author. Most recently, Paul van Helvert and John van Wyhe have produced the most updated and richest single-book resource that exists about Darwin in *Darwin: A Companion* (World Scientific Publishing, 2021).

For those wishing to go further and deeper into historical and philosophical issues, two must-read books are *The Cambridge Companion to Darwin* (Cambridge University Press, 2nd edition, 2009) edited by Jonathan Hodge and Gregory Radick, and *The Cambridge Companion to the Origin of Species*, edited by Michael Ruse and Robert J. Richards (Cambridge University Press, 2009). Another valuable, but old and perhaps difficult to find, book is *The Darwinian Heritage*, edited by David Kohn (Princeton University Press, 1985). Finally, *The Cambridge Encyclopedia of Darwin and Evolutionary Thought* (Cambridge University Press, 2013), edited by Michael Ruse, is a useful resource on almost everything about Darwin.

Not surprisingly, Darwin and religion is a topic of many books. For a historical overview, a great book is John Hedley Brooke's *Science and Religion: Some Historical Perspectives* (Cambridge University Press, 1991). For Darwin himself, Nick Spencer's *Darwin and God* (SPCK, 2009) provides a nice overview. Philosophers of science have also provided detailed accounts of the impact of Darwinism and evolutionary theory on religiosity. Particularly interesting and useful is Philip Kitcher's *Living with Darwin* (Oxford University Press, 2007). Another informative one is Michael Ruse's *Can a Darwinian Be a Christian? The Relationship between Science and Religion* (Cambridge University Press, 2001).

Index

Printed in the United States
by Baker & Taylor Publisher Services